Electrical Stimulation and
Electropathology

Electrical Stimulation and Electropathology

J. PATRICK REILLY

Applied Physics Laboratory
The Johns Hopkins University

with chapters contributed by

HERMANN ANTONI
University of Freiburg i. Br.

MICHAEL A. CHILBERT
Medical College of Wisconsin

WALTER SKUGGEVIG
Underwriters Laboratories Inc.

JAMES D. SWEENEY
Arizona State University

CAMBRIDGE
UNIVERSITY PRESS

Published by the Press Syndicate of the University of Cambridge
The Pitt Building, Trumpington Street, Cambridge CB2 1RP
40 West 20th Street, New York, NY 10011-4211, USA
10 Stamford Road, Oakleigh, Victoria 3166, Australia

First published 1992

Printed in the United States of America

Library of Congress Cataloging-in-Publication Data
Reilly, J. Patrick.
    Electrical stimulation and electropathology / J. Patrick Reilly;
with chapters contributed by Hermann Antoni ... [et al.].
        p.   cm.
    Includes bibliographical references and index.
    ISBN 0-521-41791-0 (hc)
    1. Electric shock.   2. Electric stimulation.   3. Electricity –
Physiological effect.   I. Antoni, H. (Hermann), 1929–
II. Title.
    [DNLM: 1. Electric Stimulation.   2. Electric Stimulation – adverse
effects.   3. Electric Stimulation Therapy.   WB 495 R362e]
RC87.5.R45   1992
615.8′45 – dc20
DNLM/DLC
for Library of Congress   91-37632
                                                              CIP

A catalog record for this book is available from the British Library.

ISBN 0-521-41791-0   hardback

The authors have made every attempt to provide an accurate and
balanced treatment. However, the material in this book has been
drawn from numerous sources, and in matters of critical interpretation
the authors suggest that the reader refer to the original sources.

Contents

Contributors

Hermann Antoni
Physiological Institute
University of Freiburg
Hermann Herder Str. 7
D 7800 Freiburg i. Br.
FR Germany

Michael A. Chilbert
Department of Neurosurgery
Medical College of Wisconsin
8701 Watertown Plank Road
Milwaukee, Wisconsin 53226

J. Patrick Reilly
The Johns Hopkins University
Applied Physics Laboratory
Johns Hopkins Road
Laurel, Maryland 20723

Walter Skuggevig
Underwriters Laboratories Inc.
1285 Walt Whitman Road
Melville, New York 11747

James D. Sweeney
College of Engineering and Applied Science
Department of Chemical, Bio, and Materials Engineering
Arizona State University
Tempe, Arizona 85287-6006

Preface

Today, the occasions for human exposure to electricity are numerous. A myriad of electrical products can be a source of unintended shock unless they are designed with an understanding of safety precautions and human reaction thresholds. Intense electric fields from high voltage transmission lines can be a source of electric shock to both workers and the public. Furthermore, the applications of electrical stimulation for medical purposes are becoming more numerous. Whether our interest is in accidental or in intended exposure, it is important to understand the range of potential human reactions to electrical stimulation.

This book describes human reactions to electrical stimulation, with special attention to reactions of a detrimental nature. The term "electropathology" is used here to indicate any undesirable biological reaction to electrical stimulation. A reaction may be judged undesirable in one context, such as in chance electrical exposure, but may be desirable in another context, such as in stimulation for medical reasons.

I first began to study electrical stimulation of people because of my work in electrical safety and environmental acceptability. I soon realized that much of the engineering literature on electrical acceptability seemed to ignore research from medical applications of electrical stimulation. And the literature by biomedical researchers on electrical stimulation was largely devoted to particular applications that did not include electrical acceptability in a broader context. It became clear to me that there was a need for a book that bridged the gap between engineering and medical applications of electrical stimulation—one that would give a treatment of human reactions ranging from the just-noticeable to the clearly unacceptable.

This book focuses on short-term reactions to electrical stimulation. The subject of health effects from chronic exposure to low-level electromagnetic fields has not been included. I do not wish to minimize the importance of a subject that has been much discussed in public forums. However, an appropriate treatment of chronic low-level exposure would require another book, and several excellent publications have recently appeared on this subject.

It is hoped that this book will be useful to the biomedical scientist, as well as to the engineer concerned with electrical safety. The material is directed to a reader who is familiar with basic physics and basic physiology, but who may not necessarily be an expert in either discipline. The initial chapters (2–5) are intended to provide a fundamental background by which later discussions on electropathology and electrical stimulation may be understood. Subsequent chapters (6–10) treat human reactions to electrical stimulation and are organized according to the nature of the response. A final chapter (11) treats standards and protective measures in consumer products. I hope the reader will benefit from the diversity of backgrounds of the researchers who have contributed to this book, and that this work will promote the creative and safe uses of electricity.

This book would not have been possible without the contributions of many people. I am particularly grateful for the help of those at The Johns Hopkins University Applied Physics Laboratory, and especially for the Stuart S. Janney Fellowship which supported a portion of my writing and research efforts. The Fleet Systems Department of the Applied Physics Laboratory made available the resources of their editorial group for word processing of the manuscript. That effort was carried out primarily by Rachel Mercer, Joann Choyce and Julie Dorsey. Much of the art work was done at the Laboratory's Technical Reports Group, thanks primarily to Joseph Lew and Jacob Elbaz.

A number of professional colleagues provided valuable help. Professor Gottfried Biegelmeier of Austria was most helpful in providing discussions and encouragement and also by providing an English translation of his own book *Wirkungen des elektrischen Stroms auf Menschen und Nutztiere*. Several colleagues reviewed chapters and made valuable suggestions. I thank Arnold Smoot of Underwriters Laboratories for reviewing Chapter 2 and Orest Z. Roy of the National Research Countil of Canada for reviewing Chapter 6. Chapter 7 was reviewed by Gary Rollman of the University of Western Ontario and Willard D. Larkin of the University of Maryland. I particularly wish to acknowledge Professor Larkin for the many things I learned from him during the years that we worked together researching electric shock. I am grateful to Donald McNeal of the Rancho Los Amigos Rehabilitation Center for helping me to implement a computer version of the nerve excitation model that he originated in 1976. An extended version of that model has played a key role in many of the discussions in this book. There were also many colleagues, too numerous to mention, who sent me papers and discussed research topics with me. I am grateful to them all.

Special thanks are due to those colleagues who have written sections of this book. Individual chapters were written by Hermann Antoni of the University of Freiburg (Chapter 5), James D. Sweeney of the Arizona State University (Chapter 8), Michael Chilbert of the Medical College of Wisconsin (Chapter 10), and Walter Skuggevig of Underwriters Laboratories (Chapter 11). In addition, Harry Eaton of the Applied Physics Laboratory contributed the final section of Chapter 9.

As anyone who has written a technical book undoubtedly knows, the process requires a substantial commitment over a long period of time involving many personal sacrifices that are shared by one's family. I thank my wife Lynette for her patience and understanding during this project, and for the many ways that she assisted me.

J. P. Reilly

1

Introduction

1.1 General Perspective

Electrical forces are vital for the functioning of living things, from the metabolism of individual cells to human consciousness derived from the activity of the brain. When we artificially introduce electric currents into a living organism, we can activate or modify these natural forces, with the result being either detrimental or beneficial, depending on circumstances. The term *electric shock* is generally used to describe the response of the body to inadvertent electrical exposure, and the consequences of electric shock are usually considered undesirable. However, the biological mechanisms responsible for electric shock may also be activated in a controlled fashion for beneficial medical purposes.

The fact that electricity can interact with biological processes has been known for more than 2000 years.[†] The electrical discharge of the torpedofish was reported to have been used as early as 46 A.D. to treat pain. It is tempting to scoff at early claims of beneficial uses of electricity, but when one considers modern methods of pain management by electrical means, one is led to suspect that there might have been valid reasons for some of these early beliefs.

The beginnings of quantitative bioelectric science can arguably be ascribed to the investigations of Galvani (around 1790) and later of Volta. Galvani observed motion in severed frogs' legs when he touched them with metallic wires. He supposed that he was releasing stored "animal electricity," and that this was responsible for the observed muscle activity. It was later demonstrated that the dissimilar metals of the wires used in his procedures were in fact generating the electrical forces. His efforts did, however, spark more investigations of electrical stimulation. Volta's

---

[†]McNeal (1977) provides an interesting historical account of electrical stimulation.

invention of the "Voltaic pile" gave science a battery that could be used in systematic and controlled investigations. Later biological investigations by Faraday (around 1831) demonstrated that interrupted electric current was an effective means of electrical stimulation of nerves. The term "Voltaic" stimulation came to be used to indicate direct current stimulation, and "Faradaic" to indicate pulsed or interrupted stimulation.

1.2   Electrical Exposure
Electrical Fatalities

It has been long recognized that man-made electricity can be harmful, and can even cause death. This understanding eventually led to the use of electrical executions of criminals to replace the "less civilized" means that had previously been devised. The first electrocution in the United States was administered in 1890 (Leyden, 1990). The number of electrocutions of criminals since that time is uncertain, but is estimated to number around 4100 in the United States. Today, about 900 prisoners in 14 states await execution in the electric chair.

The National Center for Health Statistics publishes data on the mortality from electrical accidents in the United States. Figure 1.1 illustrates U.S. mortality statistics for the years 1975–1987 as summarized by Smith (1990). The figure shows data for all electrical fatalities excluding lightning incidents, and for fatalities related to consumer products.

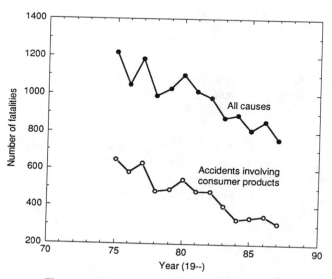

Figure 1.1 Mortality from electrical accidents in the United States for the years 1975–1987 (per data compiled by Smith, 1990).

Table 1.1. *Mortality due to electrical accidents*

	1987	1986	1985	1984
Electric current				
Total (all causes)	760	854	802	888
Domestic wiring and appliances	121	150	146	148
Power generating, distr., transmission	177	182	196	215
Industrial wiring, appliances, machinery	64	89	69	68
Other unspecified electric current	398	433	391	457
Lightning	99	78	85	--

Source: Vital Statistics of the United States, Volume II—Mortality, U.S. Department of Health and Human Services, National Center for Health Statistics, Washington, DC.

Since the mortality data do not specify the product involved in the accident, the number of fatalities related to consumer products was inferred by Smith on the basis of the reported locations of the accidents.

The downward trend of electrical mortalities is striking. And if one accounts for the increase in U.S. population over the period reported in Fig. 1.1, the trend is even more impressive—the per-capita rate of fatalities related to consumer products declined from 3.0 per million in 1975 to 1.3 per million in 1987. According to Smith, this trend could be due to a number of factors, including: (1) a steady increase in the number of residences having ground-fault current interrupters (GFCIs); (2) an increase in the number of locations having GFCIs within individual dwellings; (3) a decrease in citizen-band base-station antennas; and (4) an increase in the manufacture of double-insulated power tools. Details of some of these precautionary measures are given in Chapter 11. The downward trend in Fig. 1.1 contrasts with a slight increase in electrical mortality during the period 1960–1975 (Dalziel, 1978), a period when electrical safeguards were still being developed but were not widely implemented.

Table 1.1 provides a further breakdown of electrical mortalities. The table also enumerates deaths by lightning, of which there are nearly 100 each year in the United States. Lightning accounts for roughly 10% of the total number of electrical fatalities. It has been estimated that one-third of all lightning strikes to humans are fatal (Biegelmeier, 1986).

The worldwide experience of industrialized countries also shows a downturn of electrical fatalities (Kieback, 1988). The annual number of fatalities per million inhabitants for 1982 (the latest year reported by Kieback) ranged from a low of 0.42 for the Netherlands to a high of

7.66 for Hungary. Kieback emphasizes that such comparisons should be treated with caution because of differing reporting standards in various countries.

It is likely that the mortality data referred to above underreport the actual number of electrical fatalities. The data for the United States have been developed from death certificates. Not all electrical fatalities may have been identified in the certificates as such. Furthermore, deaths from injuries incidental to electric shock, such as falls, are probably not reported in the electrical category. Delayed fatalities, such as with burns, also may have been omitted. Even if we allow for underreporting, the number of electrical fatalities remains small in comparison with other accidental causes in the United States. Death from choking on food or other ingested objects, for example, exceeds electrical fatalities by nearly five times.

Typical Electrical Exposures

Notwithstanding the relatively small rate of electrical fatalities, the acceptability of electrical exposure by the public is a matter of growing concern. One area of concern is related to the increasing number of electrical consumer products. While potentially fatal electrical exposures are seldom an issue in most consumer products, in some cases exposures might be unpleasant, or they might lead to injuries from startle reactions. Both consumers and manufacturers hope to avoid such exposures.

High-voltage transmission lines provide another source of public exposure. Increases in population and in individual electrical demand press utilities to increase electricity generation and transmission. The trend in transmission is to higher voltages because of economies in the cost of construction and operation of transmission facilities. One environmental consequence of high-voltage transmission is the potential for electric shock arising from the electric fields that the lines inevitably produce (see Chapter 9). As far as the general public is concerned, the electric shock resulting from transmission line fields does not have the potential for injury, except possibly in certain unusual circumstances. Nevertheless, perceptible shock from transmission line fields can be unpleasant and can provoke fear in the exposed public.

Probably the most rapidly growing area of electrical stimulation is in biomedical technology. Electrical stimulation is being used increasingly as a tool for medical diagnosis, therapy, and prosthesis; examples of biomedical applications are listed in Table 1.2. Even this incomplete list is a testament to the numerous medical applications of electrical stimulation. Electrical stimulation for medical purposes may be introduced through electrodes in contact with the skin, through implanted electrodes,

Table 1.2. *Examples of electrical stimulation in biomedical applications*

Restoration of muscle function after nerve injury
Preservation of muscle tone after nerve injury
Treatment of scoliosis
Diaphragm stimulation for respiration control
Electrical stimulation of sphincter for urinary control
Correction of foot-drop
Sensory aids for the blind
Cochlear prosthesis for the deaf
Management of intractable pain
Inhibition of intractable self-injurious behavior
Diagnosis of peripheral nerve function
Diagnosis of muscle function
Functional diagnosis and mapping of the brain cortex
Stimulation of the visual cortex
Electroconvulsive therapy
Automatic cardiac pacing
Automatic sensing and reversal of fibrillation (implants)
Defibrillation in emergency aid (external)
Bone healing
Electrical diathermy

or through magnetic induction. The latter method precludes the necessity for any electrode contact whatsoever.

Electrical stimulation that might be considered detrimental in chance exposures can be beneficial when used in a controlled fashion. For example, cardiac arrhythmias caused by chance electric shock are regarded as being potentially life-threatening; these same responses can be life-saving when used in implanted pacemakers. Furthermore, a response that is considered undesirable in one medical application may be considered beneficial in another. Electrically induced pain, for example, is considered a difficulty to be avoided in procedures involving the introduction of currents via electrodes on the skin. But painful electrocutaneous stimulation has also been used in a beneficial way to inhibit otherwise intractable self-injurious behavior (Newman, 1984).

In this work, the term "electropathology" is used to indicate any undesirable biological reaction to electrical stimulation. The study of electropathology can be applied to help reduce the possibility of unacceptable exposures to electrical equipment, and to understand better how to use electrical stimulation for beneficial reasons. Regardless of our particular orientation, it is useful to understand the range of potential human reactions and their underlying mechanisms.

1.3   Scales of Short-Term Reactions

Several scales of reactions are of interest in the study of electropathology and electrical stimulation. These scales can be best appreciated by way of example. Figure 1.2 illustrates possible reactions to 60-Hz alternating current flowing between the hand and feet. It is assumed that an adult is gripping a large electrode in one hand, while a return electrode is contacting a foot or both feet. The exposure duration is assumed to be approximately 5 s. Categories of reactions are identified in the five columns in the figure, each involving a distinct continuum of reactions. Various thresholds of reaction are identified by descriptors

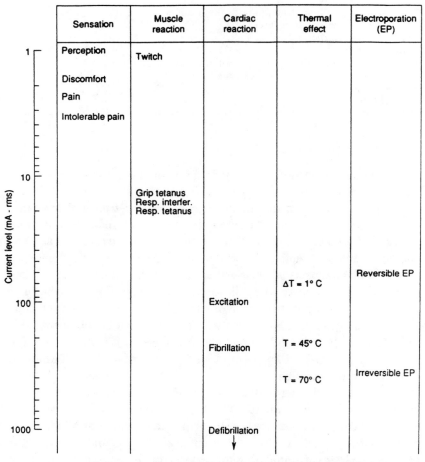

	Sensation	Muscle reaction	Cardiac reaction	Thermal effect	Electroporation (EP)
1	Perception	Twitch			
	Discomfort				
	Pain				
	Intolerable pain				
10		Grip tetanus Resp. interfer. Resp. tetanus			
				ΔT = 1° C	Reversible EP
100			Excitation		
			Fibrillation	T = 45° C	
				T = 70° C	Irreversible EP
1000			Defibrillation ↓		

Current level (mA - rms)

Figure 1.2 Potential short-term reactions to 60-Hz current. Assumed conditions: 5-s exposure; hand grip of large electrode; return electrode at feet; large adult subject; median response.

whose vertical placement indicates an approximate median value of the electrical threshold, which is read on the left-hand scale. These thresholds are only approximate, and are subject to many variables. The intent of Fig. 1.2 is to provide a rough comparison of thresholds, rather than precise numerical values. Since the indicated values are approximate medians of a statistical distribution, they would not normally be suitable as acceptability criteria unless some safety factor were applied.

The following is a commentary on each category. The commentaries call attention to particular chapters (6–10) where a detailed treatment may be found. Additional chapters (2–5) provide a foundation of bioelectric principles and models that explain the experimental data and related mechanisms described in later chapters.

Sensory Reactions (Chapter 7)

Electrical sensation involves the same pathways that are used to sense our external and internal environment. The dynamic range of electrical thresholds from perception to pain is, however, a small fraction of that for natural stimuli. The dynamic range for 60-Hz current is particularly small. While pain thresholds can be measured with a reasonable degree of repeatability, the measurement of thresholds for "intolerable pain" may vary considerably with the experimental context. The tolerance threshold indicated in Fig. 1.2 uses data described in Chapter 7 to indicate a hypothetical acceptance limit, but does not necessarily represent the most extreme limit of human endurance. Indeed, subjects participating in hand-grip experiments (the "let-go" level in the next column) willingly endured considerably higher levels.

Muscle Reactions (Chapter 8)

The quantum of muscle reaction is called a "twitch." The threshold for the smallest measurable twitch for a hand-grip electrode is expected to be nearly equal to that for sensation. For the conditions represented in Fig. 1.2, the threshold twitch (around 1.0 mA) will probably result from the small musculature of the hand. At much higher levels (around 15 mA), one encounters grip tetanus, where the hand "freezes" to the conductor (the so-called let-go threshold). The muscles responsible for grip tetanus lie much higher up in the forearm. It is probable that the small muscles of the hand would be tetanized at much lower current levels, but their effects on grip would be easily overcome by the finger extensors in the forearm. However, at current levels that cause grip tetanus, electrical stimulation of more powerful flexors located in the forearm would dominate.

Respiratory interference can also result from electrical exposure. Quantitative data supporting particular thresholds are rather sparse. Anecdotal reports gathered during grip tetanus tests indicate that respiratory interference may occur at levels somewhat above the grip tetanus threshold, and respiratory tetanus somewhat above that. The current levels indicated for respiratory effects in Fig. 1.2 are rough estimates based on these reports.

Cardiac Reactions (Chapters 5 and 6)

The cardiac excitation threshold indicated at about 100 mA in Fig. 1.2 applies to an extra beat elicited during the normally relaxed state of the heart. Such excitation is not necessarily life-threatening, but is nevertheless treated as a potentially serious effect. At a higher current (around 240 mA), it is possible to induce ventricular fibrillation, in which uncoordinated contractions of heart muscle preclude pumping. Since the human heart rarely recovers spontaneously from electrically induced fibrillation, death will soon result unless defibrillation equipment can be applied to the victim.

Defibrillation occurs at still higher current levels (several amperes). Here, the fibrillating heart is forced into a uniform state of excitation, after which normal rhythmic activity becomes possible. Consequently, fibrillation is most probable within a band of current. Below the lower limit of the band, the current is too feeble to excite the heart; above the upper limit, the heart is defibrillated. This would explain why electrical fatalities due to cardiac failure occur most often at low voltages, whereas high-voltage injuries are principally from burns (see Chapter 10).

Thermal Reactions (Chapter 10)

The greatest temperature rise due to the current from a gripped conductor would probably occur in the high-current-density region of the wrist, or possibly in the hand at the edges of the conductor. With a 60-Hz stimulus, a 1 °C temperature rise would not be sensed, because the current necessary to produce it would also produce severe pain that would mask any thermal perception.

Tissue heating depends largely on the root-mean-square value of the current, and relatively little on the specific stimulus waveform. Excitation of sensory nerves, on the other hand, is very sensitive to the stimulus waveform. If the waveform were inefficient for electrical stimulation, it would be possible for electrical perception thresholds to exceed thermal perception thresholds. Such a condition occurs if the stimulus frequency of a gripped contact exceeds 10^5 Hz. In that case, a small temperature rise will first be sensed as warmth, without electrical stimulation.

When skin or muscle tissue is heated to about 45°C for prolonged periods, thermal damage can result. At that temperature, cutaneous nociceptors would probably be stimulated, resulting in pain (the body's natural defense against heat injury). Thermal damage, however, would not likely occur at 45°C for a duration as short as 5 s. A temperature rise of somewhere around 70°C would be needed to sustain permanent heat damage at that duration. Thermal perception for current frequency above 10^5 Hz is indicated at 35 mA for a touched contact (see Fig. 7.12). A large area grip contact will require an increased current to produce the same local temperature rise. Assuming that perception occurs with a temperature rise of 1°C, the current for higher temperature rises can be determined by using the relationship that the current is proportional to $\sqrt{\Delta T}$ [see Eq. (10.3)].

Electroporation (Chapter 10)

Current flowing within a biological medium will create potential differences across a cellular membrane. This effect is strongest in elongated cells (e.g., nerve and muscle cells) that are oriented in a direction parallel to the current flow. At the lowest levels on the scale in Fig. 1.2, the alteration of membrane potential will excite nerves and result in both sensory and muscular reactions. At much higher levels, the intense electric field that develops across the cellular membrane will promote the formation of pores that interfere with the cell's normal function. This process of electroporation (EP) is reversible at formative levels of current density but becomes irreversible at higher levels, leading to cellular death.

All cells normally maintain a natural "resting" potential, which for muscle cells is around 90 mV, with the inside negative relative to the outside (see Chapter 3). In response to an external current flowing parallel to the long axis of the cell, the membrane will be hyperpolarized at the cathode-facing end and depolarized at its anode-facing end. The thresholds indicated in Fig. 1.2 are predicated on a membrane voltage of 200 mV for reversible EP and 800 mV for irreversible EP (see Chapter 10). These thresholds correspond to hyperpolarization of 110 mV for reversible EP and 710 mV for irreversible EP in the hypothetical muscle cell used in this example. The corresponding current levels indicated in Fig. 1.2 are with reference to the maximum current density (and cellular polarization) occurring in the wrist. The assignment of electroporation thresholds in Fig. 1.2 is only approximate, as a number of simplifying assumptions have been made in their derivation.

1.4   Variables Affecting Thresholds

The thresholds indicated in Fig. 1.2 are illustrative values for a particular set of conditions. Many variables strongly affect these thresholds. These variables can be grouped into categories associated with the stimulus waveform, the electrode interface, and the subject. The importance of these variables can best be expressed by way of example.

Consider the electric shock that can result from the electric fields produced near high-voltage power transmission lines (see Chapter 9). If a conducting object in the field is insulated from ground, it will have a voltage induced on it through the action of the field. If a person touches that object, there will be a transfer of charge between object and person, which can result in a perceptible shock. In a DC field, the charge transfer will occur with a single capacitive discharge; if the initial voltage is sufficiently high, the discharge occurs through a spark. If a person touches a conducting object that is energized by an AC electric field, there can exist a series of spark discharges, followed by steady sinusoidal current when direct contact is made with the object.

To illustrate the effects of sensitivity variables, consider that a single discharge from a 100-pF capacitor is conveyed to a small electrode held against the finger of an adult subject. A reasonable example value for the median threshold of perception in this case would be around 0.1 μC of charge, requiring an initial voltage on the capacitor of 1000 V (see Fig. 7.4). The average threshold of annoyance for this case is predicted to occur at a charge of about 0.23 μC; the threshold of pain is predicted at 0.35 μC (see Table 7.2).

The thresholds cited above are nominal example values that can be significantly affected by many variables. Table 1.3 illustrates variables that were studied by the author and co-workers in connection with electric shock induced by DC or AC electric fields (see Chapter 7). Variables are organized by those associated with the stimulus, with the electrode delivery or interface, and with subjective factors. Significant parameters for DC or AC electric field applications are noted by an "x." Where parameters have relatively minor importance, an "o" is indicated. One column in the table indicates the range of variables studied in the author's research. The last column lists the approximate range of threshold charge that is associated with the parameter variation, assuming that all other parameters are held fixed. Item A.1, for example, indicates that when discharge capacitance varies from 100 to 6400 pF, the threshold of perception, measured in units of charge, can vary by a factor of 4 under appropriate experimental conditions. The "x" in the DC and AC columns indicates that discharge capacitance can affect thresholds of induced electric shock in either DC

Table 1.3. *Variables for sensory effects, electric field-induced shock*

Variables	DC field	AC field	Range studied	Possible range of threshold charge	Notes
A. *Stimulus related*					
1. Capacitance	x	x	100–6400 pF	6:1	
2. Discharge time constant	x	x	0.5–1000 μs	8:1	
3. Polarity	x	o	+/–	1.25:1	
4. Source leakage conductance	o	x	$k = \frac{1}{4}$ to 1	2:1	
5. Stimulus repetition number	o	x	1–16	2:1	*a, b*
6. Waveform oscillation	x	x	$f = 80$ kHz	1.4:1	*c*
B. *Electrode delivery/interface*					
1. Force of touch	x	x	0–50 dB	4:1	
2. Speed of approach	o	x	2–16 mm/s	7:1	
3. Electrode size					
Tapped electrode	o	x	1–34 mm diam.	1.5:1	*d*
Contact electrode	x	x	0.2–40 mm diam.	10:1	*d*
4. Electrode impedance	x	x	0–2 MΩ	3:1	*e*
5. Skin hydration	x	x	Dry/elect. paste	4:1	
C. *Subject*					
1. Body location of electrodes	x	x	7 loci	4:1	
2. Skin temperature	x	x	10–41°C	2:1	
3. Body size	x	x	41–127 kg	2:1	*f*
4. Population variance	x	x	124 subjects	3:1	*g*
5. Preexisting pathology	x	x	Not studied	—	*h*

Notes: x = relevant parameter; o = not relevant or not applicable.
[a]Stimulus repetition number applies to number of half-cycles of 60-Hz supporting discharges (see Fig. 9.5).
[b]Threshold range applies to suprathreshold response.
[c]Waveform oscillation with capacitor discharges can occur when discharge path includes significant inductance. Threshold variation will depend strongly on frequency of oscillation.
[d]Electrode size is important principally when discharge is to an electrode previously in contact with the skin.
[e]Electrode impedance affects discharge time constant.
[f]Body-size relationship based on regression formula (Eq. 7.7).
[g]Population variance includes unidentified factors in addition to body-size effect.
[h]Preexisting pathology is important, but not quantified in the author's research.
Source: Adapted from Reilly and Larkin (1985a).

or AC electric fields. The polarity of discharge (item A.3) can affect thresholds by a factor of 1.25, but only in a DC field environment.

Table 1.3 demonstrates that the potential range of thresholds for electrical sensory effects is large. A large range of parametric sensitivity would also apply to the other categories in Fig. 1.2 Clearly, a single number cannot represent an electrical threshold, unless many other qualifications are also stated.

A significant part of the parametric sensitivity noted in Table 1.3 can be understood in the light of basic principles of bioelectricity. These basic laws are treated in the initial chapters of this book. Chapter 2 covers impedance and internal current distribution. In Chapter 2 the biological subject is considered for its passive electrical properties. Chapter 3 develops the principles that govern the electrical response of nerve and muscle. In Chapter 4 these principles are extended to computational models that allow one to study the excitatory effects of electrical stimulation. An "electrical cable" model is developed to help explain how electrical excitation is related to the temporal and spatial aspects of the stimulus. The electrical properties of the heart are developed in Chapter 5.

The excitation model of Chapter 4 has been extensively referred to in this book when discussing electrical excitation of the heart (Chapter 6), of sensory processes (Chapter 7), and of muscle (Chapter 8). The model is also used to derive excitation thresholds pertaining to electric and magnetic field exposures (Chapter 9).

Considering the extensive number of parameters that affect electrical sensitivity, one might wonder whether it is feasible to define "safe" or "acceptable" exposure levels. The subject of safety criteria is treated in Chapter 11, where performance criteria and electrical safeguards in consumer products are discussed. The approach for setting electrical safety standards described in Chapter 11 is to make conservative assumptions regarding parameters, such as body size, impedance, stimulus waveform, and statistical threshold variations. While this approach does not necessarily protect against the most extreme combination of sensitivity factors, these standards are intended to provide an acceptable margin of safety.

In selecting criteria for the acceptability of electrical exposure, we are confronted with questions that cannot always be answered with scientific objectivity. Frequently there are policy or judgment issues that are settled on the basis of historical precedent, or on some other basis. One example of a judgment issue is the selection of the population percentile that should be assumed for a particular threshold reaction. Should a median sensitivity for a large segment of the population be used because it is "representative" in some sense? Should a lower percentile be selected to minimize the probability of an unacceptable exposure? If a safety factor is to be applied, on what basis should it be selected? The object of this book is not to define the "correct" criteria for the acceptability of electrical exposure. Rather, we hope to present the scientific data that will aid in the process of criteria selection.

The material of this book is the short-term human reactions to electrical stimulation, with particular attention to those of a detrimental nature. In this context, the term *electropathology* is meant to include those human reactions that might be considered undesirable in some context. It is necessary to understand the full range of reactions, from just-noticeable reactions to clearly undesirable or life-threatening ones. The intent of this book is to cover fundamental principles by which such responses may be understood.

2

Impedance and Current Distribution

This chapter examines a number of issues concerning body impedance as it affects the evaluation of electrical stimulation. We are concerned generally with current levels exceeding 100 µA, and voltage levels exceeding 1 V. As a result, many of the special problems associated with characterizing the body's response to microampere currents or millivolt potentials will not be relevant. For a thorough discussion of impedance in biological measurements at very low voltages and currents, the reader is directed to the work of Geddes (1972). The reader is also directed to Chapter 11, which discusses impedance models for use in product safety analysis.

2.1   Dielectric Properties of Biological Materials

The bulk impedance properties of biological materials are important in many applied problems of electrical stimulation. They dictate the current densities and pathways that result from an applied stimulus. In order to appreciate these bulk properties, it is helpful to discuss dielectric properties in a more general context.

Conductivity and Permittivity

Figure 2.1*a* illustrates a simple measurement of the electrical resistivity (ρ) or conductivity ($\sigma = 1/\rho$) of a substance. Two electrodes of area A contact a cylindrical block of biological material having length d. The resistance between the electrodes, given by the ratio V/I, will be directly proportional to the length of the material and inversely proportional to the area of the electrodes and the material's resistivity in accordance with

$$R = \frac{\rho d}{A} = \frac{d}{\sigma A} \tag{2.1}$$

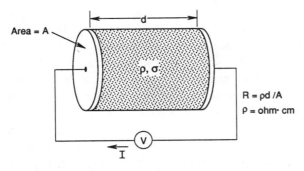

(a) Resistivity for volume material

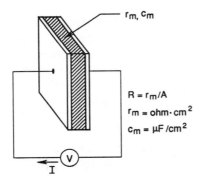

(b) Area proportional resistance of membrane

Figure 2.1   Electrical resistivity in volume and area materials: (*a*) resistivity for volume material; (*b*) area-proportional resistance of membrane.

Inversion of Eq. (2.1) yields a simple definition of the material's resistivity or conductivity:

$$\rho = \frac{1}{\sigma} = \frac{RA}{d} \tag{2.2}$$

While the illustration in Fig. 2.1*a* indicates a simple conceptual method for measuring resistivity, a more accurate and practical method would use four electrodes—two to supply a current within the medium, and two intermediate electrodes to sample the voltage drop (Ruch et al., 1963). Resistivity is ordinarily cited in units of Ωm or Ωcm; the relationship between the two units of measurements is $\rho(\Omega cm) = 100\rho(\Omega m)$. The inverse of resistivity in Ωm is conductivity, expressed in units of siemens per meter (S/m).[†]

[†]Formerly called the "mho," the siemen is now the accepted international unit of conductance.

While the simple concept of conductivity expressed in Eq. (2.1) is adequate for many calculations, a more complete description of the dielectric properties of a material is often needed. The dielectric characteristics of a material are described in complex notation and include both conductive and capacitive properties. The concept of "polarizability" will help to explain the relationship among the variables, and the remarkable electrical properties of biological materials.

Figure 2.2 illustrates a nonconductive material held between a pair of parallel-plate electrodes on which a potential of V volts is applied. The amount of charge accumulated on the electrodes is directly proportional to the product of applied voltage and capacitance in accordance with

$$Q = CV \tag{2.3}$$

where Q is charge in coulombs, C is capacitance in farads, and V is the potential in volts. Assuming that the plate separation, d, is much smaller than its linear dimensions, the capacitance of the parallel-plate arrangement is

$$C = \frac{\varepsilon_0 \varepsilon_r A}{d} \tag{2.4}$$

where ε_0 is the dielectric constant of free space (8.85×10^{-12} F/m), and ε_r is the permittivity of the material relative to that of free space (expressed in dimensionless units, with $\varepsilon_r \geqslant 1$). The relative permittivity of air is very nearly equal to unity.

The relative permittivity constant, ε_r, is a measure of a material's ability to become polarized in response to an applied electric field. In the parallel-plate example, the electric field is simply given by $E = V/d$. The hypothetical material indicated in Fig 2.2 is assumed to be nonconductive; that is, it lacks free electrons or ions. Consequently, there is no net

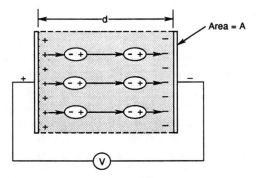

Figure 2.2   Dielectric property. Dipoles within material enhance internal field as indicated by flux lines (arrows).

current flow in response to the applied field. Nevertheless, the material is assumed to contain units of separated charge that are bound together into electrically neutral entities called *dipoles*. The displaced charge centers represent attractive forces within the dielectric medium, enhancing the internal electric field as indicated by the arrows in Fig. 2.2. Analogously, the dipoles act as if they were reducing the plate separation to an effective value of d/ε_r, and thereby increasing the capacitance.

The capacitive current is given by the change of charge versus time, dQ/dt. For a time-varying applied voltage, it follows from Eq. (2.3) that

$$I = C \frac{dV}{dt} \tag{2.5}$$

For a sinusoidal voltage, $dV/dt = j\omega V$, where $\omega = 2\pi f$, f is the frequency of oscillation, and j is the phaser operator indicating $90°$ phase shift. The capacitive current can be expressed as

$$I = j\omega\varepsilon_0\varepsilon_r \frac{AV}{d} \tag{2.6}$$

When the material contains both dipoles and free charges, its dielectric description requires complex notation. Figure 2.3 illustrates an equivalent circuit of a partially conductive material. The total current is the sum of the resistive and capacitive components $I = I_R + I_C$, which is given by

$$I = \frac{V}{R} + C \frac{dV}{dt} \tag{2.7}$$

For a sinusoidal voltage, Eq. (2.7) can be expressed as

$$I = \frac{V\sigma A}{d} + V \left(\frac{\varepsilon_0\varepsilon_r A}{d} \right) j\omega$$

or, equivalently

$$I = \frac{VA}{d} (\sigma + j\omega\varepsilon_0\varepsilon_r) \tag{2.8}$$

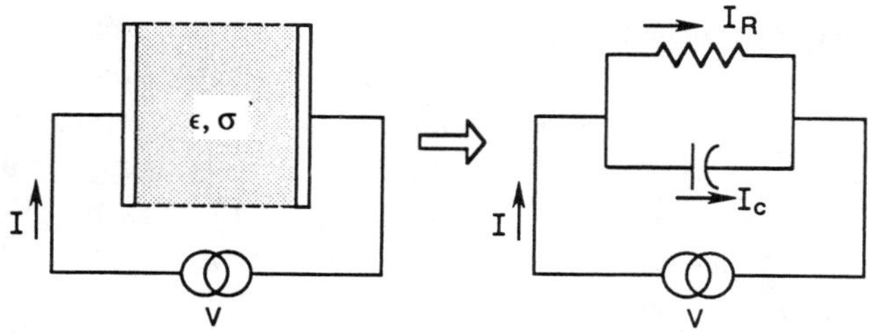

Figure 2.3 Complex permittivity. A partially conductive dielectric material is subjected to an alternating voltage V, resulting in resistive and capacitive current.

The macroscopic electric field is $E = V/d$, and the current density is $J = I/A$. Consequently, Eq. (2.8) can be written as

$$J = \sigma^* E$$

where σ^* is the complex conductivity given by

$$\sigma^* = \sigma + j\omega\varepsilon_0\varepsilon_r \tag{2.9}$$

Alternatively, Eq. (2.8) can be written as

$$I = \frac{j\omega\varepsilon_0 A}{d}\left(\varepsilon_r - \frac{j\sigma}{\omega\varepsilon_0}\right)V \tag{2.10}$$

By analogy with Eq. (2.6), the term in parentheses is the complex permittivity constant, given by

$$\varepsilon^* = \varepsilon_r - \frac{j\sigma}{\omega\varepsilon_0} \tag{2.11}$$

or, in conventional notation,

$$\varepsilon^* = \varepsilon' - j\varepsilon'' \tag{2.12}$$

in which $\varepsilon' = \varepsilon_r$, and $\varepsilon'' = \sigma/(\omega\varepsilon_0)$. Equations (2.8) and (2.11) are equivalent descriptions of the material. The form indicated by Eq. (2.11) is more often used; its symbolic representation is given as in Eq. (2.12) and is called the *complex permittivity*. Its imaginary part includes the conventional conductivity, and the real part is the relative dielectric constant.

In the above development, it is assumed that the dipoles orient themselves in response to the internal alternating field, requiring that they reverse their orientation every $\frac{1}{2}$ cycle of the applied voltage. The dipoles have a certain degree of inertia, and cannot follow the field oscillation if it is sufficiently rapid. We therefore expect that ε_r would be a maximum at low frequencies, and drop when the frequency is raised above some critical value. In the limit with very high frequencies, the dipoles would retain random orientation, and the relative permittivity of the material would approach unity.

The ability of a dielectric material to respond to an applied field can be expressed in terms of its *relaxation time constant* τ_r, or, equivalently, in terms of the relaxation frequency $\omega_r = 2\pi f_r = 1/\tau_r$. In a typical biological medium there may exist several mechanisms for producing dipoles, each having a different relaxation time constant. In addition, the presence of boundaries between regions of differing permittivity will result in an equivalent relaxation time constant due to the buildup of charges at those boundaries (referred to as "interfacial" effects).

Figure 2.4 illustrates an example of the relative permittivity of a typical biological material (Pethig, 1979; Foster and Schwan, 1986). The dips in the curve arise from different mechanisms of polarization and are termed *dielectric dispersion.* At the lowest frequency, the so-called α dispersion has been attributed to electronic bilayers in organic molecules, ionic dispersion processes in micrometer-sized particles, active membrane conductance phenomena, and other membrane effects. The β dispersion is attributed to capacitive charging of cellular membranes (interfacial effects), and dipolar relaxation of proteins. The γ dispersion represents dielectric relaxation of water molecules.

Table 2.1 lists experimental values of conductivity (Part A) and relative permittivity (Part B) for various biological material. The tabulated values are geometric means of data published by Foster and Schwann (1986). Useful summaries have also been published by others (Geddes and Baker, 1967; Schwann, 1968; Stuchly and Stuchly, 1980; Stoy et al., 1982).

The relative permittivity of biomaterials at low frequencies can be on the order of 10^6. These values are remarkably large in comparison with synthetic dielectric materials, for which relative permittivities of 5–10 are typical. Despite the appearance of such large permittivity, the biological material remains overwhelmingly resistive, as can be seen from the following example. Assume the following values: $\sigma = 0.1 \, \text{S/m}$, $\varepsilon_r = 10^6$, and f = 100 Hz. In accordance with Eq. (2.9), the complex conductivity is $\sigma^* = 0.1 + j\,5.6 \times 10^{-3}$. In this example, the capacitive component of current will be only 5.6% of the resistive component. In general, the bulk properties of biological tissue are dominantly resistive.

The conductivity of some biological preparations is markedly aniso-tropic. Skeletal muscle, for example, is shown in Table 2.1 to be about

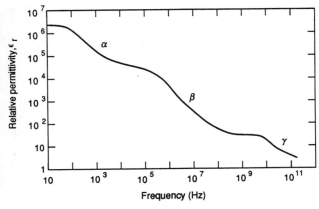

Figure 2.4   Frequency variation of dielectric permittivity typical of soft tissue.

Table 2.1. *Dielectric properties of biological materials. Part A: Conductivity (values given in S/m)*

Frequency (Hz)	Skeletal muscle Parallel	Skeletal muscle Perpend.	Skeletal muscle Nonspecific	Liver	Lung	Spleen	Kidney	Brain white matter	Brain gray matter	Bone	Whole blood	Fat
10	0.52	0.076	—	0.12	0.089	—	—	—	—	—	—	—
10^2	0.52	0.076	—	0.13	0.092	—	—	—	—	0.013	0.60	—
10^3	0.52	0.08	—	0.13	0.096	—	—	—	—	0.013	0.68	0.04
10^4	0.55	0.085	—	0.15	0.11	—	—	—	—	0.013	0.68	—
10^5	0.65	0.4	0.48	0.16	—	0.62	0.24	0.13	0.17	0.014	0.61	—
10^6	—	—	0.71	0.20	—	0.63	0.38	0.16	0.21	0.017	0.71	—
10^7	—	—	0.87	0.46	—	0.63	0.59	0.28	0.52	0.024	1.11	—
10^8	—	—	0.85	0.65	0.53	0.83	0.80	0.48	0.68	0.057	0.82	0.04
10^9	—	—	1.41	1.03	0.73	1.31	0.97	0.85	1.05	0.05	1.43	0.05
10^{10}	—	—	8.23	7.3	—	8.1	5.8	8.0	10.0	0.92	9.8	0.35

Part B: Relativity permittivity

Frequency (Hz)	Skeletal muscle Parallel	Skeletal muscle Perpend.	Skeletal muscle Nonspecific	Liver	Lung	Spleen	Kidney	Brain white matter	Brain gray matter	Bone	Whole blood	Fat
10	10^7	10^6	—	5×10^7	2.5×10^7	—	—	—	—	—	—	—
10^2	1.1×10^6	3.2×10^5	—	8.5×10^5	4.5×10^5	—	—	—	—	3800	2900	1.5×10^5
10^3	2.2×10^5	1.2×10^5	—	1.3×10^5	8.5×10^4	—	—	—	—	1000	2810	5×10^4
10^4	8×10^4	7×10^4	—	5.5×10^4	2.5×10^4	—	—	—	—	640	3300	2×10^4
10^5	1.5×10^4	3×10^4	2×10^4	1.2×10^4	—	3260	1.2×10^4	2500	3800	280	2040	—
10^6	—	—	2200	1970	—	1450	2540	670	1250	87	200	—
10^7	—	—	184	232	—	357	294	190	309	37	71	—
10^8	—	—	68	72	35	78	76	62	81	23	63	—
10^9	—	—	55	49	35	51	44	39	46	8	45	5
10^{10}	—	—	36	37	—	38	33	25	40	—	—	—

Source: Table adapted from Foster and Schwan (1986). Reprinted with permission from *CRC Handbook of Biological Effects of Electro-*

6–7 times more conductive for a current orientation parallel to the muscle fibers, as compared with a perpendicular orientation. The degree of anisotropicity of animal muscle tissue appears to vary greatly with the tested species (Chilbert et al., 1983). Anisotropic conductivity ratios ranging from about 5:1 to 10:1 apply to cardiac tissue, depending on the method of measurement (Plonsey and Barr, 1986). An anisotropic ratio of about 10:1 has been attributed to nerve bundles (Nicholson, 1965; Ranck, 1963). For a critique of anisotropic impedance measurements in cardiac and other muscle tissue, the reviews of Plonsey and Barr (1986) and Roth (1989) are recommended.

Cellular Membranes

The previous discussion has dealt with the bulk dielectric properties of composite biological materials. Characteristics of microscopic biological components, such as the cellular membrane, may also be important in applied studies of electrical stimulation. The cellular membrane consists of a bimolecular lipid structure whose impedance properties are usually expressed as area-proportional quantities—unit area resistance in $\Omega \, cm^2$, and capacitance in $\mu F/cm^2$ as in Fig. 2.1b. These units stand in contrast to the bulk resistivity units ($\Omega \, cm$) indicated in Fig. 2.1a for composite materials. The reason is that the thickness of the biological membrane cannot be subdivided without altering its basic structure, and is considered a "given." The capacity of a cellular membrane generally is in the region 0.5–$1 \, \mu F/cm^2$, resistivity in the range 10^2–$10^4 \, \Omega \, cm^2$, and relative permittivity around 2.5 (Pethig, 1979). The ionic permeability is specific to particular ionic species. For excitable membranes (nerve and muscle tissue), ionic permeability is highly dependent on the transmembrane voltage, as discussed in Chapter 3.

Skin Depth

The penetration depth of incident electromagnetic energy is often described in terms of the material's "skin depth." This concept can be understood in relationship to the penetration of an incident magnetic field. As described in Chapter 10, an incident magnetic field will set up eddy currents in a conducting material. The eddy currents, in turn, create their own magnetic field, which tends to oppose the incident field and resist its penetration into the material. Consequently, the current induced within the material will drop off in an exponential fashion from the surface. The distance at which the current density falls to e^{-1} of its surface value is known as the *skin depth*.

The skin depth of a material of arbitrary conductivity was described in 1888 by Oliver Heaviside (Nahin, 1987):

$$\delta = \frac{1}{2\pi f\{(\mu\varepsilon/2)[\sqrt{1 + (\sigma/2\pi f\varepsilon)^2} - 1]\}^{1/2}} \tag{2.13}$$

where $\mu = \mu_0\mu_r$ is the magnetic permeability of the material, $\varepsilon = \varepsilon_0\varepsilon_r$ is its dielectric permittivity, and f is the frequency of the induced current. The magnetic permeability of free space or air is $\mu_0 = 4\pi \times 10^{-7}$ H/m; for all practical purposes, the permeability of biological materials is that of free space, that is, $\mu_r = 1$. For a good conductor, $\sigma/(2\pi f\varepsilon) \gg 1$, and Eq. (2.13) reduces to

$$\delta = (\pi f\mu\sigma)^{-1/2} \tag{2.14}$$

which is the more familiar expression for the skin depth as expressed in most engineering texts.

The skin depth of biological materials for frequencies below about 10 MHz is generally much greater than any practically attainable material thickness. This can be seen by way of example. Assume that $\mu_r = 1$, $\varepsilon_r = 200$, $\sigma = 0.1$ S/m, $f = 10$ MHz. Then, from Eq. (2.13), $\delta = 0.8$ m. It is only when the frequency is well above 10 MHz that skin depth becomes a significant consideration in most cases. One interpretation of this result is that, for frequencies below 10 MHz, a magnetic field will pass readily through biological material, and the internal magnetic field differs negligibly from the external field.

2.2   Skin Impedance

In most situations involving electrical stimulation, current is introduced through metallic electrode contact with the skin. The total circuit impedance will include contributions from the source, the electrode interface, the skin, and the internal tissues of the body. Of these, skin impedance is the most difficult to characterize. It is nonlinear, time-variable, and depends on environmental and physiological factors that are usually difficult to control in an experimental setting. Yet skin impedance is often the primary factor that limits current flow in the body, particularly where the applied voltage is moderately low (< 200 V) and where the skin is undamaged. It is important to examine the electrical properties of the skin in some detail.

Detailed Structure of Skin[†]

The skin consists of various layers, as indicated in Fig. 2.5. An outer layer (the *epidermis*) overlays an inner dermis. (An additional view

[†]See also Edelberg (1971), Harkness (1971), and Tregear (1966).

of the structure of the skin is shown in Fig. 3.18.) The epidermis consists principally of keratin, derived from dead cells of the lower layers, arranged in a flattened and irregular fashion. The bonding strength of these cells is low. They are constantly flaking away naturally, and can be easily broken or removed. The germinating layer at the boundary of the epidermis contains a mixture of living and dead cells. The dermis contains living cells, and a great density of blood vessels that are related to both nutrition of the skin and to its thermoregulation. The dermis consists of bundles of *collagen* fibrils oriented in all directions, giving it strength and elasticity. The distribution of skin thickness varies greatly for different body areas. The epidermis itself ranges from about 10 to over 100 μm. It is typically 10 times as thick on the palms of the hands as compared with other body areas, and is endowed with a much greater density of sweat glands.

The corneum (the outermost layer of dead skin cells) is a relatively poor conductor when dry. But when wet or sweaty, or when bypassed (as with an injury), the conductivity of the skin can rise dramatically. The contribution of the corneum to the total impedance can be studied by stripping the skin with cellophane tape (Harkness, 1971; Lykken, 1971; Tregear, 1966; Reilly et al., 1982; Clar et al., 1975). Figure 2.6 illustrates the drop of skin resistivity when the corneum is successively stripped away, showing an ultimate drop by a factor over 300. Impedance drop with corneal stripping has also been noted with high-voltage spark discharge stimuli (refer to Sec. 2.5). When a microelectrode penetrates the corneal

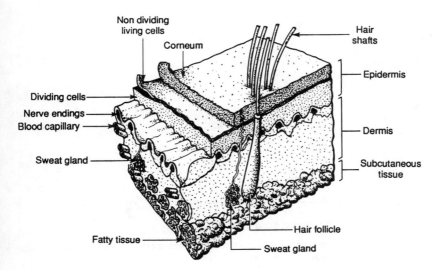

Figure 2.5   Structure of the skin.

layer, the resistance drops suddenly, as noted in the data of Suchi (1954) presented in Table 2.2.

Sweat is chiefly a 0.1–0.4% saline solution of sodium chloride, with a resistivity of $140\,\Omega\,cm$ at 37°C (for a 0.3% solution). The density of sweat ducts varies according to the part of the body being considered. On the insides of the hands and feet, it is approximately $370/cm^2$; on the forearm it is approximately $160/cm^2$. The diameter of the ducts ranges from 5 to $20\,\mu m$.

Current appears to be conducted in dry skin through discrete channels beneath a contact electrode. With a multipoint electrode, current is found

Table 2.2. *Impedance layers in the epidermis*

	Ball of thumb	Forearm
Dry, outer layer thickness (μm)	200	45
Penetration distance for impedance breakdown (μm)	350	50

Source: Suchi (1954)

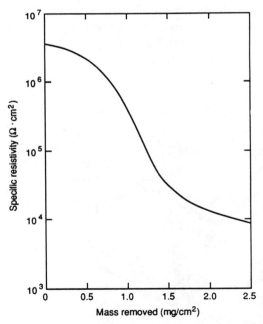

Figure 2.6   Impedance of skin as related to the mass of corneum removed dry skin; sinusoidal current at 1.5 Hz. (Adapted from Tregear, 1966.)

to be preferentially conducted at one point (Mueller et al., 1953). In experiments with electroplating on the skin surface (Saunders, 1974), the pattern of silver deposition on the arm shows that current is conducted in discrete channels at a density of about 1 channel/mm^2. This density approximately corresponds to the density of sweat ducts. The punctate nature of skin conductance was also explored with high-voltage discharges (Reilly et al., 1982). Those investigations are discussed in Sec. 2.5.

The sweat ducts form electrical weak points in the epidermis, acting as conductive tubes into the well-conducting dermis and tissues below. Suchi (1954) explored the impedance role of sweat ducts using a silver microelectrode to scan various parts of the skin. When the electrode touched a duct filled with sweat, the impedance dropped by a factor of 10 as compared with adjacent areas of the skin. If the duct was dry, the drop was approximately a factor of 2.

Equivalent Circuit Models

It would be desirable to represent the impedance of the skin and body as an equivalent circuit model that allows one to determine internal currents for a wide range of exposure conditions. Unfortunately, the impedance of the human body, and especially the skin, is not readily expressible as a simple passive circuit—the skin is a distributed electrical system having markedly nonlinear and time-variant properties complicated by electrolytic interactions at the electrode interface. Nevertheless, impedance models are often valid under a sufficient range of circumstances to be of practical use.

The simplest model that is sometimes used to represent skin impedance is a parallel network consisting of a capacitor and resistor, followed by a series resister (cf. Yamamoto and Yamamoto, 1977; Burton et al., 1974). In this model, the parallel resistor and capacitor represent the resistivity and capacity of the skin, and the series resistance the well-conducting subepidermal medium. Evidence for this simple model can be seen when measuring the current response to a constant-voltage stimulus pulse applied to the skin as illustrated in Fig. 2.7 (from Lykken, 1971). The response, illustrated in Fig. 2.7b, shows an initial current spike that is limited by the series resistance, R_s. Afterward, the current decays to the value limited by $R_p + R_s$. After the corneum has been removed by abrading, the current, illustrated in Fig. 2.7c, is approximately limited by R_s.

A more complete model considers the skin as composed of numerous layers of cells, each having capacitance and conductance, as illustrated in Fig. 2.8a (Edelberg, 1971; Lykken, 1971). The individual strings of elements are meant to represent the parallel paths beneath an electrode. In addition to the resistance and capacitance elements, the electrical model contains DC potential sources to account for the observed bulk potential of the skin of about 15–60 mV, with the surface negative relative to the underlying layers. These potentials are extremely small in comparison with the stimulus potentials typically needed for cutaneous electrical stimulation,

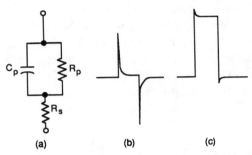

Figure 2.7   Current response of skin to square-wave constant-voltage pulses: (*a*) equivalent circuit; (*b*) response of intact skin; (*c*) response of skin with corneum removed. (Adapted from Lykken, 1971.) Copyright 1971, The Society for Psychophysiological Research. Reprinted with permission of the publisher and the author.

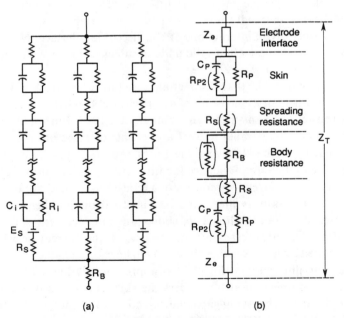

Figure 2.8   More complex impedance models: (*a*) multilayer model for skin impedance; (*b*) simplified body impedance model.

and can be omitted from further consideration. The element R_B is treated as the body impedance, exclusive of the skin.

Figure 2.8b illustrates a model of intermediate complexity. The element Z_e represents the impedance at the electrode interface. In studies of body impedance the term Z_e is seldom determined explicitly, but rather is lumped together into total impedance. The parallel RC circuit represents the epidermis, with the element R_{P2} added in the capacitive branch. It is shown in parentheses to indicate that it is frequently ignored in impedance representations. The subepidermal layer is shown as consisting of a spreading resistance element, R_s, which is inversely related to electrode size, and a body resistance element, R_B, which increases with electrode separation. In most studies of body impedance, R_s and R_B are lumped together into a single resistive term. The element R_B is generally treated as a pure resistance, although in fact it contains a small reactive component, indicated by the components in parentheses. The reactive component can be ignored for most practical applications. The circuit representation includes additional terms to account for the second electrode. The total circuit impedance is designated Z_T. The circuit models discussed here are merely approximations that are sometimes suitable; the individual terms may behave in a more complex fashion than would a laboratory component. Some of these complexities are discussed in the following sections.

Area Proportionality of Skin Admittance

From geometric considerations, it might seem reasonable to suppose that skin admittance is directly proportional to contact area, as long as the electrode diameter is significantly larger than the skin thickness. Many investigators report equivalent circuit parameters as being area-proportional. Lykken (1971), for example, notes an area-proportional admittance for DC currents with electrodes of $0.72 \, cm^2$ and larger, and for pulse stimuli with electrodes $2.4 \, cm^2$ and larger. Tregear (1966) reports that the product of impedance and area is constant for wet skin only for electrodes having areas larger than $2 \, cm^2$. For smaller areas, the impedance departs considerably from inverse area proportionality, with significant differences between wet and dry skin. Biegelmeier and Rotter (1971) evaluate equivalent circuit parameters for electrodes of 1.5 and $100 \, cm^2$ (an area difference of 67:1); they note a 4:1 change in $R_{P2} + R_S$ as defined in Fig. 2.8b, and a 25:1 change in R_p, showing that admittance rises more slowly than does contact area.

The area-dependent portion of the resistive path (R_s in Fig. 2.8b) comprises the region where the current still undergoes spherical spreading from the stimulating electrode. At the point where current no longer spreads out with distance, the resistance is no longer dependent on electrode area, but is more a function of electrode separation and body geometry. Thus, R_B in Fig. 2.8*b* is taken to be independent of electrode contact area.

Area-proportional parameters are usually determined by dividing admittance by electrode contact area. This simple calculation is confounded by the fact that the distribution of current beneath a contact electrode may be very nonuniform, as demonstrated in theoretical (Caruso et al., 1979; Rattay, 1988) and experimental (Lane and Zebo, 1967) studies. Figure 2.9 illustrates the theoretical current density for a three-layer conductivity model representing a layer of skin, fat, and muscle tissue; a two-layer model consisting of fat and muscle; and a single-layer model consisting of muscle tissue (Caruso et al., 1979). For the three-layer model, the calculated current density near the electrode edge is nearly a factor of 10 greater than that at the center. For the single-layer model, the current density from center to edge differs by a factor of about 2.2. The issue of area conductivity is further complicated by the fact that current travels

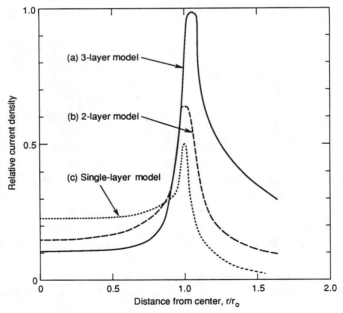

Figure 2.9   Current density beneath contact electrode; r_0 = electrode radius; vertical axis on dimensionless scale. (*a*) Three layer model with skin, fat, and muscle; (*b*) two layer with fat and muscle; (*c*) single layer with muscle. (From Caruso et al., 1979.)

Table 2.3. *Low-frequency resistivity of human and animal skin*

Species	Hydration	Body location	Frequency (Hz)	Impedance ($k\Omega$ cm^2)
Human	Dry	Arm	0	100–1000
Human	Dry	Arm	1.5–10	600–1200
Human	Dry	Fingertip	0–1	120–130
Human	Dry	Palm	30–65	60–80
Human	Wet	Forearm	1.5	880
Pig	Wet	Flank	1.5	12
Rabbit	Wet	Flank	1.5	18

Source: Adapted from Tregear (1966); data compiled from a variety of sources.

laterally beyond the confines of the electrode, such that the effective contact area may be larger than the physical contact area.

With these caveats in mind, Table 2.3 lists some of the area-proportional values reported for low-frequency (0- to 65-Hz) skin impedance (Tregear, 1966). Because·of the low frequency, the list applies primarily to the resistive component $R_P + R_S$ defined in Figs. 2.7a and 2.8b. The values are a compilation from a variety of sources, and do not necessarily represent data measured in a consistent manner.

Skin Capacitance

A variety of testing methods indicate that the skin's capacity lies in the range 0.02 to 0.06 $\mu F/cm^2$ (Edelberg, 1971); by conventional calculations, this is considered very high. Consider, for example, a corneum thickness of 10 μm, and a dielectric constant of 2.5 for biological membranes. With these values, the capacity is calculated to be about 10^{-4} $\mu F/cm^2$—a small fraction of the experimental values.

Skin capacitance will be affected by *polarization capacitance*—the phenomenon of stored charges that appear around an electrode in an electrolytic medium, forming an effective ionic capacitor that is dependent on excitation frequency (Schwan, 1966). Lykken (1971) argued that the apparent frequency-dependent property of skin capacitance is simply a consequence of the choice of equivalent circuit, and that a representation with a resistively coupled capacitor (C_P and R_{P2} in Figure 2.8b will demonstrate a fixed value versus frequency.

Much of the skin's capacity lies in the corneum. If the corneum is stripped away, the skin capacitance is reduced with each successive stripping operation (Edelberg, 1971). When the corneum is removed entirely, the capacity drops to a small fraction of its intact value (Lykken, 1971; van

Boxtel, 1977). These observations are counter to a model in which the corneum is simply a dielectric separating the electrode and the underlying conductive dermis. If that model were correct, we would expect to see an increase of capacitance as the thickness of the corneum is reduced. Biegelmeier and Miksch (1980) postulate that the skin's capacity is derived from membrane capacitance of the sweat gland duct. However, this explanation does not account for the leading-edge current spikes that are observed when constant-voltage pulses are applied to the sweat gland-free skin of Rheus monkeys (Bridges, 1985).

An alternative explanation for the skin's high capacity was provided by Tregear (1966), who treated the skin capacitance as being due to individual cell membranes as in Fig. 2.8a. If we assume that each cell's membrane may have a capacity as large as $5 \, \mu F/cm^2$, and acknowledging that each cell accounts for two membrane layers, then 200 cell layers could account for a capacitance of $0.05 \, \mu F/cm^2$. A related mechanism that might account for skin capacitance has been described as an ionic bilayer surrounding individual corneal cells (Clar et al., 1975).

Time-Variant and Nonlinear Aspects of Skin Impedance

When an electrode is placed on dry skin, impedance gradually falls with time, as noted in Fig. 2.10 (Mason and Mackay, 1976). During the first 2 min, the impedance undergoes rapid fluctuations superimposed on a relatively sharp overall drop (see insert). Thereafter, the impedance

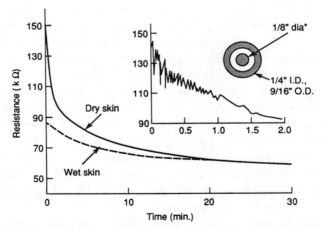

Figure 2.10   Variation of skin resistance with time. Concentric electrode as illustrated (*inner diameter not clearly described in source). Wet skin treated with tap water. Inset shows expanded scale for first 2 min. Stimulus current: 1 mA. (Adapted from Mason and Mackay, 1976, © 1976 IEEE.)

continues to drop, reaching an apparent asymptotic value after some 20 or 30 min. This effect may be explained by gradual hydration of the corneum from sweat buildup beneath the electrode. In the same experiments, skin pretreated with conductive electrode paste started with an initially lower impedance, rose somewhat, and after about 1 min, attained a final value only slightly below its initial value.

Past studies have shown that the electrical properties of the skin are nonlinear. Edelberg (1971) suggests that the linear region for current density lies in the range below $2 \, mA/cm^2$. Lykken (1971) concludes that it is applied voltage, and not current density, that determines the upper limit of linearity, and is around 2 V. Biegelmeier and Rotter (1971) note variations in equivalent circuit values of over 2:1 for voltage changes from 4 to 45 V. Stevens (1963) defines the DC current conducted in dry skin in terms of $I = aV + bV^2$, in which the V^2 term dominates beyond about 3 V.

Nonlinear phenomena in living skin was studied by Grimnes (1983a). He hypothesized that the time scale of impedance changes could be explained by either a nondielectric mechanism or a dielectric breakdown mechanism, depending on the applied voltage in the range 600–1000 V. In this work, the current was limited to a few microamperes, and the results may not necessarily apply directly to higher current densities.

A dramatic display of skin nonlinearity is seen in the sudden breakdown of impedance if the applied voltage is great enough. Dielectric breakdown of dry, excised corneum has been observed by Mason and Mackay (1976) at 600 V for a 15-μm-thick sample, and at 450 V by Yamamoto et al. (1986). Dielectric breakdown of intact skin from high-voltage spark discharges appears at similar voltages, as described in Sec. 2.5.

With intact, living skin, a sudden impedance breakdown has been observed at significantly lower voltages than that needed to break down excised corneum. It is likely that some mechanism other than dielectric breakdown is at work at these lower voltages. Pricking pain has been associated with the impedance breakdown (Nute, 1985; Gibson, 1968; Mason and Mackay, 1976; Mueller et al., 1953). During breakdown, current appears to be concentrated in discrete channels, causing a marked rise in current density in those channels. According to Mason and Mackay, microscopic examination of the electrode site reveals small blackened punctures under a dry-skin electrode, but no such evidence when the skin has been pretreated with water or electrode paste.

It is likely that the major source of skin nonlinearity lies in the corneum. Equivalent circuit values R_P and R_S were evaluated by van Boxtel (1977)

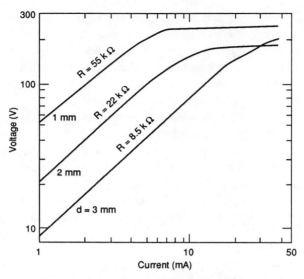

Figure 2.11   Nonlinear voltage/current relationship for 5-μs cathodal pulses applied to abdomen; resistance (*R*) applies to linear regions of curves; *d* = electrode diameter. (Adapted from F. A. Saunders, in *Conference on Cutaneous Communication System's and Devices*, pp. 20–26, reprinted by permission of Psychonomic Society, Inc.)

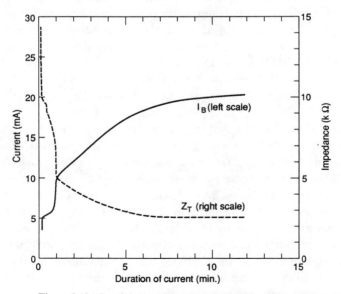

Figure 2.12   Breakdown of the skin at 50 V. Forearm-to-forearm current path; circular electrodes, 12 cm²; dry skin. (From Freiberger, 1934.)

from the transient current response to constant-voltage pulses. With intact skin, R_P varied significantly with the applied voltage, whereas R_S remained independent of the stimulus level (up to a voltage of 40 V, and 40 mA final current). With the corneum removed, both R_P and R_S remained independent of the stimulus level.

Figure 2.11 illustrates nonlinear impedance response for pulses of 5-μs duration (Saunders, 1974). At these short durations, it might be anticipated that the corneal layer is bypassed through capacitive coupling to the dermis. However, the skin exhibits marked nonlinear properties, as noted in the knee of the curves at the higher current and voltage values. The point of marked nonlinear breakdown occurs in the voltage region from about 150 to 250 V.

Freiberger (1934) describes breakdown as occurring at small contact areas within a fraction of the second for voltages above 100 V. Biegelmeier and Miksch (1980) noted breakdown above 200 V for large hand-held electrodes (refer to Sec. 2.3). With small electrodes and voltages above 100 V, breakdown may occur within fractions of a second with dry-skin contact.

Nonlinear response to sinusoidal voltage shows up as distortions from a sinusoidal current waveform. Short-term distortions appear as an instantaneous nonlinear response within individual cycles; longer-term nonlinearities appear as a gradual reduction of impedance. Figure 2.12 illustrates impedance breakdown with 50-Hz AC stimulation on a relatively long time scale (Freiberger, 1934). An initial precipitous drop occurs within a small fraction of a minute, a second drop during the first minute, and a gradual drop during the next seven or eight minutes, finally reaching a stable plateau.

2.3   Total Body Impedance: Low-Frequency and DC

The previous section showed that body impedance depends on a variety of factors, including size and location of contact electrodes, skin hydration, and applied voltage. These factors were studied in detail by Biegelmeier in a series of experiments that also demonstrated the human protection afforded by current interruption devices. Measurements consisted of the peak and root-mean-square (rms) voltage and current, dissipated power, and the phase angle between the voltage and current applied to the subject. The subject grasped cylindrical electrodes of 8 cm diameter, providing a hand-to-hand current path. With these electrodes, the contact area on each hand was estimated to be 82 cm^2. After covering the cylindrical electrodes with insulating material, smaller hand contact

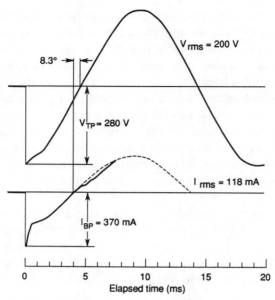

Figure 2.13   Oscillograms of applied voltage (upper trace) and body current (lower trace), with contact at the approximate peak of the voltage waveform; applied voltage = 200 V rms. (From Biegelmeir, 1985a.)

areas consisted of 12.5, 1.0, 0.1, and 0.01 cm^2 on each hand. Foot electrodes consisted of copper foil inserted into the shoes.

Figure 2.13 illustrates example current and voltage waveforms at the onset of a stimulus having a steady-state voltage and current[†] of 200 V and 118 mA, respectively (Biegelmeier, 1985a). Corresponding peak values were 280 V and 370 mA. Total body impedance (Z_T) and internal body resistance (R_B) were determined by

$$Z_T = \frac{V_T}{I_B} \tag{2.15}$$

$$R_B = \frac{V_{TP}}{I_{BP}} \tag{2.16}$$

where V_T and I_B are steady-state (rms) values of applied voltage and body current; V_{TP} and V_{BP} are the peak values of voltage and current. In the example of Fig. 2.13, $Z_T = 1.7 \, \mathrm{k\Omega}$, and $R_B = 0.75 \, \mathrm{k\Omega}$.

Figures 2.14–2.16 illustrate the relationship between impedance and the applied voltage for large-area contacts and different electrode

[†] When referring to sinusoidal voltage and current, cited magnitudes indicate rms values, unless specifically stated otherwise. This convention applies throughout this book.

placements (from Biegelmeier, 1985b). Figure 2.14 applies to hand-to-hand contacts, and indicates Z_T and R_B separately; Z_T is further subdivided into wet and dry contacts. Each measurement point is the average of six procedures with a single subject. The dry-contact data are indicated by filled circles; the error bars show the mean and extreme measurements. In Fig. 2.14, the wet-contact conditions were obtained by soaking the hands in either tap water or a saline solution prior to the measurement. The tap water treatment lowers the impedance somewhat relative to the

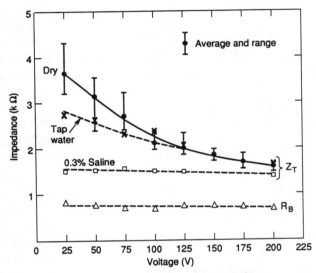

Figure 2.14   Impedance versus voltage for large-area hand-to-hand contacts. (From Biegelmeier, 1985b.)

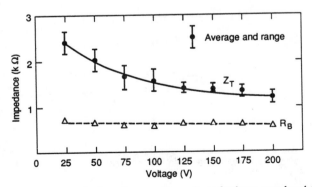

Figure 2.15   Impedance versus voltage for large-area hand-to-feet contacts. (From Biegelmeier, 1985b.)

dry condition; treatment with saline lowers the impedance much more. At 200 V there is little difference between wet and dry electrodes.

Figures 2.14 and 2.15 also show internal body resistance, R_B, for the indicated paths. The value of R_B drops only slightly as the applied voltage is raised from 25 to 200 V, dropping to 650 Ω for the hand-to-hand path, and to 550 Ω for the hand-to-foot path. These values are reasonably close to the value of approximately 500 Ω determined for R_B from high-frequency measurements (refer to Sec. 2.4), and also with high-voltage capacitive discharges (refer to Sec. 2.5).

The cited values of R_B are of the same order as the resistance of a solution of NaCl of physiological concentration, and of geometric dimensions similar to that of the body (Biegelmeier, 1986). If the resistivity of the NaCl solution is taken to be 80 Ω cm, Eq. (2.1) indicates that the resistance of a cylindrical cross-sectional area of 25 cm^2 and of length 150 cm (for two arms in series) is approximately 480 Ω. This simple calculation correlates quite well with measured values of R_B.

Figure 2.17 illustrates further results of the Biegelmeier (1985a) experiments showing the relationship between total impedance and applied voltage for various contact areas. At an applied voltage of 25 V, impedance is nearly inversely proportional to contact area. As the voltage is increased, impedance drops rapidly, and area dependence is markedly reduced. At 225 V, total impedance varies by only about 4:1 for a change in electrode contact area of 8200:1.

Impedance-versus-voltage characteristics depend very much on the body locus of skin contact. With tests on cadavers, Freiberger found impedance of dry contacts to drop precipitously for voltages above 200 V on the palmar and plantar surfaces of the hands and feet. On the forearm, the point of impedance breakdown was about 50 V.

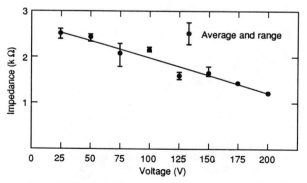

Figure 2.16   Impedance versus voltage for large-area foot-to-foot contacts. (From Biegelmeier, 1985b.)

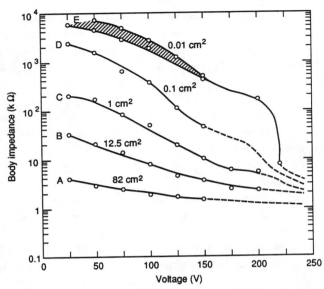

Figure 2.17   Total body impedance as a function of applied voltage for various contact areas; dry hand-to-hand electrodes. The various curves are identified by the contact area on each hand. (From Biegelmeier, 1985b.)

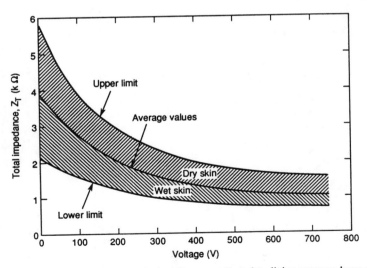

Figure 2.18   Total body impedance attributed to living persons; large-area hand-to-hand or hand-to-foot contacts. Measurements above 50 V conducted on cadavers, and corrected for living persons. (From Freiberger, 1934.)

Figure 2.18 illustrates measurements of Freiberger (1934) showing the relationship between large-area hand-to-hand or hand-to-foot contacts at 50 Hz. The illustrated values are reasonably consistent with those of Biegelmeier (Fig. 2.17) within the voltage range 25–200 V. In Fig. 2.18, impedance declines with voltage, reaching a minimum plateau somewhat beyond 500 V. Freiberger attributed the data in Fig. 2.18 to living persons, although the measurements above 50 V were actually conducted on corpses, and those below 50 V on living persons. He adjusted the measurements on corpses to living persons using a correction procedure that will be described presently.

Distribution of Current and Impedance within the Body

The pathway and density of currents within the body, as well as the total impedance presented to the stimulating device, depend critically on the size and placement of the electrodes. Figure 2.19 (from Roy et al., 1986) illustrates the current distribution measured for electrodes of various sizes within a saline-filled tank. A large (400-cm^2) "indifferent" electrode and a smaller "active" electrode were separated by 16 cm. The current density was measured as a function of distance from the active electrode; the applied current was 3 mA. The current density near the active electrode depends substantially on electrode size. But when

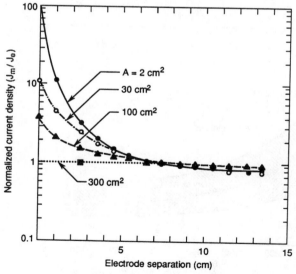

Figure 2.19   Normalized current density in medium with resistivity of 1860 Ω cm; J_m/J_e is current density in medium divided by average current density of electrode. Indifferent electrode area is 400 cm^2. (From Roy et al., 1986.)

measured farther into the medium (beyond 6 cm in this example) the current density depends but little on electrode size. This is an illustration of "spreading" resistance, mentioned in connection with Fig. 2.8. When the electrodes are widely spaced on the body, the total impedance will depend on the separation of body contact points, plus the skin and spreading resistance. For high voltages, large contact areas, and hydrated skin, the total impedance will be dominated by internal body impedance.

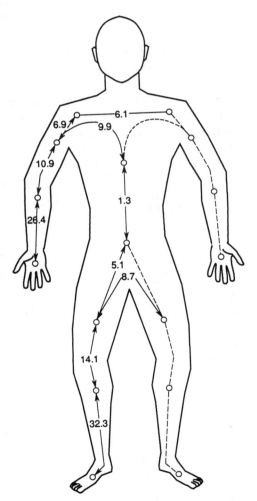

Figure 2.20   Distribution of internal impedance of the human body. Numbers indicate percentages of total internal impedance for hand-to-foot contacts. (Adapted from Freiberger, 1934.)

A substantial body of knowledge concerning body impedance was provided in early experiments by Freiberger (1934). Even today, after the passage of more than half a century, we can still learn much from his remarkable work. Much of Freiberger's research was conducted on cadavers. He was able to separate out the contribution of skin impedance by removing the cornium from the electrode site using a process of inducing heat blisters on the skin. The impedance after this treatment was called "body" impedance. Figure 2.20 illustrates the distribution of body impedance for various current paths. The numbers indicate internal impedance for various electrode placements as a percentage of the total hand-to-foot impedance. These data apply specifically to the internal component, analogous to R_B in Fig. 2.8. Roughly 50% of the internal impedance for hand-to-hand or hand-to-foot contacts resides in the wrists or ankles. The bulk of these high-impedance regions are dominated by relatively poorly conducting bone and ligament. The impedance distribution shown in Fig. 2.20 is consistent with the investigations of Taylor (1985), who used high-voltage capacitive discharges on living people (see Sec. 2.5). The internal body impedance for hand-to-hand or hand-to-foot paths as measured by Freiberger averaged $1000\,\Omega$ for 60 corpses— male and female, of various ages and body structure. This value is some-

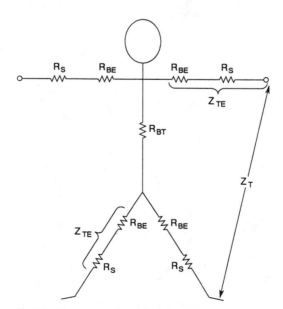

Figure 2.21 Simplified impedance model for current paths across extremities—wet skin, large-area contacts. Example values: $Z_T = 1000\,\Omega$; $Z_{TE} = 500\,\Omega$; $R_{BE} = 250\text{--}375\,\Omega$; $R_S = 125\text{--}250\,\Omega$; $R_{BT} = 50\,\Omega$.

what larger than the previously discussed measurements (500–750 Ω) in living persons.

Figure 2.21 shows a simplified representation of body impedance for evaluation of electrical accidents in which the current path is from one extremity to another. If the contact area is large and if the skin is sweat- or saline-soaked, the impedance to low-frequency currents is largely resistive—a typical example value would use $Z_T = 1000\,\Omega$. As a further approximation, we can consider the impedance in each extremity (Z_{TE}) as being equal, and the impedance of the body trunk (R_{BT}) to be negligible in comparison. The internal body impedance across both extremities lies in the range 500–750 Ω. If $Z_T = 1000\,\Omega$, then the skin impedance at each electrode must be 125–250 Ω. Using the model of Fig. 2.21, the example indicates that approximately 1000 Ω would be measured across any combination of two extremities. If a person were to grasp equipment with both hands while sitting on a conductive floor, the impedance would be reduced to 250 Ω. For other current paths, calculation of total impedance can be carried out by reducing R_B in accordance with Fig. 2.20. If the skin is damaged, the value of R_S would also be reduced.

Current Through the Heart

In an electrical accident, it is important to know the percent of current that is conducted through the heart. Freiberger applied 50-Hz currents to corpses to study this question. After opening the chest of a corpse, he encircled the heart with a ring-shaped current transformer, and measured the current flowing through its cross-sectional area (51 cm²) as a function of electrode placement on the body. Measurements made in this manner are sensitive only to current in a direction along the long axis of the heart. The percentage of current flowing through the ring transformer is shown in Table 2.4. In general, the percentage of current flowing through the heart is quite small for current paths through the limbs. The greatest percentage was 8.5% for a right-hand-to-feet current path. For a foot-to-foot path, the current through the heart does not surpass 0.4% of the total body current. It is likely that a more favorable orientation of the ring transformer was responsible for the larger currents measured from the right-hand in comparison with the left-hand contact. Tests similar to those of Freiberger have been carried out on dogs (Kouwenhoven et al., 1932) with similar results.

Applicability of Measurements on Corpses

Tests on biological tissue reveal impedance changes after death. The dielectric permittivity of frog muscle, for example, is reported to

Table 2.4. *Percentage of longitudinal heart current for various current paths*

Current path	Heart current (%)		
	Min.	Avg.	Max.
Hand to hand	1.9	3.3	4.4
Left hand to feet	1.2	3.3	5.1
Right hand to feet	4.8	6.7	8.5
Foot to foot	<0.1	—	0.4
Head to feet	4.8	5.5	5.9

Source: Data from Freiberger (1934).

decrease by a factor of 2 at 10 Hz within 2.5 h after death, but changes above 100 Hz are much less pronounced (Schwan, 1954). Permittivity changes would have little effect on tissue conductivity, since most tissue is dominantly resistive, even at low frequencies.

Freiberger measured systematic increases in body resistance with time after death. The internal body impedance increase after 5 h, for example, was about 20%. More gradual increases continued up to 30 h, at which time the increase from time of death averaged about 45%. These increases were attributed almost entirely to the cooling of body temperature to room temperature after death. Freiberger determined that the resistivity of physiological saline varies with temperatures in a manner that would account for the observed change in body impedance. Similar changes after death in the resistivity of muscle tissue of animals have been observed; resistivity changes in fatty tissue, however, are found to be minimal (Chilbert et al., 1983).

Changes in the processes of blood circulation and sweat activity after death should have the greatest influence on skin impedance. Considering that skin impedance is most significant at low voltages, we expect the impedance difference between living persons and corpses to be small when the voltage is raised above 200 V. Skin hydration should play a prominent role in the impedance changes after death. It is very difficult to measure skin hydration in a meaningful way. Freiberger instead studied the correlation of impedance with body temperature after death, and used a temperature correction factor to relate the impedance of corpses to living persons. These correction factors were applied in Fig. 2.18 to the measurements on corpses for voltages above 50 V; below 50 V, the data were derived from measurements on living persons.

Table 2.5. *Statistical impedance measurements*

Voltage (V)	Number	Total (Z_T)		Body (R_B)	
		Avg. (kΩ)	S.D. (kΩ)	Avg. (kΩ)	S.D. (kΩ)
25	100	3.52	1.40	0.78	0.11
15	50	3.72	1.12	0.64	0.10

Notes: Large area contacts, 50 Hz, dry skin, hand-to-hand.
Source: Data from Swiss and Austrian measurements, as summarized by Biegelmeier (1985c).

Statistical Distribution of Impedance

Table 2.5 provides statistical data from Swiss and Austrian measurements at 50 Hz (Biegelmeier, 1985c). Data are shown for total impedance (Z_T) with large-area contacts on dry skin, and also for the body impedance component (R_B). Freiberger also determined the statistical distribution of 50-Hz impedance on corpses above 50 V, and on living persons below 50 V. According to Biegelmeier (1986), the statistical data of Freiberger have been adapted by the International Electrotechnical Commission (IEC) to produce an impedance model applicable to living persons. Table 2.6 presents the IEC model for living persons at the 5, 50, and 95 percentile ranks, and for voltages ranging from 25 to 1000 V. The asymptotic values are the presumed values of internal body impedance (R_B).

Freiberger tested 25 subjects using direct or 50-Hz alternating voltages ranging from 5 to 50 V. The current path was hand-to-hand, with large cylindrical electrodes on dry skin. At 5 V, the average impedance was about 4.8 kΩ for DC voltages, and 3.8 kΩ for AC voltages. At 50 V, the impedance dropped to about 70% of the 5-V values. Maximum and minimum impedances at DC were typically a factor of 2 and 0.5 times the averages. At AC, the extremes were 1.5 and 0.53 times the averages. These extremes probably represent approximately the 5 and 95 percentile ranks of a cumulative distribution. The mean DC impedance exceeded the AC value by a typical factor of 1.25, although there was considerable overlap in the distributions. Lower AC values are attributed to 50-Hz capacitive coupling through the high-impedance corneal layer.

Statistical impedance data were developed by Underwriters Laboratories (UL) for DC currents and low voltages (approximately 12 V)

Table 2.6. *Statistical data for total body impedance (Z_T) adopted by the IEC for 50-Hz currents*

Touch voltage (V)	Total body impedance (Ω) at the indicated percentile rank		
	5%	50%	95%
25	1750	3250	6100
50	1450	2625	4375
75	1250	2200	3500
100	1200	1875	3200
125	1125	1625	2875
220	1000	1350	2125
700	750	1100	1550
1000	700	1050	1500
Asymptotic value	650	750	850

Note: Large area contacts, hand-to-hand, dry skin.
Source: Data from IEC (1984).

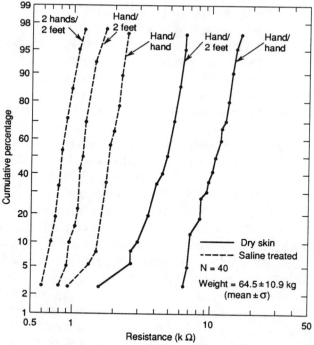

Figure 2.22   Cumulative distribution of DC resistance of adults, with various electrode paths. Hand electrodes = 2 No.10 Awg twisted wires, feet on plate, $I = 5$ mA. (Data from Whitaker, 1939.)

(Whitaker, 1939). Figure 2.22 illustrates statistical distributions for measurements on 40 adults; Fig. 2.23 applies to 46 children. Table 2.7 summarizes the data of adults and children. Adult weight ranged from 45.4 to 94.4 kg (median = 62.2 kg), and age ranged from 18 to 58 years (median = 30 years). For children, weight ranged from 14.1 to 58.1 kg (median = 31.8 kg), and age ranged from 3 to 15 years (median = 10 years). Hand electrodes consisted of two No. 10 AWG bare copper wires twisted together; the foot electrodes were large copper plates. Under "wet" conditions, the hands and feet of subjects were initially soaked in a 20% NaCl solution. Preliminary tests showed that resistance was independent of measuring current in the range 1–15 mA, provided a constant contact area and pressure were maintained, and provided that the subject's hands were wetted in the solution prior to each measurement. The test voltage was adjusted up to a maximum of 12 V to maintain a current of 1 mA with children and 5 mA with adults, except when body resistance was too great to allow these values.

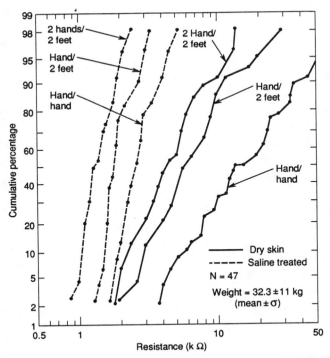

Figure 2.23   Cumulative distribution of DC resistance of children, age 3–15 yrs. Electrodes as in Fig. 2.22, $I = 1$ mA. (Data from Whitaker, 1939.)

Table 2.7. *Statistical summary of DC body impedance measurements (data listed in kΩ)*

	Adults			Children		
	5%	50%	95%	5%	50%	95%
A. *Dry conditions*						
Hand/hand	6.96	11.45	15.69	4.04	14.35	51.10
Hand/two feet	2.62	4.00	6.51	2.62	5.70	18.72
Hand/two feet	—	—	—	2.00	4.25	12.68
B. *Wet conditions*						
Hand/hand	1.28	1.86	2.45	1.70	2.55	4.47
Hand/two feet	0.93	1.20	1.67	1.43	1.80	3.02
Two hands/two feet	0.63	0.84	1.16	0.90	1.30	2.04

Notes: (a) Hand electrodes: two No. 10 Awg twisted copper wires.
(b) Voltage: 12 V DC; curent ~ 1 mA (children), ~ 5 mA (adults).
(c) Wet conditions apply to treatment with 20% NaCl solution.
(d) Children's ages: 3–15 years; adults ages: 18–58 years.
(e) Data from H. B. Whitaker (1939).

The curves plotted in Figs. 2.22 and 2.23 form approximately straight lines, which indicate the log-normal distribution on this plotting format. The slopes of the distribution curves are substantially less for wet than for dry conditions, and dry-skin slopes are greater for children than for adults. In general, children's resistance is greater than that of adults. Apparently, the shorter current paths of children is more than offset by their reduced volume. This can be appreciated by a simple calculation treating the limbs as conducting cylinders of length L and cross-sectional radius r. According to Eq. (2.1), the limb resistance would be calculated by $R = L\rho/(\pi r^2)$. If we assume that the ratio L/r is independent of body size, then it is concluded that resistance would vary inversely with r.

2.4   Impedance at Higher Frequencies

Skin impedance decreases as the frequency of the excitation current is increased, as noted in Table 2.8 (Schwan, 1968). The capacitive component of skin impedance is primarily responsible for its frequency dependence. Figure 2.24 illustrates the frequency dependence of impedance over the range 0–2000 Hz, using large hand-to-hand electrodes on dry skin (from Biegelmeier, 1986). At the highest frequency, impedance is reduced to about 750 Ω, a value approximately that for the internal body impedance with hand-to-hand contacts.

Figure 2.25 illustrates the measurements of Osypka (1963) in the frequency range 0.3–100 kHz, using low voltages (approximately 10 V) and

arge copper-cylinder hand-to-hand contacts. The test results with dry
and wet hands show that above 5 kHz, the effects of skin hydration become
negligible. Table 2.9 shows impedance measurements at 0.375 and 1 MHz
(Schwan, 1968). The hand-to-hand data are similar to the values for
internal body impedance as discussed in Sec. 2.3.

Chatterjee and colleagues (1986) measured impedance in the range of
10 kHz to 10 MHz. Adult subjects numbered 367 (170 M, 197 F). The hand
electrode was a brass rod of diameter 1.5 cm; subjects stood barefoot on
a copper plate. Prior to measurements, the subject's hand was moistened

Table 2.8. *Dry-skin   impedance   versus*
frequency

Frequency (kHz)	Magnitude (Ω cm^2)	Phase angle (deg)
1	14,000	—
5	3,000	-70
10	1,800	-65
20	1,000	-55
50	500	-30
100	300	-20
200	250	-10

Note: Phase angle given by $\tan^{-1} X/R$,
where X is capacitive reactance, and R is
resistance.
Source: Schwan (1968).

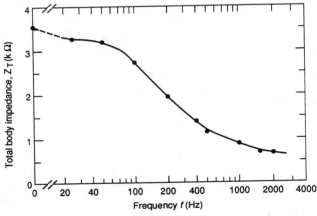

Figure 2.24   Frequency dependence of impedance with large electrodes, dry
hand-to-hand current path. Applied voltage = 25 V rms. (From Beigelmier, 1986.)

with 0.9% physiological saline. The subject's feet were not treated with saline. Figure 2.26 shows results from these experiments separately for male and female subjects. The authors assumed that impedance was related to body size, rather than sex per se, and that impedance was inversely proportional to subject's height. The phase-angle data show that the impedance is largely resistive in the indicated frequency range. The magnitude of impedance drops steadily with increasing frequency, to a

Table 2.9. *Body impedance at 0.375 and 1 MHz*

Electrode location	Area (cm²)	Impedance $f = 0.375$ MHz (Ω)	$f = 1$ MHz (Ω)
Hand to hand	90	475	460
Finger to other arm	10/65	500	470
Across left arm	32	34	21
Across elbow joint	32	37	21
Across shoulder joint	32	47	31
Across neck	32	36	18
Forehead to neck	32	82	57
Chest to back	150	31	20
Across thorax	150	29	19
Right wrist to left leg	75	248	234
Left wrist to right leg	75	274	266

Note: Area applies to each electrode, except for finger/arm, where two different electrode sizes were used.
Source: From Schwan (1968).

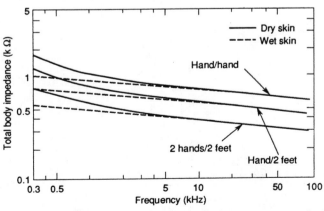

Figure 2.25 Total body impedance for large-area contacts in the frequency range 0.3–100 kHz. (From Osypka, 1963.)

value nearly equal to the internal body impedance reported in Sec. 2.3
for hand-to-feet contacts.

2.5   High-Voltage and Transient Properties

The author and colleagues studied current and voltage
relationships under high-voltage conditions with living subjects (Reilly
et al., 1983, 1984, 1985). These experiments could be safely conducted on
living subjects because the duration and energy of the stimuli were limited
by the use of capacitive discharges. These experiments provided the first
detailed account of high-voltage current response in the microsecond time
scale.

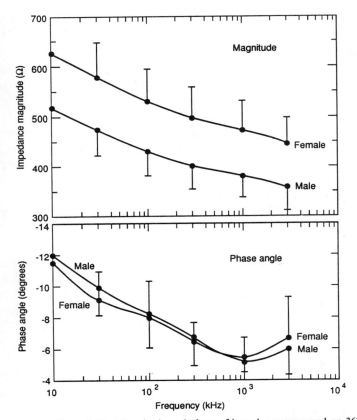

Figure 2.26   Magnitude and phase of impedance measured on 367 subjects,
hand-to-feet contact. Hand electrode: 3.5-cm rod; wet hand contacts; bare
feet on plate. Vertical bars show standard deviations. (From Chatterjee
et al., 1986 © 1986 IEEE.)

The experimental apparatus is illustrated in Fig. 2.27. A high-voltage transformer T was followed by a variable series-coupling resister R_c (variable from 1 to 128 MΩ) and a capacitor C_o (variable from 100 to 6400 pF in steps of 100 pF). An internal transformer resistance of 1 MΩ, and permanent series resistance of 2 MΩ, limited the current for human safety. The voltage at the energized electrode could be varied from 0 to ± 15 kV, in the DC mode, or 15 kV peak in the AC mode.

The stored charge on C_o could be discharged to a subject by one of several methods: by actively touching the energized electrode; by bringing to some location on the body a probe energized by C_o; or by closing a switch between C_o and a passive electrode already in contact with the body. Stimulus voltage and current could be simultaneously sampled at a maximum rate of 50 ns, and stored in a dual-channel digital oscilloscope. The digital oscilloscope was connected to a digital computer for manipulation and plotting of the data. In most cases the return electrode was silver-plated, with a contact area of 15 cm^2. It was covered with conductive electrode paste, and generally worn near the site of stimulation, such as the forearm for stimulation on the fingertip. Because of the size and treatment of the return electrode, its impedance was negligible compared with that at the stimulus site.

What may be thought to be a simple stimulus—a single capacitive discharge—actually has complexities that distinguish it from other

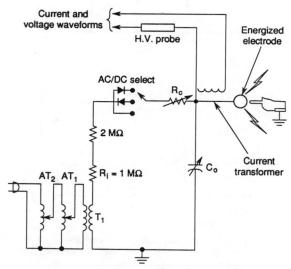

Figure 2.27   High-voltage stimulator, schematic. (From Reilly and Larkin, 1983.)

electrical stimuli discussed previously. These complexities include a separation of the discharge into spark and contact components, a plateau voltage, marked impedance nonlinearities, differences in response to positive and negative stimuli, and significant waveform differences at different body locations.

It is important to distinguish between two modes whereby charge from a capacitor can be transferred to the skin. The stimulus can be presented via a metallic contact on the body, or via a spark contact. When a subject touches an energized electrode, there can result both a spark (if the initial voltage is sufficient) and a subsequent discharge when mechanical contact is made. To understand the characteristics of the spark and contact components, we studied the factors that govern the shape and time course of discharge waveforms under a variety of stimulating conditions.

Spark and Contact Components

When a voltage is applied across the air gap between two electrodes, the free electrons and negative ions that normally exist in air will move toward the anode (positive electrode). The positive ions move toward the cathode (negative electrode). If the electric field produced by the voltage is sufficient, the free electrons will be accelerated to velocities such that their collisions with neutral atoms or molecules create more free electrons. This process of free-electron multiplication is known as an *electron avalanche*.

When one of the electrodes is the human body, the nature of the electrical breakdown will be greatly affected by the nonlinear impedance properties of the skin. The nature of the breakdown phenomenon is evident in Fig. 2.28, where a subject lightly tapped with his finger an electrode having a 1-mm-diameter tip that protruded by 0.5 mm above an insulated housing. For each set of conditions depicted in the figure, four successive stimuli were applied; the four corresponding voltage and current waveforms were then averaged.

In this sequence of four recordings, the discharge capacitance has been constant at 400 pF, but the initial voltage has been varied. In Fig. 2.28a, the initial voltage of 405 V is too small to produce a spark because of the dielectric protection of the skin. As a result, the current is conducted entirely by direct contact when the finger touches the electrode. In Fig. 2.28b, the initial voltage of 515 V is just barely at the point of a spark discharge at the initiation of the current trace as revealed by the small initial current transient at about $t = 75\,\mu$s. In Fig. 2.28c, the initial voltage has been raised to 586 V (a 14% increase over Fig. 2.28b), and the magnitude of the spark discharge grows by a factor of 10. In Fig. 2.28d,

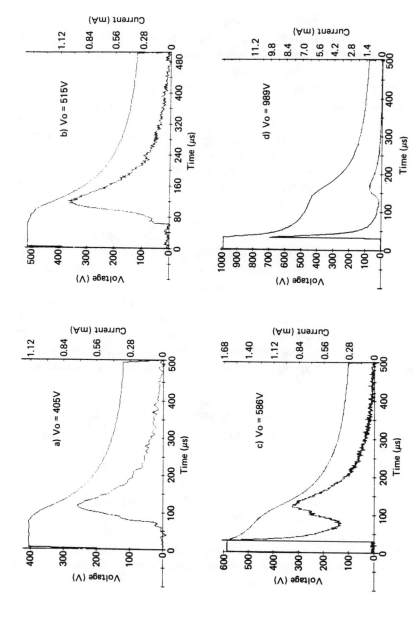

Figure 2.28 Stimulus waveforms for light tap force, capacitive discharges with $C = 400\,\text{pF}$. (From Reilly et al., 1982.)

with the initial voltage at 989 V, the spark discharge component of the stimulus has increased by another factor of 10, and we can see the convergence of voltage to a plateau around 450 V, followed by the contact at about $t = 130\,\mu s$.

In this sequence of waveforms, an initial voltage of about 500 V delineates a level above which spark discharges occur, and below which they do not. This is also approximately the plateau voltage level that is discussed in the succeeding paragraphs. The experimental data suggest that the breakdown voltage for spark discharges to the skin is at about the level of 500 V. This is consistent with the dielectric strength of the excised human corneum reported in Sec. 2.2.

Plateau Voltage

In the example illustrated in Figure 2.28, separate voltage plateaus can be seen for the spark and the contact phases of the discharge. The contact plateau is around 100 V, and the spark plateau is around 450 V. These plateaus suggest that as the voltage declines, the impedance becomes very large in both phases of the discharge. The presence of a voltage plateau reveals that a charged capacitor cannot be totally discharged through a spark to the skin. It also delineates a voltage level above which spark discharges can occur.

The plateau voltage for a spark stimulus can be studied by bringing an energized electrode slowly to the body to produce a spark discharge that is not followed by a contact. Figure 2.29 illustrates four successive

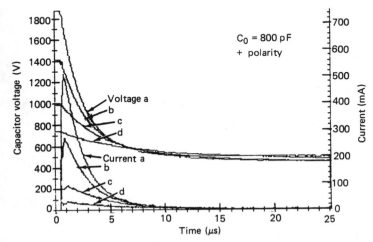

Figure 2.29   Four successive discharges to the same point (second finger, point *A*), with variable initial voltage. (From Reilly and Larkin, 1983.)

discharge waveforms for a pencil-shaped probe (0.8-mm-diameter tip) that is brought slowly to the same point on the finger pad. Discharge to the same point was ensured by the use of a dielectric mask with a 0.5-mm hole. The initial voltage has been varied from 750 to 1900 V. In each case, a plateau voltage in the range of 450–500 V is evident. The presence of this plateau indicates that the spark component does not completely discharge the capacitor and that the discharge impedance converges to a very large value in the region where the discharge current approaches zero. When tested over various subjects and body locations, the plateau voltage was found to range from about 450 to 650 V, with an average of 525 V. Stripping away the corneal layer of skin lowers the plateau voltage to about 330 V.

Discharge Impedance

Knowledge of the dynamic properties of discharge impedance is important in a number of applications related to transient high-voltage shock exposure. With capacitive discharges, for example, the impedance affects the time constant of the discharge. The time constant, in turn, significantly affects the neural excitation potency of the discharge, as discussed in Chapters 3 and 4. In other applications, knowledge about body impedance is needed to define the limiting value of current in a high-voltage exposure—such as with fault conditions in high-voltage installations.

Figure 2.30 expresses impedance data for Fig. 2.29 as the ratio voltage/current. These dynamic impedance functions reveal that the skin

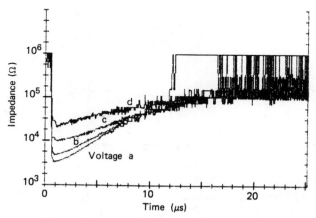

Figure 2.30   Impedance versus time for four successive discharges to the same point (second finger, point *A*) with variable initial voltage. Voltage and current waveforms as shown in Fig. 2.29. (From Reilly et al., 1982.)

impedance to a spark discharge exhibits a nonlinear relationship with voltage in which peak current is not directly proportional to the initial voltage. The variation in initial impedance with initial voltage is evident in Fig. 2.30. The impedance attains a minimum value in a fraction of a microsecond after the stimulus onset, and increases during the time course of the stimulus, during which time the stimulus voltage is decreasing. The large excursions beyond $16\,\mu s$ are due to digital quantization effects.

During a capacitive spark discharge, impedance seen by the capacitor consists of two contributions: the voltage drop across the arc itself, and the voltage drop across the subject. The author measured the arc voltage/current relationship when the human body is in direct contact with one of two metallic electrodes forming the arc gap, and is in series with the electrical path. The voltage drop across the arc for this case varied from nearly zero at the point of arc initiation to about 50–75 V at the point of arc extinction. As a result, when two metallic electrodes form the discharge path, the arc does not contribute significantly to the overall measured impedance. We cannot be certain that the same arc voltage drop will apply when the spark terminates at the human skin rather than a metallic electrode. It does not seem possible to make a more direct measurement of the arc impedance when the discharge terminates at the human skin without significantly altering the skin's electrical response. If, however, the arc voltage drop is similar to that when two metallic electrodes are used, then the arc voltage drop must be small compared with that across the human subject.

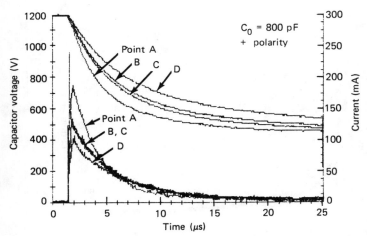

Figure 2.31   Waveforms for stimuli to four different points on the same fingertip (second, points *A–D*); each point separated by 2 mm. (From Reilly et al., 1982.)

Variation with Stimulus Location

Waveforms for discharges to different discrete body locations can differ significantly (e.g., a factor of 2 in minimum impedance), even though the stimulated points may be separated by only 2 mm. Figure 2.31 illustrates voltage and current waveforms for anodal (positive polarity) spark discharges to four points on the fingertip separated by 2 mm. In order to localize the spark stimulation, a dielectric mask with a 0.5-mm-diameter hole was placed on the finger. On the other hand, repeated positive polarity stimuli at the same point produce nearly identical waveforms. This closely spaced pattern of skin impedance may reflect the mosaic of sweat ducts.

The fact that identical nonlinear current response is seen with repeated discharges to the same point on the finger suggests that, with the tested discharge parameters, the impedance breakdown is not permanent. Perhaps ionic disassociation rather than membrane rupture is involved. Repeated discharges at similar energy levels on the dry skin of the calf on the leg, however, eventually resulted in visible erosion of the corneum.

Polarity Effects

Positive-polarity current transients appear relatively smooth and repeatable. In contrast, negative-polarity waveforms often are bistable and have lower initial impedance than the positive waveform, as illustrated in Fig. 2.32. In this example, four cathodal spark discharges are applied to point *A* (also used for the anodal discharges shown in Figs. 2.29–3.31).

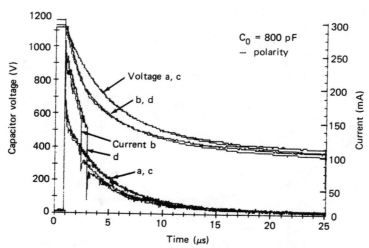

Figure 2.32   Waveforms for successive discharges to the same point (second finger, point *A*); negative polarity. (From Reilly et al., 1982.)

One possible explanation for these polarity differences is related to the ionized discharge process. Electrons move so rapidly that by the time the avalanche has reached the anode, the newly created positive ions are virtually still in their original positions, forming a positive space charge conically concentrated at the anode, and tapering or decreasing toward the cathode as shown in Fig. 2.33a (Howatson, 1965). This distribution of positive ions modifies the field near the anode as shown in Fig. 2.33b. The conical spreading of the plasma discharge from the cathode to the anode via the avalanche process results in a larger area of a spark contact at the anode than at the cathode. The lower initial minimum impedance for negative polarity may result from the greater ion contact area with the body as an anode than as a cathode. The bistable character of negative polarity discharges may result from an unstable space charge distribution and its effective area of anodal contact.

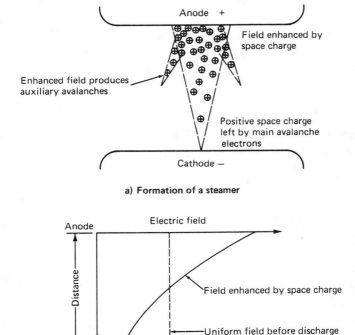

a) **Formation of a steamer**

b) **Electric field**

Figure 2.33   Spark discharge formation: (a) Formation of a steamer; (b) electric field. (Reprinted with permission from A. M. Howatson, *An Introduction to Gas Discharges*, Copyright 1965, Pergamon Press PLC.)

Stimulus Time Constants

A capacitance C discharged to an ideal linear resistance R will have an exponentially decaying current and voltage waveform with a time constant $\tau = RC$, which defines the time at which the voltage or current decays to e^{-1} of its maximum value. As evident in the previous examples, many of the discharge patterns can be approximated by exponential functions with current decaying to zero and voltage decaying to the plateau value. Although an ideal RC circuit results in an exponential discharge pattern, the converse statement should not be assumed: an exponential current decay pattern does not necessarily imply a linear resistance. This can be appreciated with reference to the discharge patterns of Fig. 2.29: both current and voltage waveforms are approximately exponential functions. However, because of the nonzero voltage plateau, the impedance varies from a minimum value at the onset of the discharge to a very large value at its extinction.

The instantaneous impedance to a capacitive discharge departs significantly from what would be expected in a linear model. Linear circuit models for skin impedance, commonly invoked in studies of low-voltage cutaneous currents, do not adequately represent the measured impedance response. It appears, however, that the measured transient response might be accounted for if the model parameters were voltage- or time-dependent.

We define time constants for capacitive discharge stimuli as the time required for the voltage to decay to $e^{-1} \Delta V$, where ΔV is the difference between the initial and plateau voltages. For example, in Fig. 29, the initial voltage for waveform (a) is 1900 V, and the final voltage is 480 V. The decay time is measured from the onset of the current waveform. The measured plateau for the touch component was generally in the vicinity of 100 V (as in Fig. 2.28), in contrast to the plateau of about 500 V for the spark component. The value of ΔV defines the amount of charge passed, in accordance with $Q = C \Delta V$.

Figure 2.34 depicts the measured time constants for the spark and contact components as a function of voltage for a single subject who discharged a capacitor by actively touching an energized electrode. Both spark and contact component time constants are shown. The discontinuity around 500–600 V occurs because spark discharges could not be produced below this range but occurred above it. The time constants decrease as the initial voltage on the capacitor is increased, and as the capacitance is decreased.

The time constants illustrated in Fig. 2.34 apply to a procedure in which a subject discharged a capacitor by actively touching an electrode energized by a charged capacitor. The resulting effective discharge contact area is very small for both spark and contact phases. We hypothesize that at the instant of a tapped electrode contact, an initially small point of conduction establishes a preferred current channel. As a result, both spark and contact phases may be thought of as utilizing a small contact area, regardless of the electrode size. In contrast, discharges to an electrode held in contact with the skin appear to utilize multiple current channels. The result is a much lower impedance at the electrode/skin interface, and significantly smaller discharge time constants. For example, discharges from a 6400-pF capacitor to a 1.3-cm-diameter contact electrode result in discharge time constants under 3 μs, even when the initial voltage is as low as 50 V.

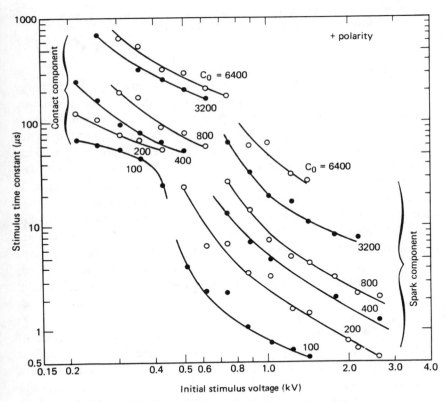

Figure 2.34   Time constants for monophasic capacitive discharge stimulation of the fingertip. Parameter shown is the capacitance in picofarads. (From W. D. Larkin and J. P. Reilly, *Perception & Psychophysics* 36 (1): 68–78, 1984, reprinted by permission of Psychonomic Society, Inc.)

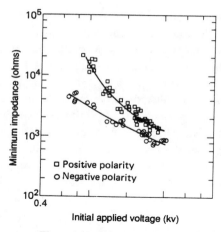

Figure 2.35   Minimum impedance of leg (volar calf) versus voltage; for spark discharges to skin; leg grounds. (From Reilly et al., 1982.)

Minimum Impedance with Spark Discharges[†]

The minimum impedance to a capacitive spark discharge occurs at the onset of the stimulus, when the applied voltage is greatest. Figure 2.35 shows the minimum impedance for the case where the spark discharge and return electrode are on the calf of the leg. The curves are second-order least-squares fits to the experimental data. Two striking features are the voltage dependence and the polarity sensitivity. The voltage dependence reflects the nonlinear nature of skin impedance as discussed in Sec. 2.2. Additionally, the diameter of spark contact area may increase with voltage—at higher voltages, the distance between the skin and the discharge electrode is greater, allowing the base area of the space charge cone (Fig. 2.33) to increase with distance. When the spark discharge terminates at an electrode already in contact with the skin, the impedance will be much lower than when the spark terminates directly on the skin.

Internal Body Impedance[†]

The internal body impedance can be estimated from transient measurements of capacitive discharge stimuli where the electrode area is large. In this case, large contact area and high voltage will minimize skin impedance. Figure 2.36 shows an example of the ratio of voltage and current during a capacitive discharge to a 3-cm-diameter cylinder electrode grasped by a subject with the left hand while standing with the bare left

[†]Adapted from the work of R J. Taylor, as reported in Reilly et al. (1982, 1983) and Taylor (1985).

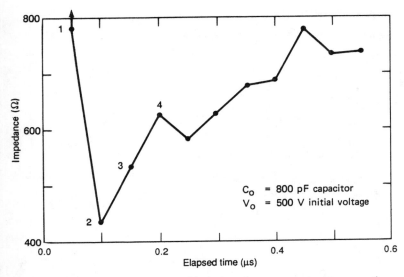

Figure 2.36 An example of body impedance measurements versus time; 3-cm-diameter electrode in left hand, left bare foot on plate. Spark discharge was initiated between 0.0 and 0.05 μs. (From Taylor, 1985.)

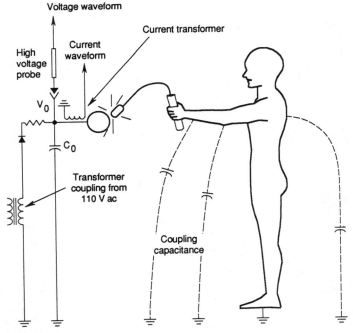

Figure 2.37 Body impedance measurement with capacitive coupling of body to ground. (From Taylor, 1985.)

Table 2.10. *Initial total body impedance* (Ω)

Path	Mean	Standard deviation
Dry skin		
Left hand/left foot	533	52
Right hand/left foot	521	37
Right hand/left hand	508	29
Treated skin		
Left hand/left foot	516	55
Right hand/left foot	507	36
Right hand/left hand	490	36

Source: Data based on measurements of seven individuals (from Taylor, 1985).

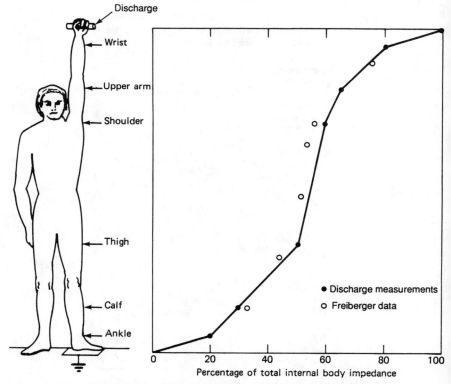

Figure 2.38   Internal body impedance between sole of foot and various locations of the active electrode along the body. Open points from Fig. 2.19. (From Taylor, 1985.)

foot on a copper plate. The discharge capacitance was 800 pF at an initial voltage of 500 V. The discharge process was essentially complete during the interval illustrated in the figure. Data point 1 is not considered meaningful because of the limited rise time of the instrumentation (0.1 μs). At data point 2, the impedance is 434 Ω (434 V and 1 A current). Data point 2 is strongly influenced by the distributed capacitance of the body to ground as shown in Fig. 2.37; and initial current surge is required to supply the body's surface charge in addition to the current that passes internally. The impedance measurement at data point 2 changed as much as 50% as the proximity between the body and grounded objects was varied. The measurements at data point 3 changed only a little (about 10%) with variations in distributed capacitance. The measurements at data point 3, termed "initial body impedance," were considered by Taylor to represent the core body impedance.

Table 2.10 shows initial body impedance for various electrode configurations and skin treatment conditions (from Taylor, 1985). The treated

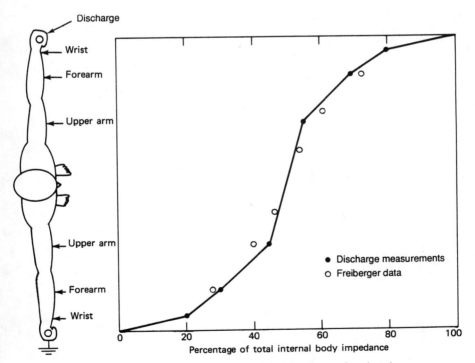

Figure 2.39   Internal body impedance from one hand to various locations of the active electrode along the arms. Open points from Fig. 2.19. (Adapted from Taylor, 1985.)

skin had conductive electrode paste applied to it. The impedance for dry and treated skin differs only a little, thus supporting the hypothesis that the contribution of skin impedance is negligible with this measurement technique. The data in Table 2.10 agree well with internal body impedance data determined by other methods (refer to Sec. 2.3).

Figures 2.38 and 2.39 illustrate the internal body impedance for different locations of the active electrode, measured as a percentage of the total hand-to-hand and hand-to-foot internal impedance. The filled circles represent the transient discharge measurements, averaged over seven adult individuals. The open circles are derived from the model of Freiberger (Fig. 2.20) The agreement between the two experiments is apparent.

3

**Electrical Principles of
Nerve and Muscle Function**

An understanding of the electrical basis of nerve and muscle function
will be a valuable foundation for later discussions of electrical excitation.
This chapter begins with some basic electrical properties of biological
cells. We then examine the function of a special class of cells that have
electrically excitable membranes. We show how these properties are
organized for sensory and muscle function. This information will help
explain how externally applied currents can modify or interfere with
normal function.

Consider a familiar example that points out several pertinent electrophysiological functions. You touch a hot object with your finger—after
a brief delay, you feel pain, and you jerk your hand away. This sequence
of events involves a number of electrical functions of nerve and muscle.
First, the skin's temperature rise is converted to an electrical potential by a
specialized transducer or *receptor* in the skin. An example of a sensory
receptor is shown in Fig. 3.1. The receptor's voltage response, called a
generator potential, initiates a nerve impulse, called an *action potential*
(AP). The AP travels along an electrical cable known as *a nerve axon*.
The axon runs from the receptor to the spinal column, where connections,
called *synapses*, are made with additional nerve cells; some of these
ultimately carry information to the brain. The receptor, axon, and synaptic
terminus comprise a single nerve cell (also called a *neuron*). Bundles of
neurons are commonly called *nerves*. In our example, the length of the
nerve cell, from the fingertip's receptor to the spinal synapse, will be
about 1 m.

Thus far, we have discussed an *afferent* nerve cell, i.e., one that carries
information from the body's sensory system to the spinal cord and then

For background material, see Ruch et al. (1968), Stein (1980), and Plonsey (1969).
Sections 3.1 and 3.2 substantially adapt material from Reilly (1988).

to the brain. Our example sensory nerve cell is also a slowly conducting type in which the conduction velocity may be only a few meters per second. Because of the delay in the conduction and synaptic processes, a substantial fraction of a second may pass before the brain is notified of the finger's temperature rise, and pain in consciously registered. Even before that, however, a reflex action may be set in motion within the spinal column. In the reflex action, sensory inputs, and possibly other inputs from the brain, are processed within the spinal column, and, if the inputs satisfy threshold criteria, AP signals are sent to the muscles of the hand and arm. There signals travel along *efferent* (motor) neurons, i.e., ones that carry signals from the central nervous system (CNS) to the muscles (Fig. 3.1*a*). When the efferent APs reach the neuron terminus at the muscles, they initiate a series of events leading to electrical excitation and contraction of muscle fibers. The result is that the hand pulls away in a reflex action.

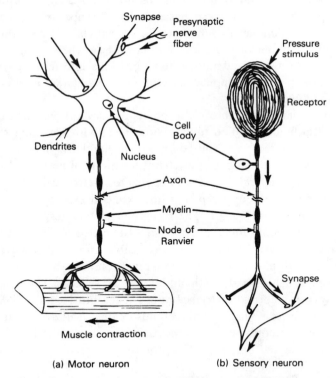

Figure 3.1   Functional components of (*a*) motor and (*b*) sensory neurons. Arrows indicate the direction of information flow. Signals are propagated across synapses via chemical neurotransmitters and elsewhere by membrane depolarization. Synapses are inside the spinal column. The sizes of the components are drawn on a distorted scale to emphasize various features.

Myelin

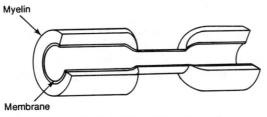

Membrane

Figure 3.2   Detail of myelinated fiber at a node.

The mechanisms responsible for normal function will also respond to externally applied currents. With appropriate control, electrical stimulation can provide medical assistance and diagnostic benefits. Beneficial uses of electrical stimulation include muscle control, pain relief, sensory prosthesis, and diagnosis of nerve and muscle pathology. However, in chance encounters with electric currents (electric shock), the result can be pain, hazardous muscle reactions, heart disturbances, and ultimately death.

Figure 3.1 illustrates several functional components of a sensory and a motor (muscle) neuron. In this example, the neuron is *myelinated*, i.e., covered with a fatty layer of insulation called *myelin*, and has exposed *nodes of Ranvier*. Other neurons are unmyelinated. The conducting portion of the neuron is a long, hollow structure known as the axon (illustrated in Fig. 3.2). The axon plus myelin wrapping is frequently referred to as a nerve *fiber*. The arrows in Fig. 3.1 indicate the direction of information flow. For the motor neuron in Fig. 3.1a, APs propagate from synapses in the spinal column to the terminus at the muscle. For the sensory neuron in Fig. 3.1b, APs originate from one of a variety of specialized receptors (a *pacinian corpuscle* is illustrated) and proceed to a synapse in the spinal column. Communication across the synapses is accomplished through chemical substances known as *neurotransmitters*.

3.1   Cellular Membranes

Cells are the basic building blocks of both plant and animal life. The functional boundary of the cell is a thin (about 10-nm) bimolecular lipid and protein structure. One role of the membrane is to regulate chemical exchange from within the cell (the *plasm*) to its surroundings (the *interstitial fluid*). The electrochemical forces across the membrane are intimately involved in the regulation.

The plasm and interstitial fluids are composed largely of water containing ions of different species. The concentration of ions inside and outside the cell is not the same, and this difference leads to the electrochemical forces across the cell membrane. The membrane is said to be *semipermeable*; i.e., it is basically a dielectric insulator that allows some ionic interchange. Figure 3.3 represents a membrane as a barrier, with pores that permit the passage of ions; the individual pores may be very selective with respect to the ionic species that are allowed to pass. Typical

Table 3.1. *Example cellular ionic concentrations*

Species	Concentration ($\mu M/cm^3$)		Nernst potential
	Inside	Outside	(mV)
A. Mammalian muscle cells			
Na^+	12	145	66
K^+	155	4	-97
Cl^-	4	120	-90
Resting potential			-90
B. Squid axon			
Na^+	50	460	59
K^+	400	10	98
Cl^-	40–100	540	-45 to -69
Resting potential			-60

Source: Data from Ruch et al. (1968) and Katz (1966).

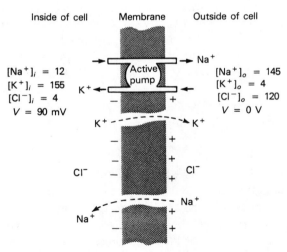

Figure 3.3   Schematic of a typical cell membrane. The pores allow the passage of ions. Numerical values indicate approximate steady-state concentrations ($\mu mol/cm^3$) for typical mammalian muscle cells. An active metabolic pump drives Na^+ out of the cell and K^+ into the cell. The transmembrane potential difference is about $-90\,mV$; the inside is negative relative to the outside.

concentration inside and outside a cell are shown in Table 3.1 for Na^+, K^+, and Cl^{-1} ions. Although other species.are also present, confining our attention to these ions is sufficient for our purposes. The concentrations indicated in Table 3.1 are markedly different inside and outside the cell. The differences lead to two forces that tend to drive ions across the membrane: a concentration gradient and a voltage gradient. In order to understand these forces, first consider an environment where only one ionic species, substance S, is present.

The Nernst Equation

In a solution where the concentration varies from one region to another, there will be a net flux from the region of higher concentration to the lower. The concentration potential energy difference, W_c, is work required to move a mole of S against the gradient. This quantity is proportional to the logarithm of the concentration difference in accordance with

$$W_c = RT(\ln[S]_i - \ln[S]_o) \qquad (3.1a)$$

$$= RT \ln \frac{[S]_i}{[S]_o} \qquad (3.1b)$$

where $[S]_i$ and $[S]_o$ represent the concentrations of S inside and outside the cell, R is the gas constant, and T is the absolute temperature in degrees Kelvin. The product RT has units of energy per mole.

If S is ionized, an electrical potential difference will occur between the two regions of differing concentration. The electrical potential energy, W_e, is given by

$$W_e = ZFV_m \qquad (3.2)$$

where Z is the valence of S, F is the Faraday constant (the number of coulumbs per mole of charge), and V_m is the potential difference across the membrane.

The total electrochemical potential difference is the sum of the concentration and electrical potentials:

$$\Delta W = W_c + W_e \qquad (3.3)$$

Substituting the quantities from Eqs. (3.1b) and (3.2) results in

$$\Delta W = RT \ln \frac{[S]_i}{[S]_o} + ZFV_m \qquad (3.4)$$

When $\Delta W = 0$, S is at equilibrium across the membrane, i.e., there is no net force in either direction, and the net flux across the membrane is zero. Under conditions of equilibrium, the membrane will attain the potential:

$$V_m = \frac{RT}{FZ} \ln \frac{[S]_o}{[S]_i} \qquad (3.5)$$

Equation (3.5) is known as the *Nernst equation*. It is a statement of the membrane potential for an ionic substance in electrochemical equilibrium. Using the values $R = 8.31 \, \text{J/mol K}$, $T = 310 \, \text{K}$ (37°C), $F = 96,500 \, \text{C/mol}$, and $Z = +1$ (for a monovalent cation), converting to the base 10 logarithm, and expressing V_m in millivolts, we obtain:

$$V_m = 61 \log \frac{[S]_o}{[S]_i} \quad \text{(mV)} \tag{3.6}$$

For a system with more than one permeable ionic species, the equilibrium voltage will depend on the concentration and relative permeability of the individual ions. For a system consisting of K^+ and Na^+, for example, the expression is

$$V_m = 61 \log \frac{P_K[K^+]_o + P_{Na}[Na^+]_o}{P_K[K^+]_i + P_{Na}[Na^+]_i} \tag{3.7}$$

where P_K and P_{Na} are the permeabilities (expressed in units of centimeters per second) for K^+ and Na^+, respectively. An alternate expression for Eq. (3.7) uses the ratio $q = P_{Na}/P_K$ to obtain

$$V_m = 61 \log \frac{[K^+]_o + q[Na^+]_o}{[K^+]_i + q[Na^+]_i} \tag{3.8}$$

One can get a feel for the changes in V_m during excitation by considering the simplified circuit diagram in Fig. 3.4. The membrane permeability is

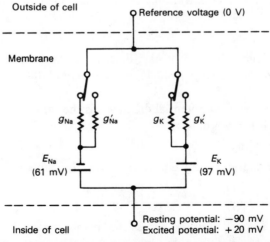

Figure 3.4   Circuit diagram representing membrane conductance for Na^+ and K^+ ions. In the resting condition, the inside of the cell is at a potential of $-90 \, \text{mV}$. In the excited state, the inside of the cell is at a potential of $20 \, \text{mV}$.

represented by conductances g_{Na} and g_K, and the electrochemical gradients as potential sources E_{Na} and E_K. For an excitable membrane in the resting state, $g_{Na} \ll g_K$ and the membrane potential approaches the Nernst potential for K^+, as indicated by Eq. (3.8). In the excited state, $g_{Na} > g_K$, and the switches in Fig. 3.4 would be connected in the alternate positions, forcing the membrane to move toward the Nernst potential for Na^+.

Consider now the individual Nernst potentials for the ionic species listed in the right-hand column of Table 3.1. The equilibrium potential for Na^+ (66 mV) is far removed from the membrane potential (-90 mV), K^+ is slightly out of equilibrium, and Cl^- is essentially in equilibrium. The magnitudes and signs of the potentials show that a strong electrochemical force tends to drive Na^+ into the cell and a relatively weaker force tends to drive K^+ out of the cell. Given that the membrane is at least somewhat permeable to the ions discussed here, these forces ought eventually to bring the species into equilibrium. Clearly, another force is working to maintain the system in disequilibrium. The responsible force is the so-called sodium pump, an active system that pumps Na^+ out of the cell, and K^+ into the cell. The energy for the pump is derived from the cell's metabolism. A dead cell would eventually reach equilibrium potential.

The electrical forces on the membrane are quite large. Considering the membrane potential ($\sim 10^{-1}$ V), and thickness ($\sim 10^{-8}$ m), the electric field developed across the membrane is about 10^7 V/m. Conductivity properties of the excitable membrane are intimately tied to the membrane electric field; disturbances from the resting condition can lead to profound changes in the membrane's electrical properties. These changes ultimately initiate and sustain the functional responses of nerve and muscle.

3.2   The Excitable Nerve Membrane

Nerve and muscle cells possess membranes that are excitable, such that an adequate disturbance of the cell's resting potential can trigger a sudden change in the membrane conductance. The resulting membrane voltage change will affect adjacent portions of the membrane, and in a nerve, will propagate as a nerve impulse. The response of the excited membrane is an *action potential* (AP).

In order to illustrate the properties of excitability and propagation consider the experiment illustrated in Fig. 3.5, in which a small stimulating electrode (SE) is near a nerve fiber, and two small recording microelectrodes (RE) pierce the membrane; the return electrodes are assumed to be immersed in the conducting medium some distance away. The stimulating electrode is connected to a current source. In Fig. 3.5, the small arrows represent the distribution of current flow if SE is a cathode, showing cationic flow toward the cathode (or, equivalently, anionic flow in the opposite direction). The voltage disturbance due to the stimulating electrode will tend to reduce the membrane potential (depolarization) near the cathode, and increase the potential (hyperpolarization) elsewhere along the axon.

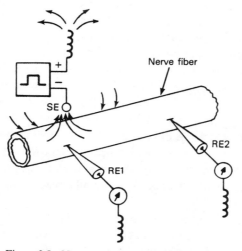

Figure 3.5   Nerve excitation and measurement arrangement. Excitation is initiated near the cathode of the stimulating electrode. Arrows indicate direction of conventional (electronic) current flow—actual anionic flow is in opposite direction.

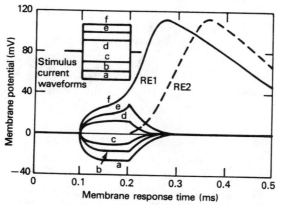

Figure 3.6   Excitation of a nerve fiber by an applied current, as in Fig. 3.5.

Figure 3.6 illustrates the response of the membrane to the rectangular current pulses shown in the upper part of the figure. Six possible current magnitudes labeled *a–f* are shown. Pulses *a–c* apply if SE is an anode; pulses *d–f* apply if SE is a cathode. The membrane response is shown as measured by RE1 and RE2, where 0 V represents the resting potential. Responses *a–c* are in the direction of hyperpolarization, and responses *d–f* are in the direction of depolarization. Responses *a–d* exhibit the characteristic of a linear *RC* network (see Chapter 4). Response *e* is slightly below the excitation threshold, and response *f* illustrates a fully developed action potential. The signal at RE2 is a delayed version of the AP, demonstrating conduction along the axon. The membrane response is often referred to as "all-or-nothing," because the peak AP response is not normally graded—the membrane is either excited or it is not.

To develop a quantitative model of excitation of the neuron by externally applied currents, nonlinear membrane models and electrical cable theory must be used. Let us begin with an understanding of membrane events associated with the production of an AP. In Fig. 3.3, the membrane can be treated as a lossy capacitor. The membrane itself acts as the dielectric, and the ions on either side of the membrane act as the conductive plates of the capacitor. The nature of the leakage channels in the dielectric distinguishes the excitable membrane from the ordinary cellular membrane.

The Hodgkin-Huxley Membrane

A detailed description of the electrical properties of the excitable membrane was developed by Hodgkin and Huxley (1952). With a series of ingenious experiments that eventually lead to a Nobel Prize in Physiology, they provided the first detailed description of the electrical properties of the excitable membrane of unmyelinated nerve cells. This work was later extended by Frankenhaeuser and Huxley (1964) to describe the myelinated nerve membrane. For brevity, we shall refer to the Hodgkin-Huxley and Frankenhaeuser-Huxley work as HH and FH equations, respectively. The system of HH and FH equations is largely empirical.

Figure 3.7 illustrates the HH membrane schematically. The electrical model consists of membrane capacitance, nonlinear conductances for Na^+ and K^+, and a linear leakage element. For a parallel combination of capacitance and conductance, the current through the membrane is related to the capacitive and leakage currents by

$$J_m = c_m \frac{dV}{dt} + \left(J_K + J_L \right) \tag{3.9}$$

where J_m is the membrane current density, c_m is the membrane capacity, V is the membrane voltage, and $J_{Na}, J_K,$ and J_L are the ionic current densities. The ionic terms are expressed by

$$J_{Na} = g_{Na}(V - V_{Na}) \tag{3.10}$$

$$J_K = g_K(V - V_K) \tag{3.11}$$

$$J_L = g_L(V - V_L) \tag{3.12}$$

where $g_{Na}, g_K,$ and g_L are the ionic conductances, and $V_{Na}, V_K,$ and V_L are the ionic Nernst potentials. The g_L conductance is a linear conductance; the other two conductances are more complex nonlinear functions of the form

$$g_{Na} = \bar{g}_{Na} m^3 h \tag{3.13}$$

$$g_K = \bar{g}_K n^4 \tag{3.14}$$

where $\bar{g}_{Na}$ and $\bar{g}_K$ represent the maximum conductance values; and $m, n,$ and h are so-called activation and deactivation variables that modulate the maximum conductances. The $m, n,$ and h variables are governed by the first-order differential equations:

$$\frac{dn}{dt} = \alpha_n(1 - n) - \beta_n n \tag{3.15}$$

$$\frac{dm}{dt} = \alpha_m(1 - m) - \beta_m m \tag{3.16}$$

$$\frac{dh}{dt} = \alpha_h(1 - h) - \beta_h h \tag{3.17}$$

The α and β terms in Eqs. (3.15)–(3.17) are functions of the membrane voltage. At the experimental HH temperature of 6°C, the α and β constants are

$$\alpha_n = \frac{0.01(10 - \Delta V)}{\exp[(10\,\Delta V)/10] - 1} \tag{3.18}$$

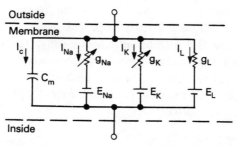

Figure 3.7   Hodgkin–Huxley membrane model.

$$\beta_n = 0.125 \exp\left(-\frac{\Delta V}{80}\right) \tag{3.19}$$

$$\alpha_m = \frac{0.1(25 - \Delta V)}{\exp[(25 - \Delta V)/10] - 1} \tag{3.20}$$

$$\beta_m = 4 \exp\left(-\frac{\Delta V}{20}\right) \tag{3.21}$$

$$\alpha_h = 0.07 \exp\left(-\frac{\Delta V}{20}\right) \tag{3.22}$$

$$\beta_h = \left[\exp\left(\frac{30 - \Delta V}{10}\right) + 1\right]^{-1} \tag{3.23}$$

In Eqs. (3.18)–(3.23), ΔV is the change in membrane potential, expressed in millivolts, relative to the resting potential. Solutions to Eqs. (3.15)–(3.17) can be obtained in the form:

$$n(t) = n_\infty - (n_\infty - n_0)e^{-t/\tau_n} \tag{3.24}$$

$$n_\infty = \frac{\alpha_n}{\alpha_n + \beta_n} \tag{3.25}$$

$$\tau_n = \frac{1}{\alpha_n + \beta_n} \tag{3.26}$$

where $n(0) = n_0$, and $n(t \to \infty) = n_\infty$. Expressions similar to Eqs. (3.24)–(3.26) are obtained for $m(t)$ and $h(t)$. The m, n, and h variables are constrained between 0 and 1, and can be regarded as the fraction of ion gates that is open at any one time. As indicated by Eqs. (3.13) and (3.14), these gates modulate the maximum conductance of Na and K. The τ_m, τ_n, and τ_h variables determine the rate at which the gates can open and close. The asymptotic values of the m, n, and h variables, and the associated time constants, are all functions of membrane depolarization voltage as illustrated in Fig. 3.8 (after Stein, 1980).

Depolarization of the membrane is necessary for excitation. Figure 3.9 illustrates the membrane events accompanying excitation of the HH membrane. The upper part shows the AP voltage waveform along with the membrane conductance. The lower part shows membrane current—the positive axis refers to ionic current influx. An initial surge of Na^+ influx serves to further depolarize the membrane; this influx is followed by K^+ efflux, which repolarizes the membrane.

The Frankenhaeuser-Huxley Membrane
The FH equations for the myelinated membrane use four ionic terms, in contrast with the three for the HH unmyelinated membrane.

The FH ionic current densities are

$$J = J_{Na} + J_K + J_p + J_L \qquad (3.27)$$

The terms J_{Na}, J_K, and J_L have the same interpretation as they do with the HH membrane. The term J_p was described as a "nonspecific" ionic component that responds to the concentration gradient of Na^+. Whether J_p represents a second type of Na^+ channel, or another ionic species, was left unresolved by the FH researchers.

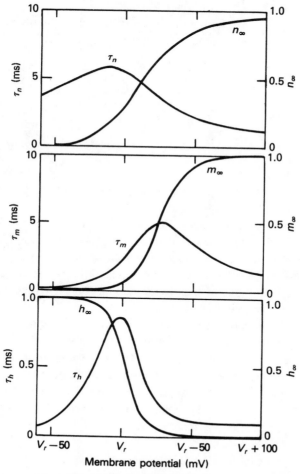

Figure 3.8   Relationship of m, n, and h constants to the membrane voltage. V_r represents the resting potential. (Adapted from Stein, 1980.)

In the FH membranes, the individual ionic current densities have the expressions:

$$J_{Na} = \bar{P}_{Na}\, hm^2 \left(\frac{EF^2}{RT}\right) \frac{[\text{Na}]_o - [\text{Na}]_i \exp(EF/RT)}{1 - \exp(EF/RT)} \tag{3.28}$$

$$J_K = \bar{P}_K n^2 \left(\frac{EF^2}{RT}\right) \frac{[\text{K}]_o - [\text{K}]_i \exp(EF/RT)}{1 - \exp(EF/RT)} \tag{3.29}$$

$$J_p = \bar{P}_p p^2 \left(\frac{EF^2}{RT}\right) \frac{[\text{Na}]_o - [\text{Na}]_i \exp(EF/RT)}{1 - \exp(EF/RT)} \tag{3.30}$$

$$J_L = g_L(V - V_L) \tag{3.31}$$

where $E = V - V_r$, V is the membrane potential, and V_r is the resting potential. The variables $m, n, h,$ and p are defined by differential equations

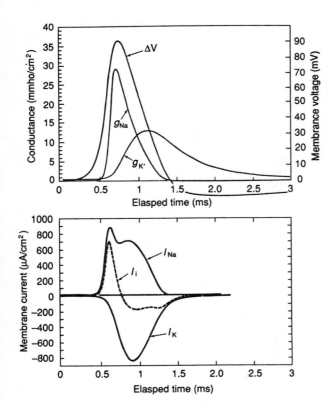

Figure 3.9   Membrane events during propagating action potential. The top figure shows the membrane voltage change (ΔV) and the conductances g_{Na} and g_K. The bottom figure shows the sodium current (I_{Na}), potassium current (I_K), and total ionic current (I_i). The positive axis indicates the ionic current influx. (Adapted from Hodgkin and Huxley, 1952.)

Table 3.2. *Constants for FH equations*

Constant	Value	Description
$\bar{P}_{Na}$	8×10^{-3} cm/s	Sodium permeability constant
$\bar{P}_K$	1.2×10^{-3} cm/s	Potassium permeability constant
$\bar{P}_p$	0.54×10^{-3} cm/s	Nonspecific permeability constant
g_L	30.3 mS/cm^2	Leakage conductance
V_L	0.026 mV	Leakage equilibrium potential
$[Na]_o$	114.5 mM	External sodium concentration
$[Na]_i$	13.7 mM	Internal sodium concentration
$[K]_o$	2.5 mM	External potassium concentration
$[K]_i$	12 mM	Internal potassium concentration
F	96,514 C/g mol	Faraday constant
R	8.3144 J/K mol	Gas constant
T	295.18 K	Absolute temperature
V_r	-70 mV	Resting potential
c_m	2μF/cm^2	Membrane capacitance per unit area

Note: Initial conditions: $m(0) = 0.0005$, $h(0) = 0.8249$, $n(0) = 0.0268$, $p(0) = 0.0049$.
Source: Frankenhaeuser and Huxley, 1963.

of a form identical to Eqs. (3.15)–(3.17). The α and β constants for the FH equations are given by

$$\alpha_m = 0.36(V - 22)\left[1 - \exp\left(\frac{-V + 22}{3}\right)\right]^{-1} \tag{3.32}$$

$$\beta_m = 0.4(13 - V)\left[1 - \exp\left(\frac{V - 13}{20}\right)\right] \tag{3.33}$$

$$\alpha_h = 0.1(-10 - V)\left[1 - \exp\left(\frac{V + 10}{6}\right)\right]^{-1} \tag{3.34}$$

$$\beta_h = 45\left[1 + \exp\left(\frac{45 - V}{10}\right)\right]^{-1} \tag{3.35}$$

$$\alpha_n = 0.02(V - 35)\left[1 - \exp\left(\frac{35 - V}{10}\right)\right]^{-1} \tag{3.36}$$

$$\beta_n = 0.05(-V + 10)\left[1 - \exp\left(\frac{V - 10}{10}\right)\right]^{-1} \tag{3.37}$$

$$\alpha_p = 0.006(V - 40)\left[1 - \exp\left(\frac{40 - V}{10}\right)\right]^{-1} \tag{3.38}$$

$$\beta_p = 0.09(-25 - V)\left[1 - \exp\left(\frac{V + 25}{20}\right)\right]^{-1} \tag{3.39}$$

Specific constants for use in the FH equations are given in Table 3.2.

Differences between the FH and HH equations reflect the specific properties of the myelinated and unmyelinated membranes. The activation variables have different powers, and the FH equations have an additional ionic term. The ionic terms appear more complex in the FH membrane as compared to the HH membrane. Despite the seemingly greater complexity of the FH membrane, the electrical properties associated with the myelinated membrane's AP development are very similar to the HH membrane. One observed difference includes a somewhat longer AP duration for the HH membrane. In addition, the HH equations respond to a prolonged current stimulus by producing multiple APs, whereas the FH membrane produces a single AP.

Other differences are seen between myelinated (A fiber) and unmyelinated (C fiber) response when the overall excitatory behavior of the neuron is considered. Some of these differences include faster conduction rates and lower thresholds to external currents for A fibers (Ruch et al., 1968). The A fiber is an ideal one to model for electrical stimulation studies; because of its lower excitation threshold for external current stimulation, the A fiber class will generally determine the limiting value for threshold currents. Nevertheless, it is important to consider the potential role of both fiber classes in electrical stimulation, and their functional properties in sensation and muscle response.

Species Dependence of Action Potential Dynamics

The HH and FH equations were developed from experiments on nonmammalian species—the HH experimenters used the giant axon of a squid, and the FH experimenters used the myelinated fiber of a toad. These species were chosen primarily for the large size of their axons, and relative ease of experimentation. The principles of membrane dynamics developed in that work have been found to be applicable to other species, including mammals. The specific details of the membrane electrodynamics may, however, depend on the species being considered. A quantitative description of membrane currents in rabbit myelinated nerve was developed by Chiu et al. (1979) using the HH experimental approach and mathematical framework. They showed that, as in squid and frog nerve, a transient inward sodium current was responsible for the initially rapid membrane depolarization. However, in contrast to squid and frog nerve, repolarization of the rabbit nerve was due to outward flow of only a single passive ionic component, and potassium current outflow was virtually absent. Despite the simpler description of the rabbit nerve, its calculated AP was very close to that of the frog nerve in both amplitude and time course.

Propagation of Nerve Impulses

The processes supporting AP propagation can be understood by referring to Fig. 3.10. Consider that point *A* on the axon is depolarized. The local point of depolarization causes ionic movement between adjacent points on the axon, thus propagating the region of depolarization. If depolarization were initiated from an external source on a resting membrane at point *A*, an AP would propagate in both directions away from the site of stimulation. Normally, however, an AP is initiated at the terminus of the axon, and propagates in only one direction.

After the membrane has regained the resting potential, it cannot be re-excited until a recovery period has passed. This period is termed the refractory state of the membrane. Before full recovery, the membrane becomes partially refractory; i.e., it requires a stronger depolarizing force to become excited. The refractory property is principally due to the prolonged decrease of the sodium deactivation variable, *h* (Fig. 3.8), which effectively turns off the membrane's sodium conductance until the passage of a recovery period that extends beyond the repolarization process.

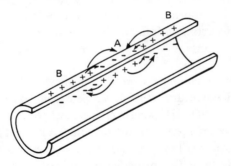

Figure 3.10   Spread of the depolarization wavefront. Depolarization occurring in region *A* results in charge transfer from the adjacent regions.

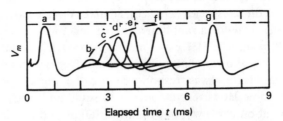

Elapsed time *t* (ms)

Figure 3.11   Illustration of the refractory period in frog nerve. The initial stimulus was applied at $t = 0$, resulting in response *a*. Subsequent stimuli were applied at various time delays, resulting in responses *b* through *g*. (Adapted from B. Katz, *Nerve, Muscle, and Synapse*, 1966, reproduced with permission of McGraw-Hill, Inc.)

Refractory behavior in frog nerve is illustrated in Fig. 3.11 (adapted from Katz, 1966). The figure shows the membrane response to an external current stimulus. An initial stimulus, applied at $t = 0$, results in response a. Successive stimuli applied at various subsequent times lead to responses $b-g$. The absolute refractory period exists from the AP spike to somewhat within the negative afterpotential. During this period, a second stimulus, no matter how strong, fails to produce a response. In a mammalian nerve at body temperature, the neuron is absolutely refractory for typically about 0.5 ms. Afterwards, for a period of several milliseconds, the nerve is relatively refractory. During this period, an increased stimulus is needed to produce a response, and that response will initially be feeble, as noted by responses $b-e$. After several milliseconds, the neuron fully recovers from its less excitable state.

The refractory recovery period sets an upper limit on the number of APs per second that the membrane can support. With externally applied electrical stimuli, an AP rate cannot be produced beyond about 2000 per second. In natural conditions in the body, the repetition rate rarely exceeds 500 per second and is more typically in the range from 10 to 100 per second (Brazier, 1977).

The AP velocity depends on the rate at which electrical charge is transferred from the locus of excitation to the region of membrane ahead of the AP. The charge-transfer rate, in turn, depends on the membrane capacity and the longitudinal resistance of the axon. Depending on the assumptions made, theoretical arguments suggest that conduction velocity ought to vary either as the square root of, or linearly with, fiber diameter (Paintal, 1967). Experimental evidence demonstrates that conduction velocity indeed increases with fiber diameter, and experimental data are frequently represented in terms of the ratio of conduction velocity to fiber diameter (Paintal, 1973).

The myelinated fiber is nature's means of obtaining fast conduction velocity without requiring unduly large fibers. The effective area requiring capacitive charging is limited mainly to the myelin-free internodes. Furthermore, the depolarization process is *saltatory*; i.e., it jumps from node to node. As a result, propagation can proceed at a much faster rate than would be the case with an unmyelinated fiber of the same diameter.

Table 3.3 lists some general characteristics of A and C fibers. The myelinated A fibers are sometimes subdivided into diameter classes designated A_δ, A_β, and A_α. The A_δ fibers are typically related to cutaneous pain and temperature sensation, the A_β fibers are related to mechanoreception, and the A_α fibers are related to proprioception and contraction of striated muscle (Li and Bak, 1976). Figure 3.12 illustrates the distribution of afferent

Table 3.3. *Characteristics of A and C fibers*

	Fiber class	
	A	C
Fiber diameter (μm)	1–22	0.3–1.3
Conduction velocity (m/s)	5–120	0.6–2.3
AP duration (ms)	0.4–0.5	2.0
Abs. refractory period (ms)	0.4–1.0	2.0
Velocity/diameter ratio (m/s μm)	6	~1.7
Myelinated	Yes	No

Source: Adapted from Ruch et al. (1968).

A fibers (Kandel and Schwartz, 1981). Unmyelinated fibers have a range more typically from 0.3 to 1.3 μm. The distribution of efferent neurons shows prominent clusters from 12 to 20 μm, and from 2 to 8 μm, with a pronounced nadir in the range from 8 to 12 μm (Ruch et al., 1968).

3.3   Action Potential Models for Cardiac Tissue

The electrical properties of excitable cardiac tissue have been defined using the experimental techniques and mathematical formalism

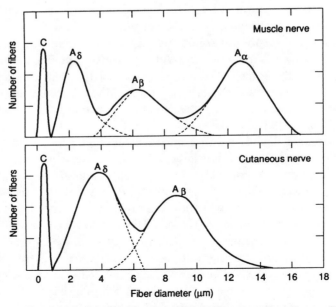

Figure 3.12   Distribution of peripheral afferent, myelinated nerve fibers: solid, cutaneous; broken, muscle. (Adapted from Martin, 1981.)

that lead to the HH model. The membrane electrodynamics are specialized for different cardiac tissue. One common feature is a prolonged excited state, as compared with a much shorter period in nerve tissue. The following describes several systems of cardiac membrane dynamics. Chapter 5 provides further details, and describes the role of the membrane dynamics in the overall function of the heart.

Cardiac Purkinje Model

The HH experimental techniques and mathematical formalism were applied to cardiac Purkinje fibers by Noble (1962). The Noble model used three ionic components, including a nonlinear sodium term similar to that in the HH equations, a nonlinear potassium term that exhibited a rectification process, and a nonspecific linear "anion" term. This model was later modified and extended by McAllister, Noble, and Tsien (1975) (referred to as the MNT model). The MNT Purkinje model is much more complex than the Noble Purkinje model—it includes nine separately described ionic components, along with five activation and deactivation variables. Some of the ionic terms in the MNT model are analogous to the HH nonlinear components; others are analogous to the HH linear leakage terms. The MNT model was modified and adapted for more efficient computer simulation by Drouhard and Roberge (1982b).

Further modifications and elaborations of the MNT model were made by DiFrancesco and Noble (1985) (DN model). This model included no less that 11 separate ionic components. A further feature of the DN model consisted of nonlinear differential equations governing the intracellular and extracellular ionic concentrations. None of the other models, including

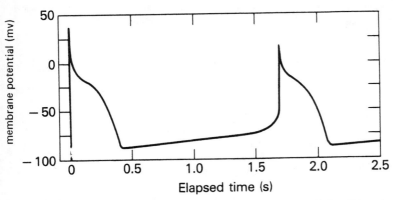

Figure 3.13   Computed action potential of cardiac Purkinje cell based on DiFrancesco-Noble model. (From DiFrancesco and Noble, 1985.)

the various nerve membrane models, includes this feature. Figure 3.13 illustrates the simulated AP of the DN model, showing pacemaker activity.

A review of the ionic processes in the Purkinje fiber and their role in the development of computational models was presented in a tutorial paper by Noble (1984). He draws attention to the growth over time of the complexity of Purkinje computational models, and remarks: "It is sincerely to be hoped that this is not a case of indefinite exponential growth, or, by the year 2000, we shall all have great difficulty in explaining cardiac excitation to ourselves, let alone to students with formidable memories" (p. 42).

The various Purkinje models may be regarded as representing differing degrees of precision in the treatment of ionic processes. From the point of view of electrical stimulation studies, it is not clear how much of this precision is required. At a minimum, we would like to reproduce the pacemaker AP process, and to simulate evoked extrasystolic excitation. Even the simplest model (Noble, 1962) gives a good reproduction of the Purkinje AP and its pacemaker action. Evoked extrasystolic excitation has been demonstrated in both the Noble (1962) and MNT (McAllister

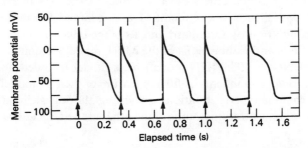

Figure 3.14   Effect of repetitive stimulation on the computed AP in the Noble Purkinje model. Arrows indicate application of stimulus at the rate 3/s. (From Noble, 1962.)

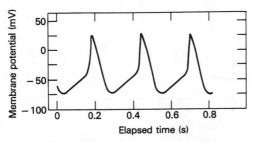

Figure 3.15   Pacemaker activity of sino-atrial node model. Response shown is for peripheral cell of frog heart. (From Noble and Noble, 1984.)

et al., 1975) papers. Figure 3.14 illustrates the response of the Noble model to repeated stimulation at a rate of 3/s. The alteration in the duration of the evoked APs seen in Fig. 3.14 is similar to experimental observations (Trautwein and Dudel, 1954; Geddes et al., 1972).

Sino-Atrial Model

The DN equations described above were modified by Noble and Noble (1984) to include electrical activity in the sino-atrial node. The modifications consisted of changes in values of particular equation parameters, such as membrane conductivity, time constants, and ionic current magnitude. Otherwise, the framework of the DN model was preserved. Figure 3.15 illustrates the response of the sino-atrial model for a peripheral cell with parameter values adjusted to reproduce the pacemaker activity in the frog heart. Slight modifications of the model were used for application to central cells, leading to a hyperpolarization potential somewhat reduced from that shown in Fig. 3.15.

Myocardial Model

Beeler and Reuter (1977) used the HH formalism to develop a mathematical model for the action potential of ventricular myocardial fibers (referred to as the BR model). Four individual components of ionic current were formulated, including potassium terms with inward-going rectification properties. The BR model included three voltage-dependent activation variables, and three inactivation variables. The authors argued that the additional ionic processes included in the MNT Purkinje model were not present to a significant degree in the myocardial AP mechanisms. In the

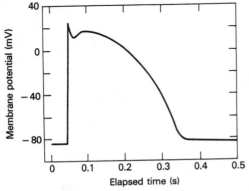

Figure 3.16.   Simulated action potential of ventricular myocardial fiber. (From Beeler and Reuter, 1977.)

BR model, a slow-inward current due to calcium involves a reversal potential that is not constant, but is computed from a differential equation involving the internal calcium ionic concentration.

Figure 3.16 illustrates the myocardial action potential as simulated by the BR model. An interesting property of the model is its ability to produce oscillatory potentials when depolarized by steady outward current, as seen in Fig. 3.17. The authors point out that although the electrical properties of the myocardium seldom produce spontaneous activity, numerous experimental studies with steady depolarizing currents have demonstrated oscillatory AP behavior such as seen in Fig. 3.17.

An Improved BR model was studied by Mogul and colleagues (1984). The authors demonstrated a foreshortening of the AP duration for stimulation at rates above 0.5 Hz. At 2 Hz, for example, the AP was shortened by nearly 20%—a result consistent with experiments. Foreshortening of the AP duration was also noted above for the Purkinje model. This property allows the cardiac system to support extrasystolic excitation at a rate higher than what might be supposed on the basis of a normal AP duration.

The BR model includes *ad hoc* assumptions about the kinetics of the sodium channel, and these assumptions lead to sodium conductance and constants of activation and deactivation that differ substantially from the

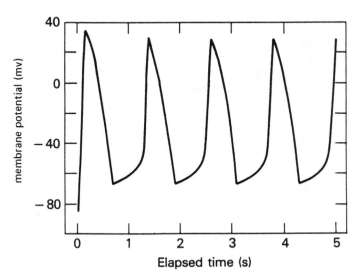

Figure 3.17   Oscillatory AP response of Beeler–Reuter model for ventricular myocardial fiber in response to steady outward current. (From Beeler and Reuter, 1977.)

HH squid model (Ebihara and Johnson, 1980). Using their own data on Na kinetics, Ebihara and Johnson developed Na conductance equations that follow the HH format, and that produce conductance terms and activation and deactivation constants much more in conformance with the squid model.

3.4 Sensory Transduction[†]

The body is equipped with a vast array of sensors (receptors) for monitoring its internal and external environment. The receptor converts a stimulus to an electrical potential that initiates a propagating AP. Receptors are specialized to respond most efficiently to a specific type of stimulus, although they may also respond to a variety of stimuli if the intensity is great enough. We shall be concerned here with the *somatosensory* system, i.e., the system of receptors found in the skin and internal organs. There are other specialized receptors in the visual and auditory systems, chemical receptors that monitor taste and smell, and other special chemical receptors by which neurons communicate with one another.

The somatosensory receptors can be classified into four general categories: *mechanoreceptors, thermoreceptors, chemoreceptors,* and *nociceptors.* Numerous specializations of mechanoreceptors respond to specific attributes of mechanical stimulation. Thermoreceptors are specialized to respond to either heat or cold stimuli. Nociceptors are unresponsive until the stimulus reaches the point where tissue damage is imminent, and are usually associated with pain. Many nociceptors are responsive to a broad spectrum of noxious levels of mechanical, heat, and chemical stimuli (Ruch and Patton, 1979). Both myelinated and unmyelinated nociceptors have been identified.

Figure 3.18 illustrates several cutaneous mechanoreceptors. The *pacinian corpuscle* is a so-called rapidly adapting receptor because it responds to the onset or termination of a pressure stimulus. As such, it transduces acceleration of skin displacement (Schmidt, 1978). The *Meisner corpuscle* is a less rapidly adapting receptor, and is most responsive to the velocity of skin displacement. The *Rufini ending* and *tactile disks* are slowly adapting, and respond most efficiently to steady pressure. Figure 3.18 also shows hair receptors that respond to displacement of hair follicles. Other mechanoreceptors, such as stretch receptors and muscle spindles, respond to movement and position of the muscles, and serve to monitor and regulate body posture and movement. An additional class of

---

[†]The following are useful references for this section: Kandel and Schwartz (1981); Lamb et al. (1984); Ruth et al. (1968), Ruch and Patton (1979).

receptors, known as free nerve endings, appear as branching structures on the end of the neuron. Free nerve endings are involved in a variety of specialized transduction processes in both nociceptive and non-nociceptive classes. It is not always clear from morphological evidence what distinguishes the transduction specializations of free nerve endings.

Intensity Coding

When a receptor is stimulated, it produces a voltage change called a *generator potential* at the terminal ending of its axonal connection. Unlike the all-or-nothing response of the axon, the generator potential is graded: if you squeeze a pacinian corpuscle, it produces a voltage at the axon terminus; if you squeeze it harder, it produces a greater voltage.

The generator potential initiates one or more APs that propagate along the axon. The AP repetition rate will depend, in part, on the intensity of the stimulus. Figure 3.19 illustrates the response rate of a slowly adapting (Fig. 3.19*a*) and an intermediately adapting (Fig. 3.19*b*) receptor (from Schmidt, 1978). The slowly adapting receptor responds to a constant pressure stimulus, although the AP rate is greatest at the onset of the stimulus. The intermediately adapting receptor AP rate is more or less constant during a constant-velocity pressure stimulus. For a rapidly

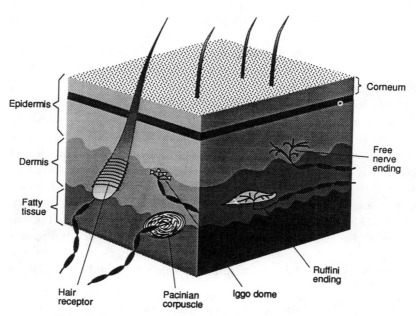

Figure 3.18   Morphology of several mechanoreceptors in the hairy skin.

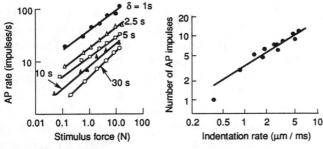

(a) Slowly adapting receptor   (b) Intermediate adapting receptor

Figure 3.19   Response of (a) slowly adapting receptor and (b) intermediate
adapting receptor to constant force stimulus. Part (a) shows AP rate of
various time delays (δ) after onset of pressure stimulus. Part (b) indicates total
number of APs during 0.5-s constant velocity stimulus. (Adapted from
Schmidt, 1978.)

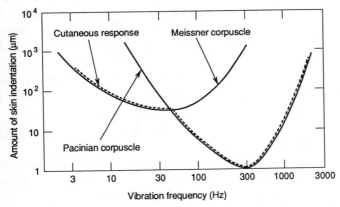

Figure 3.20   Threshold response for vibratory mechanical stimulus applied to
the skin. (Reprinted by permission from Martin, 1985, Elsevier Publishing Co. Inc.)

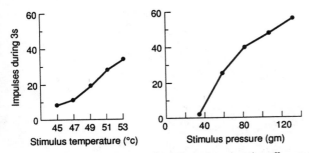

Figure 3.21   Response of myelinated nociceptive afferents innervating
monkey hand. Stimulus duration is 3 s for both types of stimuli. (From
Campbell et al., 1979.)

adapting pacinian corpuscle, a single AP is produced at the onset or termination of a pressure stimulus. However, for a vibratory stimulus, a train of APs is produced at the rate of the vibration frequency. The intensity of the threshold amplitude of the sinusoidal mechanical stimulus on the skin is shown in Fig. 3.20. Sensitivity to vibratory mechanical stimuli on the skin is greatest in the 200–300-Hz range. Anesthetization of the skin elevates thresholds only for frequencies below about 50 Hz. This can be explained by two modes of transduction: low-frequency sensations are due to more superficial hair follicle receptors or Meissner corpuscles; high-frequency vibration is detected by pacinian corpuscles located in deeper strata (Martin, 1985).

Figure 3.21 illustrates AP rate versus stimulus intensity for unmyelinated nociceptors (Campbell et al., 1979), showing a monotonic relationship between the average AP rate and intensity of both heat and pressure stimuli. Other studies (LaMotte el al., 1984) demonstrate a monotonic relationship between heat intensity and both AP rate and pain rating in nociceptive C fibers. Although AP rate is important in sensory encoding of stimulus intensity and painfulness, there are a number of other important factors as well. These include the spatial and temporal patterns of stimulation, and the number of neurons brought to excitation (recruitment). Chapter 7 provides additional discussion of the role of these factors in electrical stimulation.

3.5   Muscle Function

The previous section discussed the role of afferent neurons in conveying information from the body's sensory system to the central nervous system (CNS). Efferent neurons carry information from the CNS to the muscles to effect contraction. Figure 3.22 illustrates structural

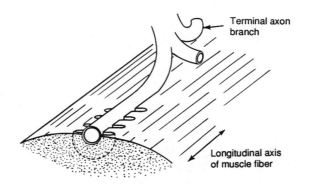

Figure 3.22   End-plate region of frog muscle fiber. (After Birks et al., 1960.)

features of the efferent neuromuscular junction at the *end plate* of the muscle (after Birks et al., 1960; Woodbury et al., 1966). When the AP reaches its terminus at the end plate, a chemical neurotransmitter is released across the nerve/muscle gap, which causes depolarization of the muscle cells. The result is that the muscle membrane is excited, and a depolarization wave is propagated in the muscle away from the end-plate region.

Figure 3.23 illustrates the anatomical structure of muscle, showing progressively smaller systems of structural elements. When the muscle is excited, the individual fibrils (shown in the lower part of the figure)

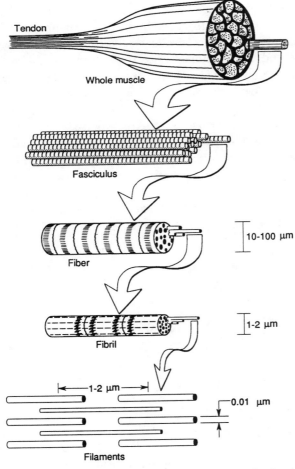

Figure 3.23   Skeletal muscle and filament structure of striated muscle.

slide together, resulting in muscle contraction. The muscle illustrated in Fig. 3.23 is called *striated* because of the microscopically visible striations that arise from the arrangement of contractile elements. Striated muscle is found in the skeletal system. Skeletal muscle is under voluntary control, and is innervated by the somatic nervous system. *Smooth muscle* differs from striated muscle in that the characteristic cross-striation pattern is absent. Smooth muscle is involuntary, and is found in vessel artery walls, air passages of the lungs, and in various tissues of the intestines and reproductive system.

An action potential launched along a motor neuron results in a single contractile quantum called a *twitch*. When a succession of APs is produced on the motor neuron, the individual twitch quanta fuse as illustrated in Fig. 3.24. In this example, maximum muscle tension is achieved at an AP rate of about 80/s, leading to a condition of maximum fusion termed *tetanus*. Gradation of muscle tension results from fusion of individual twitch quanta, and by recruitment (the excitation of additional neurons). Normally, the APs in individual motor neurons are asynchronous, and muscle tension can be finely graded. However, with stimulation by externally applied currents of a repetitive or oscillatory nature, the twitch quanta of various muscle fibers may be synchronized. Experimental data show that tetanus of the muscle group occurs at current levels that are only moderately above the twitch threshold value (Oester and Licht, 1971).

Externally applied electric currents can stimulate muscle by exciting motor neurons, or the muscle fibers themselves. Direct stimulation of muscle fiber requires much higher currents than does stimulation of enervated muscle (Harris, 1971; Walthard and Tchicaloff, 1971). As a result,

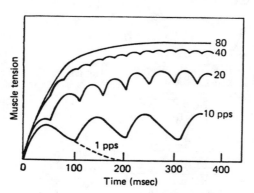

Figure 3.24   Effects of AP rate on muscle tension. [Adapted from McNeal and Bowman (1985a). Reprinted by permission from *Neural Stimulation*, vol. II. CRC Press Inc., Boca Raton, FL.]

stimulation of muscle by external electric currents will usually take place most efficiently through neural excitation. See Chapter 8 for details of electrical stimulation of muscle and motor neurons.

The propagating excitation wave along the muscle, when recorded through cutaneous electrodes, is called an *electromyogram* (EMG). The EMG signal has been used in various experimental and clinical applications as an indicator of muscle activity, and can often be more easily detected than direct observation of muscle movement or tension.

3.6 The Spinal Reflex

Muscle movement can be electrically stimulated via excitation of motor neurons, or via direct excitation of muscle fibers. Reflex activity represents a third mechanism whereby muscle movement may result from electrical stimulation. Here, stimulation may be initiated at cutaneous or muscle sensory afferents, with muscle movement ensuing through the reflex arc. When we touch a hot object, for example, our hand jerks away before we have consciously appraised the situation. Afterward, deliberate action takes place. The initial response, in this case, takes place outside of conscious control through the *spinal reflex*.

Figure 3.25 illustrates organizational features of the spinal reflex arc. Sensory inputs, originating from cutaneous or muscle receptors, communicate with muscle efferents through synaptic processes in the spinal column. Intermediate neurons (*interneurons*) may also exist within the

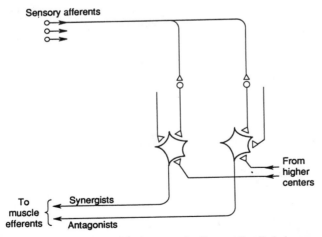

Figure 3.25 Organizational features of reflex arc. Small circles represent receptors triangles represent synaptic terminals. Synaptic summation occurs at dendritic processes within spinal column.

reflex arc. Inputs from higher centers may also be included. These modify the reflex action based on central processes—for example, we can deliberately inhibit the reflex action to maintain contact with a hot object. The synaptic summation may combine inputs as either inhibitory or excitatory, such that the presence of a synaptic output will depend on a weighted summation of inputs, thus forming a computing system within the spinal column. The output of this system consists of excitatory and inhibitory signals to muscle groups to cause coordinated, patterned movement.

The knee-jerk reflex is a simple example of a *monosynaptic* type in that it omits the interneurons such that the synapses of sensory inputs directly activate motor neurons. The motor neurons, in turn, produce AP signals that cause contraction in synergistic muscle groups, and inhibit AP signals to cause relaxation in antagonistic muscle groups.

4

Excitation Models

4.1   Introduction

Variations in the stimulus waveform or in the electrode arrangement can result in vastly different stimulation thresholds. To understand the factors responsible for electrical stimulation, computational models may be used that connect features of the stimulus current with properties of excitable tissue. A complete analysis model must account for the nonlinear membrane properties described in Chapter 3, as well as the spatial distribution of stimulus current. However, a simpler approach based on a linear, spatially limited membrane can provide insight into underlying electrical processes, and also provide closed-form mathematical expressions for some excitation relationships. This chapter begins with simple linear analysis models, and proceeds to a more complex, spatially extended nonlinear model.

The models treated here are essentially electrical cable representations of single fibers. Clearly, electrical stimulation normally involves a much more complex macroscopic system. Both sensory and motor responses will depend on the spatial and temporal patterns of stimulation; cardiac responses involve a three-dimensional dynamic system. Nevertheless, the response due to large-scale excitation is a result of the individual excitable elements. As will become apparent in subsequent chapters, electrical responses attributable to individual fibers can explain many observed macroscopic responses. Thus, single-fiber excitation models provide powerful analytic tools for evaluating a wide range of electrical stimulation properties.

This chapter substantially adapts material from the following references: Reilly et al. (1983, 1985); Reilly and Bauer, (1987); Reilly (1988).

4.2   **Linear Strength-Duration Model**

A simple analysis model treats an isolated segment of an excitable fiber as a linear electrical circuit as shown in Fig. 4.1. The membrane is assumed to consist of capacitance c_m and resistance r_m. The simplified analysis circuit is presumed to apply to a small patch of excitable membrane. The stimulus current flowing across the membrane is depicted as originating from a current source. The resting potential is treated as 0 V, and depolarization is analyzed as the voltage rise across the electrical network. In a simplified analysis, r_m is treated as a constant up to the threshold of the action potential. In actuality, because of the nonlinear membrane properties described in Chapter 3, r_m is relatively constant only up to about 80% of the action potential (AP) threshold (McNeal, 1976).

The equations governing the membrane response of the *RC* model to current injection are

$$i(t) = i_c(t) + i_R(t) \tag{4.1}$$

$$\frac{1}{c_m} \int i_c(t)\, dt = i_R(t)R \tag{4.2}$$

where $i(t)$ is the total current, $i_c(t)$ is the capacitive displacement current, and $i_R(t)$ is the current flowing in the resistance R. Consider that this simple circuit is excited by a step current pulse having the form

$$i(t) = \begin{cases} I & t \geqslant 0 \\ 0 & t < 0 \end{cases} \tag{4.3}$$

where I is the peak current. Equations (4.1) and (4.2), in connection with Eq. (4.3), may be readily solved (such as with Laplace transform methods)

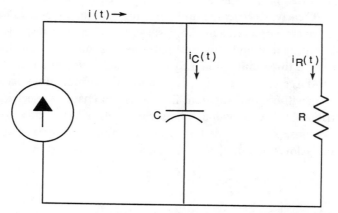

Figure 4.1   Simplified linear circuit model for isolated patch of excitable membrane.

as

$$v_m(t) = i_R(t)R = IR(1 - e^{-t/\tau_m})$$ (4.4)

where $\tau_m = r_m c_m$; τ_m is called the *membrane time constant*.

Assume that there is a single depolarization voltage, V_T, needed for excitation. This is a simplification, since the threshold depolarization voltage rises for very short pulses (e.g., $< 10\,\mu s$) (Dean and Lawrence, 1983), or for biphasic stimuli (Reilly et al., 1985). For a pulse of duration t, the maximum depolarization voltage is evaluated from Eq. (4.4). The threshold current (I_T) required to drive the transmembrane potential to V_T can be derived from Eq. (4.4) as

$$I_T = \frac{V_T/R}{1 - e^{-t/\tau_m}}$$ (4.5)

The threshold current attains a minimum value (I_0) for an infinitely long pulse. As $t \to \infty$, $I \to I_0 = V_T/R$, and Eq. (4.5) may be expressed as the normalized relationship

$$\frac{I_T}{I_0} = \frac{1}{1 - e^{-t/\tau_m}}$$ (4.6)

The threshold charge, Q_T, necessary for achieving the depolarization voltage, V_T, is

$$Q_T = I_T t = \frac{I_0 t}{1 - e^{-t/\tau_m}}$$ (4.7)

The threshold charge attains a minimum value, Q_0, for short-duration pulses. As $t \to 0$,

$$Q_T \to Q_0 = \frac{I_0}{\tau_m}$$ (4.8)

The expression for normalized depolarization charge is

$$\frac{Q_T}{Q_0} = \frac{t/\tau_m}{1 - e^{-t/\tau_m}}$$ (4.9)

The energy dissipated by the rectangular stimulus is given by $I^2 Rt$ (or, equivalently, IQR), where R is the resistance in the current path. It follows that the threshold energy, E_T, is given in normalized form as

$$\frac{E_T}{I_0^2 R} = \frac{t}{(1 - e^{-t/\tau_m})^2}$$ (4.10)

As demonstrated by Pearce and colleagues (1982), the stimulus energy attains a minimum value, E_0, when $t = 1.25\tau_e$; this corresponds to the minimum stimulus energy

$$E_0 = 2.46 I_0^2 R \tau_e$$ (4.11)

Figure 4.2 illustrates the normalized threshold current, charge, and energy versus the normalized duration (t/τ_m).

A parallel treatment has been derived (Blair, 1932a, 1932b; Reilly et al., 1983) for exponentially decaying current pulses of the form

$$i(t) = \begin{cases} Ie^{-t/\tau} & \text{for } t \geqslant 0 \\ 0 & \text{for } t < 0 \end{cases} \tag{4.12}$$

In Eq. (4.12), τ represents the time constant of the exponential current pulse. Such exponential currents are encountered with capacitive discharge stimuli (see Chapter 2). The solutions for normalized threshold current and charge are

$$\frac{I_T}{I_0} = p^{1/p-1} \tag{4.13}$$

$$\frac{Q_T}{Q_0} = p^{p/p-1} \tag{4.14}$$

where $p = \tau/\tau_m$. When rectangular and exponential stimulus criteria are compared, Eqs. (4.9) and (4.14) both converge to the same minimum charge

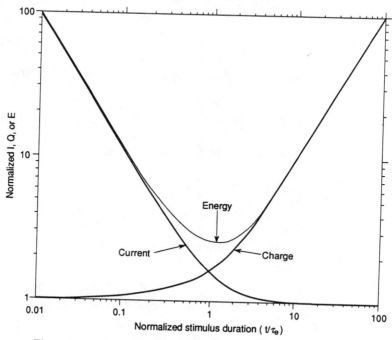

Figure 4.2   Calculated strength-duration relationships for square-wave monophasic current.

thresholds for short durations, and Eqs. (4.6) and (4.13) converge to the same minimum peak current threshold for long durations (Reilly and Larkin, 1983). This observation is equally valid when a more complete nonlinear model is used to represent the excitable fiber (see Fig. 4.14).

The simple linear model points out some important features concerning electrical thresholds for monophasic currents. For short-duration stimuli, the amount of charge transferred is the fundamental quantity defining stimulus potency. For brief monophasic transients, the details of the waveshape are relatively unimportant—both rectangular and exponential stimuli converge to the same minimum threshold charge. For long-duration monophasic currents, the peak current is the main determinant of electrical thresholds. Stimulus energy alone is not a meaningful index of excitability, as has been supposed by some in the past.

Empirical Strength-Duration Relationships

An empirical strength-duration (S-D) relationship for excitation of nerves was first derived by Weiss (1901). We can marvel today at the ingenuity of this early researcher, who, in those days of only crude electronics and measurement devices, was able to perform delicate and precise experiments that retain validity today. An expression equivalent to the Weiss formulation can be stated as

$$I_T = I_0\left(1 + \frac{\tau_e}{t}\right) \tag{4.15}$$

and

$$Q_T = Q_0\left(1 + \frac{t}{\tau_e}\right) \tag{4.16}$$

where I_T and Q_T are threshold current and charge, respectively, I_0 is the minimum threshold current for long pulses ($t \to \infty$), Q_0 is the minimum threshold charge for short pulses ($t \to 0$), t is the duration of the pulse, and τ_e is an experimental parameter related to the time response of the tissue being excited. Equation (4.16) is the expression derived by Weiss; Eq. (4.15) follows from the relationship $Q_T = I_T t$. Equation (4.16) plots as a straight line on linear paper, with a minimum value Q_0 at $t = 0$. Equation (4.15) is sometimes referred to as a "parabolic" relationship.

Lapicque (1907) studied the data of Weiss and proposed what he called a "logarithmic" relationship. Lapicque's formulation can be equivalently expressed as

$$I_T = I_0\left[1 - \exp\left(\frac{-t}{\tau_e}\right)\right]^{-1} \tag{4.17}$$

Equation (4.17) is now referred to as an "exponential" relationship.

Equation (4.17) is the same as that derived for the simple linear model [Eq. (4.6)], with the exception that Eq. (4.17) uses the experimental time constant, τ_e, rather than the membrane time constant, τ_m. The value of the time constant derived from the simple product of membrane resistivity and capacity, $r_m c_m$, is very different from the value determined by fitting experimental data to Eq. (4.17). More will be said about experimental time constants presently.

Lapicque introduced the term *rheobase* to describe the minimum threshold current achieved for very long pulses. He also defined *chronaxie* as the duration of a threshold current having a magnitude twice rheobase. These terms have become standard lexicon today. In both the parabolic and exponential formulations,

$$\text{Rheobase} = I_0 \tag{4.18}$$

Chronaxie may be related to τ_e in the two formulations as

$$\text{Chronaxie} = \tau_e \quad \text{(parabolic formula)} \tag{4.19a}$$

and

$$\text{Chronaxie} = \tau_e \ln 2 = 0.693\tau_e \quad \text{(exponential formula)} \tag{4.19b}$$

In both formulations, the parameter τ_e may be determined by the ratio of minimum charge to minimum current:

$$\tau_e = \frac{Q_0}{I_0} \tag{4.20}$$

Equation (4.20) suggests a simple and practical means of determining the parameter τ_e in both the parabolic and exponential relationships. It suggests that one may determine the experimental time constant without having to trace out the entire S-D curve—rather, it is sufficient to evaluate the stimulus threshold at only two durations (one very long and one very short).

Figure 4.3 compares the parabolic and exponential expressions for current thresholds. A third curve applies to a myelinated nerve model, which will be described presently. Both parabolic and exponential forms of the S-D relationship have been used to represent experimental data. Experimental curve fits for the two formulas have been compared for nerve (Bostock, 1983) and cardiac (Mouchawar et al., 1989) excitation. Over the range of stimulus durations studied, the two formulas provide reasonably good curve fits, although one or the other may be marginally preferred in specific instances. One attraction of the exponential formulation is that it can be related to the linear *RC* network model (Fig. 4.1), which provides a heuristic, albeit crude, physiological model of neural excitation. A more accurate analysis model would include the nonlinear and spatially extensive properties of the excitable membrane as described in the next section.

Experimental values of τ_e can depend strongly on the spatial distribution of stimulus current. The apparent time constant generally becomes longer as the membrane current is distributed in a more gradual fashion. This is clearly demonstrated in Sec. 4.7, and also by Jack et al. (1983, p. 33) using a linear cable model. These studies show that the apparent time constant with membrane current injected at a discrete point may be less than half the value observed when the current is distributed more gradually along the fiber.

Large variations in τ_e are seen for different types of excitable tissue. An experimental average for nerve excitation is around 0.27 ms, with a wide experimental range (see Table 7.1). The experimental average for cardiac excitation is around 3 ms, also with a wide experimental range (see Table 6.6).

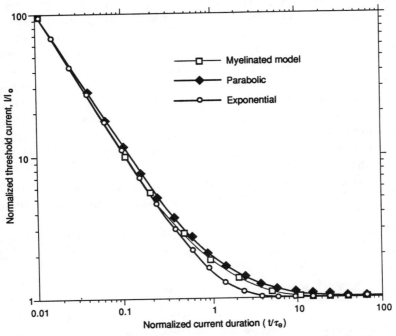

Figure 4.3   Strength-duration curves for neural excitation by rectangular monophasic current pulses. Parabolic and exponential curves are from Weiss and Lapicque formulations; thin curve applies to myelinated nerve model.

4.3   Electrical Cable Representations

While the simple linear model of the previous section provides useful insight into the neural excitation process, it is of limited value in attempting to describe the full range of interrelations between the excitable membrane and stimulus currents. In particular, the simple model does not account for the effects of the spatial distribution of stimulus current, differential sensitivity to anodal versus cathodal currents, response to biphasic currents, or response to sequences of pulsed stimuli. In order to study these factors adequately, we need a more complex model involving the spatial and temporal interrelationships of the excitable membrane.

One-Dimensional Cable Models

Excitation of nerve and muscle cells can be profitably analyzed using electrical cable theory. The cable equations to be applied to this problem were originated by O. Heaviside in 1876 in connection with the analysis of the first trans-Atlantic telegraphy cable (Nahin, 1987). This section briefly introduces cable theory as applied to the excitable membrane. More extensive treatments of this subject can be found in other publications (Rall, 1977; Jack et al., 1983; Plonsey and Barr, 1988).

An assumption of cable theory is that at any longitudinal position x along the cable, the inside and outside potentials do not vary with the position along the cable's circumference. Considering the small diameters of the cells to be studied ($< 20\,\mu m$), this assumption is a reasonable one for biological applications, and it permits the cable to be modeled in a one-dimensional fashion, as in Fig. 4.4. The illustration shows the inside as having a resistance per unit length r_i (Ω/cm), a distributed membrane resistance from inside to outside having unit length value r_m ($\Omega\,cm$), and a distributed length-proportional capacity c_m ($\mu F/cm$). I_e and I_i are the currents flowing internally and externally to the cable, V_e and V_i are the external and internal potentials, r_e is the external resistance per unit length of the cable due to the surrounding medium, and i_m is the membrane current.[†]

The currents and voltages are assumed to vary with the longitudinal distance along the cable. The cable equation is expressed in differential form as

$$\frac{\partial^2 V_m}{\partial x^2} = (r_i + r_e)i_m \tag{4.21}$$

[†]Unit length resistances are defined such that the internal and external resistances increase with length, L, in accordance with $R_i = r_i L$ and $R_e = r_e L$, and leakage resistance decreases with length as $R_m = r_m/L$ (both R_i and R_m have units of ohms). The capacitance between inside and outside over the length L is given by $C_m = c_m L$.

where V_m in the membrane voltage defined by $V_i - V_e$. An additional relationship is that membrane current consists of both ionic and capacitive components; that is,

$$i_m = I_i + c_m \frac{\partial V_m}{\partial t} \tag{4.22}$$

For a linear cable, $I_i = V_m/r_m$, where r_m is treated as a constant, independent of V_m. when the cable equations are applied to an excitable membrane at or near the excited state, the ionic term is instead described by the nonlinear equations introduced in Chapter 3.

An alternative form of cable equation is obtained by combining Eqs. (4.21) and (4.22), and expressing in normalized coordinates as

$$\frac{\partial^2 V_m}{\partial X^2} - V_m - \frac{\partial V_m}{\partial T} = 0 \tag{4.23}$$

where $X = x/\lambda$ and $T = t/\tau_m$ are dimensionless space and time variables scaled to the space constant λ and time constant τ_m of the cell membrane. The scaling constants are given by $\lambda = [r_m/(r_e + r_i)]^{1/2}$, and $\tau_m = r_m c_m$. Since r_m has units of $(\Omega \, cm)$, and r_e and r_i have units of (Ω/cm), it follows that λ has units of centimeters; τ_e has units of seconds. λ is also called the *electrotonic distance* of the membrane; it defines the distance along the membrane that a steady-state voltage disturbance due to point current injection will decay to e^{-1} of the value at the disturbed location. τ_m similarly defines the time response of an isolated piece of membrane when a step voltage is applied. In general, $r_i \gg r_e$ and it follows that $\lambda \approx \sqrt{r_m/r_i}$. Taking an example from Jack et al. (1983, pp. 23–24), for a 20-μm-diameter fiber with $r_i = 30 \, M\Omega/cm$ and $r_m = 1.6 \, M\Omega \, cm$, the space constant is calculated to be 0.23 cm.

Consider a cable of finite length $2L$ placed in a longitudinal DC field

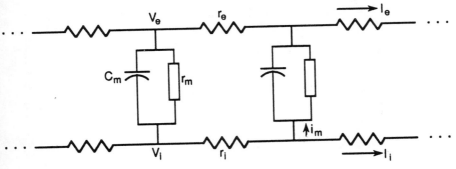

Figure 4.4   Electrical model for linear cable.

of strength E. The steady-state solution for membrane voltage in response to a nonfluctuating field is given by (Sten-Knudsen, 1960)

$$V_m(X) = -E\lambda\left(\frac{\sinh X}{\cosh L/\lambda}\right) \tag{4.24}$$

where $X = 0$ is taken as the center of the fiber, and the ends are at $\pm L$. Transient solutions of the membrane voltage are presented by Rall (1977). Figure 4.5 illustrates Eq. (4.24) as a function of distance for several cable lengths. V_m has odd-valued symmetry about $X = 0$, and only one-half of the function is displayed in Fig. 4.5. The symmetrical property means that the fiber is hyperpolarized along its anode-facing half, and is depolarized along its cathode-facing half. The maximum membrane voltage is attained at the ends of the fiber, and has the value

$$V_m = -E\lambda\tanh\frac{L}{\lambda} \tag{4.25}$$

The maximum possible value of membrane voltage is $E\lambda$, and is obtained

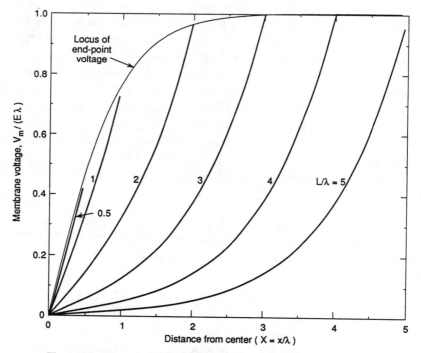

Figure 4.5   Normalized membrane voltage of a finite cable immersed in a DC field of strength E. Cable length $= 2L$. Voltage has odd-function symmetry about $X = 0$.

as $L \to \infty$. But even for fibers of only modest length, the maximum value is closely approached. For example, with $L/\lambda = 2$ (total length $= 4\lambda$), the membrane voltage at the two ends is $\pm 0.964E\lambda$.

Figure 4.6 gives a physical interpretation of the distribution of current flow around an elongated cell that is placed in a medium having a uniform electric field (or, correspondingly, a uniform current density). The fiber is presumed to be oriented parallel to the direction of the undisturbed field. The flux lines indicate that the current flow through the membrane is greatest near the ends of the fiber. At a sufficient distance from the ends, internal and external fields become equal, and there is no further current crossing the membrane.

Results from cable theory show that a longitudinal electric field (or current flow) is required to excite a nerve fiber. Other theoretical studies (McNeal, 1976) demonstrate that current flow oriented perpendicular to the long axis of a nerve fiber is relatively ineffective compared with a longitudinal orientation. These expectations are verified by a variety of experimental studies (Ranck, 1975) showing that a neuron is very much less sensitive to a transverse field, as in Fig. 4.7a, as compared with a longitudinal field, as in Fig. 4.7b.

Nonlinear Models

Various computational models have been used to study the excitation properties of nerve fibers. Excitation properties have been studied for a spatially isolated segment of membrane, in which the current density crossing the membrane is the driving force. In these models, the membrane conductance is modeled by the Hodgkin–Huxley or Frankenhauser–Huxley equations that were introduced in Chapter 3. Despite simplifications, isolated membrane models have provided substantial insights into neuronal excitation by externally applied currents

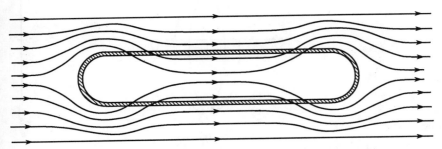

Figure 4.6 Representation of current flow around elongated cell placed in a medium having a uniform electric field (i.e., uniform current density). The membrane is assumed to be semipermeable to current flow.

(Bütikoffer and Lawrence, 1978; Dean and Lawrence, 1983, 1985; Motz and Rattay, 1986).

More complete representations include assumptions about mutual interactions among adjacent segments of the excitable membrane. One such model, developed by Cooley and Dodge (1966), uses a discrete cable model to represent an unmyelinated fiber, with stimulation by an intracellular electrode. In this model, membrane conductance is governed by the Hodgkin–Huxley equations. Myelinated fiber models studied by Fitzhugh (1962) and Bostock (1983) include in a cable model the passive circuit role of the myelin internode, in addition to the active FH conductances at the nodes. These models have also been configured to study excitation by current injection at a single point on the axon.

A difficulty with the aforementioned nonlinear models is that they presume knowledge of the current waveform and density crossing the membrane. In electrical stimulation problems we may be able to calculate the current within the medium containing the neuron, but not necessarily the current crossing the membrane. The waveform of current crossing the membrane can differ substantially from that in the surrounding medium

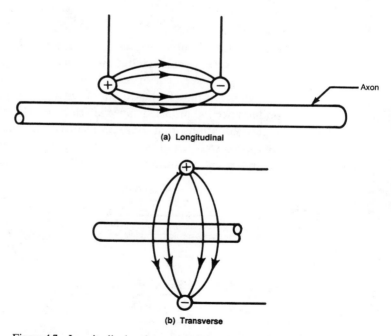

(a) **Longitudinal**

(b) **Transverse**

Figure 4.7   Longitudinal and transverse current excitation. An excitable cell is much less sensitive to a transverse field (*b*) as compared with a longitudinal field (*a*).

(McNeal, 1976). Furthermore, the force driving current into the membrane is the external field distribution along the axon, which cannot be described by the current density at a single point. These difficulties have been removed in the myelinated fiber model of McNeal, described in the following section.

4.4   Myelinated Nerve Model

Myelinated A fibers are distinguished from unmyelinated C fibers by faster conduction rates, shorter action-potential (AP) durations, and lower electrical thresholds (Ruch et al., 1968). Because of the lower thresholds, the myelinated fiber is a good choice for electrical stimulation studies.

Figure 4.8 illustrates electrical stimulation of a myelinated nerve fiber. The current emanating from the stimulus electrode through the conducting medium causes external voltage disturbances ($V_{e,n}$) at the nearby nodes. These disturbances force current across the membrane.

Figure 4.9 illustrates the representation of the myelinated nerve as formulated by McNeal (1976). The individual nodes are shown as circuit elements consisting of capacitance (C_m), resistance (R_m), and a potential source (E_r), which maintains the transmembrane resting potential. The voltages $V_{e,n}$ are the external nodal voltages as in Fig. 4.8. In the McNeal representation, the myelin internodes are treated as perfect insulators. This framework could be expanded to include passive myelin properties;

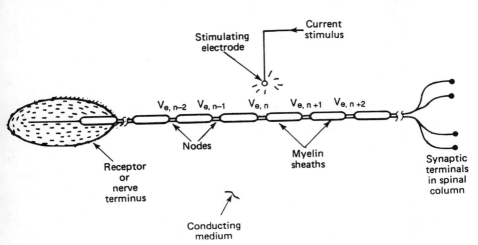

Figure 4.8   Representation of electrical stimulation of myelinated nerve. The current stimulus results in voltage disturbances (V_e) at the individual nodes. These in turn cause a local depolarization of the nerve membrane. (Adapted from Reilly and Larkin, 1984.)

however, such expansion would add significantly to the complexity of the model. The model is therefore a compromise between relatively simple single-node models and a more complete node-plus-myelin representation. Despite the compromise, the model is able to account for a variety of sensory and electrophysiological effects.

In Fig. 4.9, the current emanating from the nth node is the sum of capacitive and ionic currents, and is related to internal axonal currents by

$$C_m \frac{dV_n}{dt} + I_{i,n} = G_a(V_{i,n-1} - 2V_{i,n} + V_{i,n+1}) \tag{4.26}$$

where C_m is the membrane capacitance of the node, V_n is the trans-membrane potential at the nth node, $I_{i,n}$ is the internal ionic current flowing in the nth node, and $V_{i,n}$ is the internal voltage at the nth node. In this expression, V_n is taken relative to the resting potential, such that $V_n = 0$ applies to the resting potential and positive V_n applies to membrane depolarization.

Further relationships are given by

$$G_a = \frac{\pi d^2}{4\rho_i L} \tag{4.27}$$

$$G_m = g_m \pi d W \tag{4.28}$$

$$C_m = c_m \pi d W \tag{4.29}$$

where d is the axon diameter at the node, ρ_i is the resistivity of the axoplasm, L is the internodal gap, g_m is the subthreshold membrane conductance per unit area, c_m is the membrane capacitance per unit area, and W is the nodal gap width.

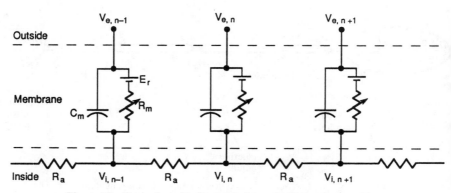

Figure 4.9   Equivalent circuit models for excitable membranes. The response near the excitation threshold requires that the membrane conductance be described by a set of nonlinear differential equations. (After McNeal, 1976.)

In Eq. (4.26), V_n is the voltage difference across the membrane:

$$V_n = V_{i,n} - V_{e,n} \tag{4.30}$$

where $V_{i,n}$ and $V_{e,n}$ are the internal and external nodal voltages, respectively, with reference to a distant point within the conducting medium outside the axon. Substituting Eq. (4.30) into Eq. (4.26) results in

$$\frac{dV_n}{dt} = \frac{1}{C_m}[G_a(V_{n-1} - 2V_n + V_{n+1} + V_{e,n-1} - 2V_{e,n} + V_{e,n+1}) - I_{i,n}] \tag{4.31a}$$

Equation (4.31a) may be analogously expressed in continuous form as

$$\tau_m \frac{dV}{dt} - \lambda^2 \frac{\partial^2 V}{\partial x^2} + V = \lambda^2 \frac{\partial^2 V_e}{\partial x^2} \tag{4.31b}$$

where V and V_e are membrane voltage and external voltage respectively at longitudinal position x, $\tau_m = C_m/G_m$, and $\lambda^2 = r_m/r_i$. Equation (4.31b) connects continuous and discrete spatial derivatives by: $\partial^2 V/\partial x^2 \approx V_{n-1} - 2V_n + V_{n+1}$, and $\partial^2 V_e/\partial x^2 \approx V_{e,n-1} - 2V_{e,n} + V_{e,n+1}$. An additional relationship used in (4.31b) is $I_{i,n} = VG_m$. A form of Eq. (4.31b), with the constants τ and λ explicitly expressed for a myelinated fiber, has been presented by Basser and Roth (1991). The main way that Eq. (4.31b) differs from the cable equation (4.23) is the inclusion of the right-hand term, in which the external field within the biological medium acts as a driving force on the membrane voltage. One conclusion that can be drawn from Eqs. (4.31a) and (4.31b) is that a second spatial derivative of voltage (or equivalently a first derivative of the electric field) must exist along the long axis of an excitable fiber in order to support excitation. Recently, the second spatial derivative of the external field has been included in an "activation function" in order to emphasize this essential aspect of stimulation (Rattay, 1986). In a typical analysis, the fiber is considered long and straight. However, excitation is nevertheless possible where the fiber is terminated or where it bends within a locally constant electric field (see additional discussion in Sec. 4.5). The orientation change or termination creates the equivalent of a spatial derivative of the applied field. Stimulation at "ends and bends" can, in fact, be the dominant mode of excitation in many cases.

The ionic current term in Eq. (4.31a) can be expressed for either a linear or a nonlinear membrane:

$$I_{i,n} = \begin{cases} G_m V_n & \text{(linear)} & (4.32a) \\ \pi d W (J_{Na} + J_K + J_L + J_p) & \text{(nonlinear)} & (4.32b) \end{cases}$$

Equation (4.32a) is a simple statement of Ohm's law for a linear conductor.

Equation (4.32*b*) applies to the nonlinear ionic current expressions for the FH membrane given by Eqs. (3.28)–(3.31).

Equation (4.26) is equivalent to the cable equation (4.23). In Eq. (4.31*a*), the $V_{e,n}$ values are specified from knowledge about the stimulus current distribution, and the V_n values are unknowns for which solutions must be found. Procedures for evaluating Eq. (4.31*a*) typically consider the $V_{e,n}$ voltages as being independent of the membrane currents, i.e., the current crossing the membrane is assumed not to perturb the voltages on the exterior of the fiber. For subthreshold conditions this is a reasonable assumption, but in an excited state, the membrane currents can be large enough to perturb the external voltages. As a consequence, the model will be less accurate in the excited state. Despite this limitation, the model accurately reproduces a wide range of excitation phenomena, including excitation and propagation, of action potentials.

The most accurate representation would treat all the nodes in the model as nonlinear. However, such treatment can result in excessive computing time, which can be reduced by limiting the number of nonlinear nodes. In McNeal's original work, he studied an 11-node array with one central node as nonlinear and all the others as linear. In his study, excitation current was introduced via an electrode near the central nonlinear node. He defined excitation as occurring when the nonlinear node reached a peak depolarization value of 80 mV. For the range of stimuli studied by McNeal, this arrangement was entirely satisfactory. However, for a more general range of stimulus parameters, some modifications are required.

The model described in this chapter is an extension of the one published by McNeal, with extensions that include FH nonlinearities at each of several adjacent nodes. Additional modifications include a test for excitation based on AP propagation,[†] the ability to model arbitrary stimulus waveforms, representation of stimulation at the neuron terminus, and representation of stimulation by uniform electric fields. This modified representation of McNeal's model as been referred to as a *spatially extended nonlinear nodal* (SENN) model (Reilly et al., 1985). Unless specifically noted, myelinated nerve model parameters used to obtain the data in this chapter are as noted in Table 4.1. The listed parameters are those originally used by McNeal.

To exercise the model, the spatial distribution of voltage along the axon as a result of the stimulating current must be specified. For example,

---

[†] The use of a single depolarization voltage is not always an adequate indicator of excitation when brief oscillatory stimuli are used. In that case, a threshold test based on propagation is needed (Reilly et al., 1985).

Table 4.1. *Base-case parameters for SENN model*

Fiber diameter (D)	[a]
Axon diameter (d)	0.7 D
Nodal gap (G)	2.5 mm
Axoplasmic resistivity (ρ_i)	100 Ω cm
External medium resistivity (ρ_e)	300 Ωcm
Membrane capacity (c_m)	2 μF/cm^2
Membrane conductivity (g_m)	30.4 mS/cm^2
Internodal distance (L)	100 D

[a]For point electrode stimulation, most data in this chapter apply to $D = 20\,\mu$m; in Section 4.1, $D = 10\,\mu$m is used for the base case.
Source: McNeal, 1976.

consider the voltage distribution for isotropic current propagation from a point electrode placed in a uniform medium. The "indifferent" electrode is taken to be in the conducting medium, far from the axon. The voltage at a radial distance, r, from the electrode is given by

$$V(r) = \frac{\rho I}{4\pi r} \tag{4.33}$$

where ρ is the resistivity of the medium. In the general case, I and V are functions of time that follow the stimulus waveshape.

The examples presented in this section are for a point electrode stimulating a 20-μm-diameter fiber having an internodal spacing of 2 mm; the electrode is assumed to be located one internodal distance (2 mm) from the fiber, and centered above the central node. For fibers smaller than 20 μm, excitation thresholds (at a given electrode distance) would be greater as noted in Sec. 4.7. Section 4.5 presents additional results for excitation by a uniform electric field within the conducting medium.

Threshold Criterion

The membrane response of the myelinated nerve model to a rectangular current stimulus is illustrated in Fig. 4.10. The example is for a small electrode that is 2 mm radially instant from a 20-μm fiber. The transmembrane voltage, ΔV, is scaled relative to the resting potential. The solid curves show the response at the node nearest the stimulating electrode. Responses to three different cathodal pulse magnitudes, all of the same duration (100 μs), are depicted. Response a is for a pulse at 80% of the threshold current, stimulus pulse b is at threshold, and pulse c is 20% above threshold. The threshold stimulus pulse in this example has an amplitude (I_T) of 0.68 mA. Figure 4.10 also shows the membrane response to a threshold pulse at the node nearest the electrode and at the next three adjacent nodes (broken lines). The time delay from node to node implies a propagation velocity of 43 m/s.

It is possible for a developing AP to be reversed and entirely abolished by the phase reversal in a biphasic waveform (Bütikoffer and Lawrence, 1978). For this reason, a simple membrane voltage test for AP excitation can be misleading, particularly for stimulus durations less than 100 μs. The difficulty is illustrated in Fig. 4.11, which shows the response to a single cycle of a sinusoidal stimulus having a period of 20 μs and an initial cathodal phase. Stimulus pulse *a*, having a peak amplitude of 9.1 mA, is above threshold: a propagated AP develops after about 0.3 ms. Stimulus pulse *b*, having a peak amplitude of 8.5 mA, is below threshold: the membrane response decays rapidly to the cell's resting potential (0 mV).

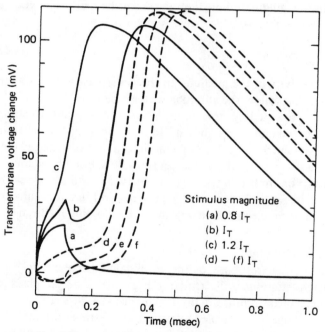

Figure 4.10   Response of myelinated nerve model to rectangular monophasic current of 100 μs duration, 20-μm-diameter fiber, point electrode 2 mm from central node. Solid lines show response at node nearest electrode for three levels of current. I_T denotes threshold current. Broken lines show propagated response at next three adjacent nodes for a stimulus at threshold. (From Reilly et al., 1985.)

Both responses, however, exceed a depolarization voltage of 80 mV, at a time corresponding to the peak of the cathodal phase of the stimulus.

In the SENN nerve model, excitation is tested on the basis of unambiguous AP propagation. The computer algorithm recognizes this condition by testing for adequate depolarization (80 mV) that propagates to the third node beyond the point of initial excitation. The threshold current value is determined by iterating between a level causing excitation with a level not causing excitation. The iteration is continued until the threshold and no-threshold levels differ by no more than 1%.

Position and Number of Nonlinear Nodes

The most general form of the nerve model would invoke the FH equations at each of a large number of nodes. However, to keep computations within feasible limits, it is necessary to minimize the number that are nonlinear, i.e., those governed by the FH equations. The necessary number of linear and nonlinear nodes depends on the stimulus waveform, the electrode/neuron geometry, and the extent to which AP propagation is modeled.

In previous work, McNeal (1976) evaluated a cathodal point electrode placed one-half an internodal distance radially from the axon. Using one

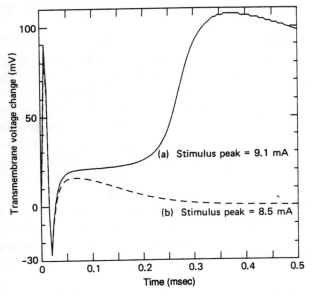

Figure 4.11   Nerve model response to one cycle of a sinusoidal pulse with initial cathodal phase and a period of 20μs. Stimulus (*a*) is suprathreshold, stimulus (*b*) is subthreshold. (From Reilly et al., 1985.)

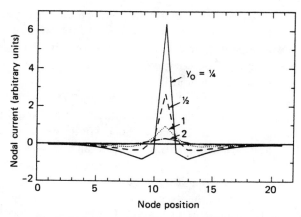

Figure 4.12   Steady-state nodal current distribution for myelinated nerve model. Stimulating electrode centered over node 11. For cathodal stimulation, positive nodal current indicates conventional current efflux (anionic flux is in opposite direction). Y_0 denotes distance between electrode and axon as a multiple of internodal distance. (From Reilly et al., 1985.)

nonlinear node closest to the electrode and five linear nodes on either side, he found that the excitation threshold could be determined within a percent or so relative to a much longer axon. However, this arrangement is not acceptable for anodal or biphasic currents, or for more distant electrode placements.

To understand the number of nodes required in the model, it is useful to examine the distribution of membrane current that arises from a stimulus. Figure 4.12 shows a spatial pattern of nodal current in response to a cathodal stimulus. The current distribution plotted here represents a steady-state condition, i.e., the convergent response to a long stimulus pulse. Current efflux[†] is represented by the positive current axis, and current influx by the negative axis. For an anodal stimulus, the figure would be inverted. An action potential results only from the depolarizing effect of nodal current efflux, represented by the positive direction of the vertical axis.

In Fig. 4.12, maximum current efflux with cathodal stimulation occurs at the node nearest the electrode. This point of maximum response is called the excitation node, i.e., the node where an AP would be initiated with a threshold-level stimulus. By inverting Fig. 4.12, it can be seen that

---

[†]In this chapter, unless otherwise specified, "current" flow refers to electronic current according to conventional engineering usage. Ionic current flow is, however, in the opposite direction.

there are two potential excitation nodes for an anodal stimulus; these move to more distant locations along the axon as the electrode is positioned farther away. In addition, the nodal current distribution becomes more gradual. In modeling anodal stimulation, it is necessary to ensure that nonlinear nodes are included at the excitation location.

Care must also be exercised in the selection of the total number of nodes. The number needed for modeling anodal stimulation is greater than with cathodal stimulation. If there are not enough, the nodal current distribution can be altered relative to a longer axon, and significant errors can be induced. These distortions can be illustrated by calculating the polarity sensitivity ratio (P), defined as the ratio of threshold current for anodal to that for cathodal stimulation (absolute values). Figure 4.13 illustrates P as a function of the electrode placement x_0 measured in a longitudinal dimension from the terminus of the node array, for various radial distances y_0 from the axon. The array is assumed to be terminated at a node. The parameters x_0 and y_0 are expressed as multiples of the internodal spacing. With this normalization, P is only weakly dependent on the ratio of nodal to axonal resistivity (R_m and R_a in Fig. 4.9), and is

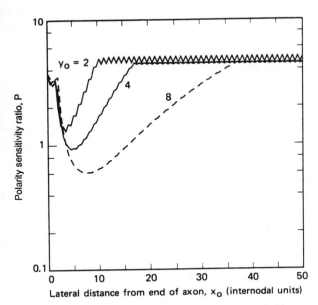

Figure 4.13   Ratio of anodal to cathodal threshold current when stimulating electrode is near the terminal end of a 91-node axon model. Y_0 denotes distance between electrode and axon as a multiple of internodal distance. Computations were done at multiples of half the internodal distance, resulting in a sawtooth pattern. (From Reilly et al., 1985.)

otherwise independent of parameters related to fiber size. The pulsewidth is assumed to be long; this incurs no loss of generality relative to other pulsewidths.

The value of x_0 at which P oscillates about fixed limits can be interpreted as the array length at which the axon model appears to be infinitely long. This point occurs when x_0 is 9, 13, and 37 nodes from the axon terminus for values of y_0 equal to 2, 4, and 8, respectively. Since these distances represent one-half the necessary extent of the axon model, we form the rule of thumb that, for general applications, the model should be at least nine times longer than the radial distance to a longitudinally centered electrode. For the computations reported in this chapter, the array length is at least 10 times y_0 but not less than 21 nodes. To study AP propagation or related phenomena (such as AP collision effects), the model requires nonlinear nodes at least as far as propagation is to be simulated.

4.5   Response to Monophasic Stimulation
Strength-Duration Relations

Figure 4.14 illustrates thresholds evaluated with the myelinated nerve model for anodal and cathodal rectangular stimuli with pulse durations of from 1 μs to 10 ms. Figure 4.13 also shows the S-D curve for

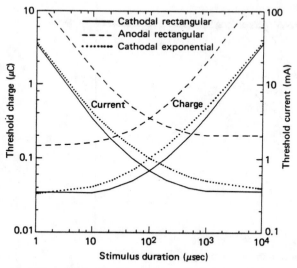

Figure 4.14   Myelinated nerve model strength-duration curves for monophasic stimuli. The left vertical axis indicates threshold charge for AP initiation. The right vertical axis indicates threshold current. The horizontal axis represents pulse duration for a rectangular stimulus or decay time constant for an exponential stimulus. (From Reilly et al., 1985.)

a cathodal exponential stimulus having the form $Ie^{-t/\tau}$. The curves in Fig. 4.14 are similar to those for the simplified linear model, described by Eqs. (4.6) and (4.9) for rectangular stimuli, and by Eqs. (4.13) and (4.14) for exponential stimuli. The threshold charge reaches a minimum at small stimulus durations for both exponential and rectangular stimuli; for long-duration stimuli, the peak current (equivalent to rheobasic current) is minimized. The exponential and rectangular stimuli have the same minimum charge threshold, indicating that, at brief durations, sensitivity is not affected by the fine structure of the monophasic pulse waveform.

The shape of the S-D curve for a linear RC network model of a single node can be characterized by the membrane's exponential time constant, given by the product $R_m C_m$ as noted in Sec. 4.2. The myelinated nerve model also has a response time that depends on both linear and nonlinear membrane properties and internodal resistance. Consequently, no single linear circuit parameter specifies the curves in Fig. 4.14. It is therefore useful to define an equivalent strength-duration time constant (τ_e) as the RC time constant of the linear single-node model having an S-D curve shape that best matches the shape of a given empirical curve. The value $\tau_e = 92.3 \, \mu s$ is obtained using a linear least-squares fit of Eq. (4.6) to the rectangular current threshold curve in Fig. 4.14. This value differs from the simple product of membrane capacitance and resistance in the subthreshold region of linear response, for which the product $R_m C_m$ is only 66 μs. As suggested by Eq. (4.20), we can more simply estimate τ_e as the ratio Q_0/I_0, which results in $\tau_e = 92 \, \mu s$. This estimate agrees quite well with the value obtained by a least-squares fit of the entire S-D curve.

The value of τ_e cited above applies to a point electrode whose radial distance from the axon is equal to an internodal space. As noted in Sec. 4.3, τ_e is expected to increase as the membrane current is distributed more gradually along the axon. We would therefore expect τ_e to be larger at increased electrode distances. This expectation is verified with the model results reported in Sec. 4.7. When the distance to a point electrode is varied from $\frac{1}{2}$ to 4 internodal spaces, the observed value of the τ_e ranges from 56 to 128 μs. The theoretical value of τ_e applying to uniform field excitation is discussed under the heading "Current Density and Electric Field Relationships."

Values of τ_e obtained with the myelinated model agree quite well with in-vivo experimentation of animal nerve (Reilly et al., 1985), and is within the range of experimental time constants reported for electrocutaneous nerve stimulation (see Table 7.1). Nevertheless, experimental time constants vary over a large range, many being more than twice the values predicted with the nerve model. It is hypothesized that electrocutaneous stimulation may originate at neural end structures (receptors, free nerve endings, and motor neuron end plates). In order to better explain experimental data, it may be necessary to develop specialized electrical models for neural end structures.

Polarity Sensitivity

The vertical separation between anodal and cathodal stimulation in Fig. 4.14 reflects the polarity sensitivity ratio (P) introduced in Fig. 4.13. In Fig. 4.14, P ranges from 4.2 at a pulse duration of 1 μs to 5.6 at a pulse duration of 10 ms. As noted in Chapter 7, polarity ratios for electrocutaneous stimulation are closer to 1.3.

It is difficult to reconcile P values from sensory experiments with those obtained with the myelinated nerve model as long as the stimulating electrode is positioned far from the end of the model axon. As seen in Fig. 4.13, the minimum value of P is about 4.5 for a variety of electrode/neuron geometries when the electrode is distant from an axon terminus. But as the electrode moves closer to the terminal end, the model predicts smaller polarity ratios, and even predicts ratios less than unity in some cases.

Ranck (1975) discusses a qualitative model for stimulation near the terminus of a CNS neuron, and shows that it is possible for anodal excitation to occur at lower stimulus levels than with cathodal excitation. This author has not encountered polarity ratios less than 1 in electrocutaneous experiments. Nevertheless, Ranck's analysis and Fig. 4.13 suggest that the modest polarity ratios found in electrocutaneous experiments may occur because the principal sites of stimulation are near terminal structures (receptors, free nerve endings). Figure 4.13 shows that P ranges from about 4.5 to less than unity, depending on the electrode/neuron configuration. With cutaneous stimulation, there will be a mix of orientations of excitable structures in the skin beneath the electrode with a spectrum of polarity ratios. The modest ratios observed in sensory experiments may represent a macroscopic average of this spectrum. A further analysis of this issue is presented in Sec. 4.7.

Current Density and Electric Field Relationships

Recall from cable theory (Sec. 4.3) that a longitudinal electric field is necessary to support excitation. The electric field must also have a spatial gradient, as can be appreciated with reference to Eq. (4.31), which shows that *second differences*[†] of the external voltages are the driving

---

[†]A second difference is calculated as $(V_{n-1} - V_n) - (V_n - V_{n+1}) = V_{n-1} - 2V_n + V_{n+1}$, where V_n is the voltage at the nth node.

forces for changes in membrane potential. If the electric field were uniform and the axon were infinitely long in both directions, there would in theory be zero net current transfer at every node. However, the field within the biological medium is never uniform. Furthermore, an effective field gradient will be realized if the orientation of the axon changes with respect to a locally uniform field, or if the axon is terminated in the field (as with receptors, free nerve endings, or nerve connections at muscle fibers).

The myelinated nerve model format can be used to analyze stimulation by a uniform electric field (i.e., uniform current density) oriented in a direction longitudinal to the fiber. Referring to Fig. 4.9, the nodal voltages to be used in the myelinated nerve model are

$$V_{e,n} = V_{e,1} + ELn \qquad (4.34)$$

where $V_{e,1}$ is a reference voltage at the terminal node, L is the internodal space, and n is the node number. In Eq. (4.34) it is assumed that the first node of the array is oriented toward the cathode of the current source. Results presented here apply to a model with 21 nodes, the first seven of which use the nonlinear FH equations. In each case examined, excitation was initiated at the terminal node nearest the cathode, and the threshold was independent of the initial reference voltage, $V_{e,1}$. This finding is consistent with cable theory predictions [Eq. (4.24)].

Strength-duration curves were obtained for the conditions of stimulation mentioned above (Reilly, 1989). The threshold curves for uniform field stimulation are similar to the point-electrode stimulation curves in Fig. 4.14, except that the units on the vertical axes must be interpreted differently. Instead of stimulus current, thresholds are expressed in terms of the normalized longitudinal electric field within the biological medium; instead of threshold charge, thresholds are expressed in terms of the product of longitudinal field amplitude and pulse duration. Alternatively, the uniform field stimulation threshold values can be expressed in terms of current density (J) or charge density (q) by multiplying the previously mentioned values by the conductivity of the medium. A linear least-square fit of the uniform field excitation thresholds to Eq. (4.6) yields an S-D time constant of $\tau_e = 120\,\mu s$ for uniform field excitation. In comparison, a value of $140\,\mu s$ was measured under experimental conditions closely approximating uniform field stimulation, in which an internal E-field was induced by time-varying magnetic field exposure over a large area of the torso (Schaefer et al., 1991).

Table 4.2 expresses the minimum threshold values used as normalizing factors for the uniform field S-D curves. For long pulses ($t \geqslant 1$ ms), the minimum threshold is expressed in Part A of Table 4.2 in terms of the

Table 4.2. *Minimum stimulus thresholds with uniform field excitation; single monophasic stimuli*

		Fiber diameter (μm)		
		5	10	20
A. *Field strength criteria*				
1. E_{min}	(V/m)	24.6	12.3	6.2
2. $(Et)_{min}$	(Vs/m)	2.98×10^{-3}	1.49×10^{-3}	0.75×10^{-3}
B. *Current density criteria*				
1. J_{min}	(A/m^2)	4.92	2.46	1.23
2. q_{min}	(C/m^2)	6.0×10^{-4}	3.0×10^{-4}	1.5×10^{-4}

Notes: Thresholds A.1 and B.1 apply to long pulses ($t \geqslant 1$ ms). Thresholds A.2 and B.2 apply to short pulses ($t \leqslant 5\,\mu$s). Current and charge density determined for conductivity $\sigma = 0.2$ S/m. A theoretical S-D time constant of 120 μs applies to these results.
Source: From Reilly (1988).

induced field. For short pulses ($t \leqslant 5\,\mu$s), minimum threshold is given in terms of the product of field strength and pulse duration.

The longitudinal separation of nodes in a myelinated fiber is directly proportional to fiber diameter. Therefore, the absolute threshold in a uniform field is inversely proportional to fiber diameter. This relationship can be seen by referring to Eq. (4.31a): If the nodal separation is increased, the external nodal potential differences within a uniform electric field will increase proportionately. Table 4.2 expresses thresholds for fiber diameters of 5, 10, and 20 μm. These values effectively cover the diameter spectrum of myelinated fibers for motor and sensory neurons (Fig. 3.12).

The electric field in the medium is the primary force governing stimulation. However, current density is perhaps a more frequently cited stimulation parameter. Table 4.2 also expresses thresholds in units of current and charge density. These quantities are determined by multiplying E and Et by the conductivity of the medium, σ. In Table 4.2, the quantities in Part B were obtained from those in Part A by assuming an example conductivity of $\sigma = 0.2$ S/m, which roughly represents the bulk conductivity of muscle tissue (see Table 2.1). The current density and electric field thresholds listed in Table 4.2 bracket experimentally determined values if appropriate allowances are made for waveform and geometric factors (refer to Table 10.7 and related discussion).

4.6   Response to Biphasic and Repetitive Stimuli

Strength-Duration Relationships

The current reversal of a biphasic pulse can reverse a developing AP process that was excited by the initial phase. As a result, a biphasic pulse may have a higher threshold than a monophasic pulse. Figure 4.15 shows S-D curves from the myelinated nerve model for three types of stimuli: a monophasic constant current (rectangular) stimulus, a symmetric biphasic rectangular stimulus, and a sinusoidal stimulus. The data apply to stimulation via a point electrode 2 mm radially distant from a 20-μm-diameter fiber. The biphasic stimuli consist of a single stimulus cycle with an initial cathodal phase followed by an anodal phase of the same duration and equal magnitude. The phase duration indicated by the horizontal axis is that for the initial cathodal half-cycle. Stimulus magnitude is given in terms of peak current on the right vertical axis, and in terms of the charge in a single monophasic phase of the stimulus on the left vertical axis. The charge is computed by $Q = It_p$ for the rectangular

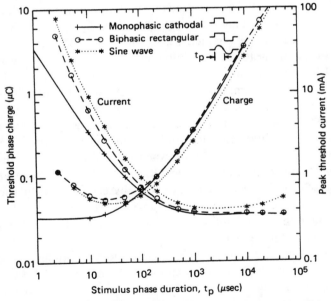

Figure 4.15   Strength-duration relationships derived from the myelinated nerve model: current thresholds and charge thresholds for single-pulse monophasic and for single-cycle biphasic stimuli with initial cathodal phase, point electrode 2 mm distant from 20-μm fiber. Threshold current refers to the peak of the stimulus waveform. Charge refers to a single phase for biphasic stimuli. (From Reilly et al., 1985.)

waveforms and from $Q = (2/\pi)I_p$ for the sinusoidal waveforms (I is threshold current amplitude and t_p is phase duration).

The effect of current reversal in the biphasic waveforms is to increase the threshold requirement for a propagating AP. This situation can be seen in Fig. 4.15 by comparing the monophasic and biphasic rectangular stimulus thresholds. For long durations, the threshold current is the same for the two stimuli. However, as the stimulus duration becomes short relative to the S-D time constant (about 100 μs), the biphasic current has an elevated threshold. The degree of elevation is magnified as the stimulus duration as reduced.

The thresholds illustrated in Fig. 4.15 apply to stimuli with an initial cathodal phase. Thresholds are greater if the initial phase is anodal, but only if the phase duration is less than 100 μs. For a single cycle of a sine wave, the model shows that if the initial phase of the stimulus is anodal, thresholds are greater than initial cathodal thresholds by 5% at a phase duration of 100 μs, 10% at 50 μs, 45% at 10 μs, and 60% at 5 μs.

Figure 4.16 illustrates threshold multipliers (M) based on the myelinated nerve model for biphasic rectangular pulses with uniform field excitation (Reilly, 1988). The vertical axis gives the threshold multiplier for a double pulse relative to a single pulse. The portion of the figure above $M = 1$ applies to a biphasic pulse doublet, where the current reversal has the same magnitude and duration as the initial pulse. The portion of Fig. 4.16 below $M = 1$ is for a monophasic pulse doublet. Figure 4.16 applies if the initial pulse is cathodal. Stimulation is also possible with an initial anodal pulse, but the thresholds are elevated.

According to Fig. 4.16, biphasic thresholds are elevated by an amount that depends on the pulse duration and the time delay before current reversal. Thresholds are most elevated when the pulse is short and the current reversal immediately follows the initial pulse. If the phase reversal is delayed by 100 μs or more, there is little detectable effect on the threshold. An implication of the results shown in Figs. 4.15 and 4.16 is that the membrane integrates the stimulus over a duration roughly equal to the S-D time constant. This integration is nonlinear: the biphasic waveforms all inject zero net charge, but have finite threshold magnitudes. Various features observed with biphasic stimulation of the nerve model are also seen experimentally in cardiac (Chapter 6) and nerve excitation (Chapters 7 and 8), as long as appropriate adjustments are made for experimental time constants.

Sinusoidal Stimuli

The sinusoidal stimulus is encountered in many applied problems. It is useful to focus on this special case of biphasic stimulation. Figure 4.15

illustrates excitation thresholds for sinusoidal stimuli consisting of a single cycle. The excitation threshold also depends on the number of stimulus cycles. Figure 4.17 illustrates the response of the myelinated nerve model to threshold-level sinusoidal stimuli at a frequency of 5 kHz (point electrode 2 mm distant from 20-μm fiber). The stimulus is composed of a single cycle in example *a*, two cycles in example *b*, and three cycles in example *c*. The threshold current requirement is reduced according to the insert in the figure as the number of stimulus cycles is varied from one to three. Figure 4.18 illustrates the relationship between threshold and stimulus duration for a sinusoidal current having frequencies of 5 and 50 kHz. When evaluated at half-cycle multiples, there is an oscillating threshold with minima at odd numbers of half-cycles and maxima at even multiples. The full-cycle thresholds reach a minimum at 8 cycles of 5-kHz stimulation, and at 64 cycles of 50-kHz stimulation. At a lower oscillation rate of 500 Hz, thresholds were nearly independent of the number of stimulus cycles.

The thresholds in Fig. 4.18 oscillate because a sinusoidal stimulus that has an even number of half-cycles is charge balanced (zero net charge), and one with an odd number of half-cycles will transfer the maximum net charge. The gradually falling aspect of the threshold is a consequence

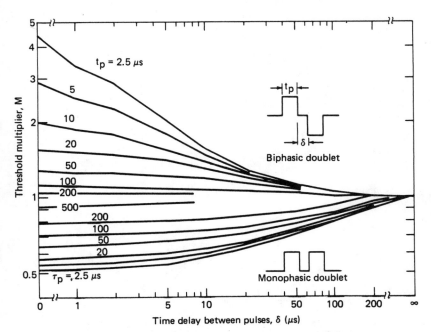

Figure 4.16   Threshold multipliers for biphasic and monophasic pulse doublets: Uniform field excitation of truncated axon. (From Reilly, 1988.)

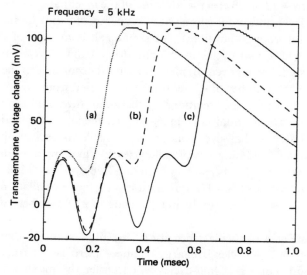

Figure 4.17   Response of myelinated nerve model to sinusoidal threshold-level stimuli at a frequency of 5 kHz. Stimulus durations: (*a*) one cycle, (*b*) two cycles, and (*c*) three cycles. Peak values of threshold currents are listed above figure. (From Reilly et al., 1985.)

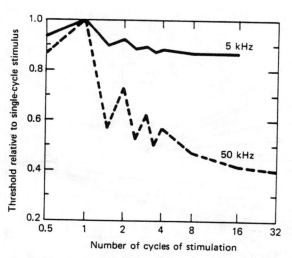

Figure 4.18   Threshold current as a function of the duration of sinusoidal stimulation. Stimulus duration was stepped in half-cycle increments out to four cycles, and in full-cycle increments beyond that point. (From Reilly et al., 1985.)

of the nonlinear conductance of the membrane. Recall from Chapter 3 that as the membrane depolarization approaches the excited state, ionic conductance for inward- and outward-going currents differs. As a consequence, an oscillating stimulus will build up a bias voltage on the membrane that increases with each successive cycle of stimulation. Experimentally these properties are found in biphasic stimulation of both nerve and cardiac tissue (see Chapters 6–8).

The stimuli discussed in previous sections of this chapter produce only a single AP. An oscillating stimulus has the ability to produce a train of APs, which can greatly enhance the intensity of the electrical response in nerve (Chapter 7), cardiac (Chapter 6), and skeletal muscle (Chapter 8) responses. Repetitive responses can be predicted from the nonlinear models described in Chapter 3 for both the myelinated and unmyelinated nerves.

Figure 4.19 illustrates the response of the myelinated nerve model to continuous sinusoidal stimulation. The axon response is shown two nodes distant from the excitation node in order to verify AP propagation. The stimulus levels indicated are multiples of the single-cycle threshold. The AP repetition rates are 100, 250, and 500 Hz for threshold multiples of 1.0, 1.2, and 1.5, respectively. In each case, AP development is synchronized

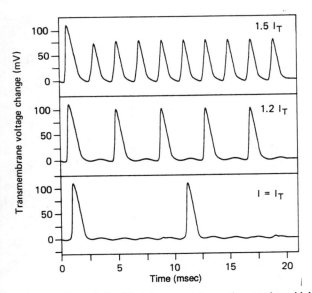

Figure 4.19 Model response to continuous sinusoidal stimulation at 500 Hz. The lower panel depicts the response to a stimulus current set at threshold level (I_T) for a single-cycle stimulus. Upper panels show response for stimulation 20% and 50% above the single-cycle threshold. (From Reilly et al., 1985.)

with the stimulus. Similar phase locking has been observed in the neural responses of sensory systems to periodic stimulation (Kiang, 1965). Responses were also studied with the myelinated nerve model for continuous sinusoidal stimulation at 5 kHz. The AP repetition rates were 320, 470, and 540 Hz at threshold multiples of 1.0, 1.2, and 1.5, respectively. At the 1.5 stimulus multiple, there was a 4.4-ms refractory period after the first AP during which excitation did not occur. After this initial refractory period, a steady AP rate of 540 Hz was produced. At both frequencies there is a reduction in magnitude of the AP spike as the AP repetition rate increases. The maximum AP rate and reduced depolarization voltage are consistent with the experimental A-fiber response illustrated in Fig. 3.11.

Sinusoidal threshold response can be represented by strength-frequency (S-F) curves as shown in Fig. 4.20. The horizontal axis in Fig. 4.20 is the inverse of twice the phase duration in Fig. 4.15. Figure 4.20 also shows a threshold curve for continuous sinusoidal stimulation.

Figure 4.20 includes experimental threshold curves for human perception and muscle contraction (Dalziel, 1972; Anderson and Munson, 1951). The experimental curves have been arbitrarily scaled on the vertical axis

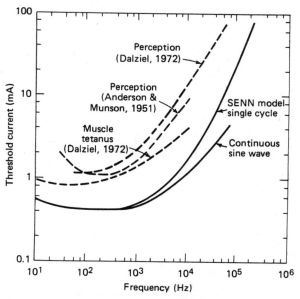

Figure 4.20   Strength-frequency curves for sinusoidal current stimuli. Dashed curves are from experimental data. Solid curves apply to myelinated nerve model. Experimental curves have been shifted vertically to facilitate comparisons. (From Reilly, 1988.)

to facilitate comparison of the curve shapes. The shapes of the experimental data and continuous stimulation model results correspond reasonably well considering that continuous stimulation was used in the cited experimental studies. Below 40 Hz, thresholds rise for sinusoidal stimulation. At low frequencies, the slow rate of change of the sinusoid prevents the membrane from building up a depolarizing voltage because membrane depolarization is counteracted by membrane leakage. Square-wave biphasic stimuli, in contrast, do not have a rate of change that depends on frequency. Consequently, there is no increase in thresholds at low square-wave frequencies.

To a first approximation, a functional S-F relationship might be obtained from a monophasic S-D curve by considering that a sinusoidal half-cycle corresponds to a single monophasic pulse. This simplification would, however, fail to account for the increased sinusoidal thresholds at both high and low frequencies relative to monophasic stimulation. An empirical fit to S-F data from the myelinated nerve model is given by

$$I_t = I_0 K_H K_L \qquad (4.35)$$

where I_t is the threshold current, I_0 is the minimum threshold current, K_H and K_L are high- and low-frequency terms, defined respectively as

$$K_H = \left[1 - \exp\left(-\frac{f_e}{f} \right) \right]^{-a} \qquad (4.36)$$

and

$$K_L = \left[1 - \exp\left(-\frac{f}{f_0} \right) \right]^{-b} \qquad (4.37)$$

where f_e and f_0 are constants that determine the points of upturn in the S-F curve at high and low frequencies, respectively. An upper limit of $K_L \leqslant K_{DC}$ is assumed in Eq. (4.37) to account for the fact that excitation may be obtained with finite DC currents. Equations (4.36) and (4.37) have the asymptotic forms

$$I_t \approx I_0 \left(\frac{f}{f_e} \right)^a \qquad \text{for } f \gg f_e \qquad (4.38)$$

and

$$I_t \approx I_0 \left(\frac{f_0}{f} \right)^b \qquad \text{for } f \ll f_0 \qquad (4.39)$$

Sinusoidal thresholds were obtained with the myelinated nerve model for stimulation by a point electrode positioned 20 mm radially distant from a 20-μm-diameter fiber. An empirical fit of Eqs. (4.35)–(4.37) to the model thresholds indicates that $a = 1.45$ for a single-cycle stimulus, and $a = 0.9$ for a continuous stimulus; $b = 0.8$ regardless of stimulus duration.

Frequency constants for the nerve model were $f_e = 5400\,\text{Hz}$ and $f_0 = 10\,\text{Hz}$. Experimentally derived values of f_e and f_0 encompass a wide range, as noted in Chapters 6 and 7. A lower transition frequency in the range 10–50 Hz is observed for nerve stimulation, and around 10 Hz for cardiac stimulation. Experimental values of f_e reported in Chapter 7 for nerve stimulation have a geometric mean of 500 Hz; those reported in Chapter 6 for cardiac stimulation have a geometric mean of about 115 Hz. Differences in model and experimental values reflect the dynamic membrane responses of the different types of tissue being stimulated, as well as the distribution of stimulating current in an experimental situation.

In characterizing S-D curves, the parameter τ_e can be simply determined as the ratio of minimum threshold charge to current [Eq. (4.20)]. It would be desirable to have a similar simple procedure for estimating the parameter f_e from S-F plots. A rough estimate of f_e can be determined from Eq. (4.36) as the upper frequency at which I/I_0 is 1.95 if $a = 1.45$; or the frequency at which I/I_0 is 1.51 if $a = 0.9$; from Eq. (4.37), f_0 is estimated as the lower frequency at which I/I_0 is 1.44.

Equations (4.35)–(4.37) have features in common with an S-F expression proposed over half a century ago by Hill and colleagues (1937), which can be equivalently stated as

$$\frac{I}{I_0} = \left[\left(1 + \frac{f_0^2}{f^2}\right)\left(1 + \frac{f^2}{f_e^2}\right)\right]^{1/2} \tag{4.40}$$

The expression of Hill et al. involves upper and lower transition frequencies, as do Eqs. (4.35)–(4.37), and has the asymptotic forms of Eqs. (4.38) and (4.39) with $a = b = 1$. The lower transition frequency was described by Hill et al. as arising from the neural property known as *accommodation*, i.e., the adaptation of a nerve to a slowly varying or constant stimulus.

Repetitive Stimuli

Repetitive stimuli can be more potent than a single stimulus through threshold reduction or through response enhancement due to multiple AP generation. In both cases an integration effect of the multiple pulses occurs. In the first case, the integration takes place at the membrane level. In the second case, response enhancement takes place at higher levels within the central nervous system for neurosensory effects, and at the muscle level for neuromuscular effects. The following considers membrane effects using the myelinated nerve model. Other integration effects are described in Chapters 7 and 8.

The lower section of Fig. 4.16 ($M < 1$) illustrates multiple-pulse threshold effects for two pulses of the same polarity. The figure gives the threshold multiplier when there are two pulses of stimulation relative to that for a single pulse. The effects are most pronounced for short pulses and short interpulse delays. (In this representation, a delay of zero is really the same as a single pulse of twice the duration.) Even at a delay of $200\,\mu s$, the integrative effects of a second pulse reduce the threshold by about 10% for $t_p \leqslant 20\,\mu s$. The nerve model was also exercised to evaluate the threshold modification for sequences of pulses, N_p, numbering 1, 2, 4, 8, 16, 32, 64, and 128. Figure 4.21 illustrates the results for $t_p = 10$, 50, and $100\,\mu s$ and for $\delta = 10$, 50, 100, 200, and $500\,\mu s$. As in Fig. 4.16, the vertical axis gives the threshold multiplier relative to a single pulse. The curve labeled $\delta = 0$ applies to continuous stimulation over a time period corresponding to N_p. Pulse integration reduces thresholds by an amount that increases as the pulse duration and delay time are shortened. At a delay of $500\,\mu s$, the nerve model shows no measurable pulse integration effect.

4.7   Parameter Variation Effects

In order to further explore mechanisms for electrocutaneous stimulation, and to reconcile data from model and experiments, a parameter variation study was conducted with the myelinated nerve model (Reilly and Bauer, 1987). The membrane parameters varied were: fiber diameter (d), nodal gap width (W), internal axonal conductivity (g_a), membrane capacity (c_m), and membrane conductance (g_m). These were individually varied by factors of $\frac{1}{2}$ or 2 times the base-case value (see Table 4.3), while holding all other base-case parameters constant. For this study, the base-case fiber diameter was $10\,\mu m$; other parameters were as in Table 4.1. Geometric factors were also varied relative to the base case. These were radial distance from the central node and longitudinal distance between the electrode and the fiber terminus.

Four aspects of model response were evaluated in the parameter variation study:

1. Minimum charge (Q_0)—the AP initiation threshold in units of charge of a short duration pulse ($1\,\mu s$). Charge is determined by the product of pulse width and current magnitude.
2. Minimum current (I_0)—the AP initiation threshold in units of peak current for a long pulse (2 ms).
3. S-D time constant (τ_e)—determined by a least-squares fit of the threshold currents for AP initiation to the theoretical S-D curve for an ideal linear membrane (see Reilly and Larkin, 1983).
4. Polarity selectivity ratio (P)—the absolute value of the AP initiation thresholds for anodal versus cathodal stimuli. P was determined separately for stimulus durations of 10 and $1000\,\mu s$.

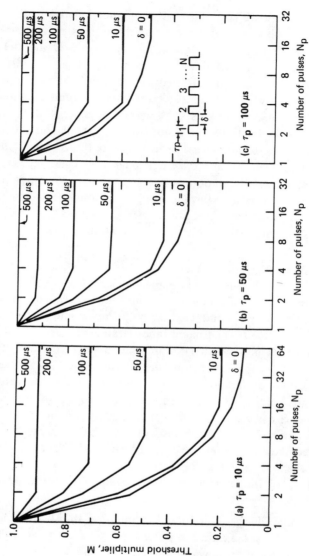

Figure 4.21 Threshold multiplier for repetitive pulse sequences—thresholds evaluated at $N_p = 1, 2, 4, 8, 16, 32$, and 64. Uniform field excitation. (From Reilly, 1988.)

Table 4.3. *Summary of parameter variation effects*

	Q_0	I_0	τ_e	P for $t = 0.01$ ms	P for $t = 1.0$ ms
Base case	15.9 nC	0.18 mA	92.3 μs	4.66	5.53
A. Variation of membrane parameters					
$D \times \frac{1}{2}$	0.50	0.50	1.00	1.00	1.00
$\times 2$	2.00	2.00	1.00	0.99	1.00
$G \times \frac{1}{2}$	0.76	0.82	0.93	0.98	1.10
$\times 2$	1.45	1.28	1.13	0.95	0.96
$r_i \times \frac{1}{2}$	0.70	0.99	0.71	1.00	1.00
$\times 2$	1.51	1.11	1.50	0.99	0.98
$c_m \times \frac{1}{2}$	0.70	0.99	0.71	1.00	1.00
$\times 2$	1.51	1.11	1.50	0.99	0.98
$g_m \times \frac{1}{2}$	1.27	1.12	1.13	0.98	0.98
$\times 2$	0.82	0.87	0.93	0.99	1.01
B. Variation of radial distance—electrode centered along axon					
$y_0 = \frac{1}{4}$	0.12	0.14	0.61	1.06	1.24
$y_0 = \frac{1}{2}$	0.31	0.34	0.90	1.06	1.18
$y_0 = 1$	1.00	1.01	1.00	1.00	1.00
$y_0 = 2$	4.28	3.58	1.17	0.92	0.83
$y_0 = 4$	23.0	15.9	1.39	1.03	0.94
C. Variation of longitudinal distance—electrode near truncated end					
$x_0 = -1, y_0 = 1$	1.78	1.65	1.08	0.77	0.87
$x_0 = 2, y_0 = 1$	1.12	1.24	0.93	0.53	0.70
$x_0 = 5, y_0 = 1$	1.00	1.01	0.99	0.77	0.48
$x_0 = -1, y_0 = 2$	4.48	3.79	1.17	0.95	0.89
$x_0 = 2, y_0 = 2$	5.38	4.92	1.12	0.69	0.56
$x_0 = 5, y_0 = 2$	4.31	3.67	1.15	0.32	0.25

Note: First row lists values in absolute units. Other table entries list values as a multiple of the base-case datum. x_0 and y_0 are dimensionless quantities (normalized to node spacing).
Source: From Reilly and Bauer (1987).

Table 4.3 summarizes results from the parameter variation study. The first row lists data for the base case in absolute units. The remaining entries in the table are expressed as multiples of the base-case values. Part A lists results for variations of fiber parameters that were set to $\frac{1}{2}$ or 2 times the base case value. In varying D, we kept the electrode distance equal to one internodal space. As a result, electrode separation, expressed in absolute distance units, varied with D. Part B lists results for variations in the radial distance (y_0) between the electrode (positioned over the center node) and the fiber; y_0 is expressed as a multiple of internodal units and is a unitless quantity. Part C lists results for variations in longitudinal distance with respect to the truncated end of the axon; $n_0 = -1$ means that the electrode was positioned one internodal unit beyond the terminus; $n_0 = 2$ and 5 means that the electrode was positioned above the second and fifth nodes, respectively, from the end.

In Part A of Table 4.3, we see that P is quite insensitive to the choice of membrane parameters. And except for c_m, the membrane parameter variations have only a modest effect on τ_e. In Part B of Table 4.3, we see that the radial separation of the electrode from a central node has a significant effect on τ_e. The fact that τ_e falls with reduced distance is consistent with the theoretical expectations discussed in Sec. 4.3. In Part C of Table 4.3, we see that the value of P can be significantly lowered when the electrode is placed near the truncated end of the axon. The degree of this reduction places the model values of P much more in conformance with experimental data from electrocutaneous sensory experiments.

Only passive membrane parameters were varied in the study. In an earlier study, Bostock (1983) also examined the effects of variations in active membrane parameters. That study used a model for current injection at a single point on the model axon, rather than excitation through an external electrode as in the myelinated nerve model. As a result, it is difficult to compare Bostock's results with the myelinated nerve model. Despite differences in model assumptions, Bostock's results provide some guidance on the potential sensitivity of τ_e to active membrane parameter variations. Of the parameters examined by Bostock, the sodium activation rate constant had the greatest effect: for a factor of 2 reduction in the activation rate constant, there resulted at most a 21% increase in the value of τ_e.

We note that stimulation near the end of the truncated model axon results in values of P that are more nearly in line with experimental data. A possible hypothesis for experimental observations is that the principal site of transcutaneous sensory stimulation may be at or near neural end structures. Although useful models for depicting axonal stimulation are available, the author is not aware of any electrical models that specifically take into account features of end-structure morphology and function. Future investigation in this direction may be fruitful.

5

Electrical Properties of the Heart

HERMANN ANTONI

5.1 Cardiovascular System: General Anatomical and Functional Aspects

The heart provides the main driving force for the movement of blood through the vessels. Additional forces are exerted by the muscular and the respiratory pumps as well as by gravity. The heart is composed of two hollow organs—its right and its left half—with muscular walls (Fig. 5.1a). Each half comprises an atrium (Ra, La) and a ventricle (Rv, Lv). The right half receives oxygen-depleted blood from the body and expels it via the pulmonary artery (Pa) to the lungs, where it is reoxygenated. Then the blood returns to the left half of the heart and is thence distributed via the aorta (Ao) to the organs of the body (Fig. 5.1b). The movement of the blood from the right to the left heart, by way of the lungs, is called the *pulmonary circulation*. Its distribution to, and return from, all the rest of the body is the *systemic circulation*. Strictly speaking, of course, the two constitute a single pathway of blood movement, with the propulsive force provided at two points by the two halves of the heart (Fig. 5.1b).

The pumping action of the heart consists of a rhythmic sequence of relaxations (*diastole*) and contractions (*systole*) of the chambers. During diastole the ventricles fill with blood, and during systole they expel it into the large vessels (aorta and pulmonary artery). Backflow from these arteries to the ventricles is prevented by the valves at their openings. Before entering the ventricles, the blood passes from the large veins into the associated atria, which act as booster pumps to help fill the ventricles. An additional pair of valves between the atria and the ventricles prevents backflow of blood during the ventricular systole.

The distinction between arteries and veins is based on the direction of blood flow within them. Veins carry the blood to the heart, and arteries carry it away. Hence, in the systemic circulation the arteries carry oxygenated blood, and in the pulmonary circulation the oxygenated blood is carried by the veins.

Because the demands made on the circulating blood vary with time, the heart must be able to adjust its activity over a wide range. For example, the volume of blood expelled per minute by one ventricle (cardiac output) is about 5 l for an adult person at rest and rises to almost 30 l during hard physical work. Because the two ventricles are arranged in series, their outputs must be nearly the same at each beat. This requires a mechanism for precise adjustment of the outputs of the two ventricles. Moreover, when the resistances to flow in the systemic or pulmonary circulation increase—for instance, because of extensive vasoconstriction—the ventricles quickly adapt to the changed conditions by contracting more strongly and raising the pressure sufficiently to propel the same volume of blood. Likewise, changes in venous return and diastolic filling are compensated by adjustment of the cardiac output. This astonishing adaptability of the heart arises from two mechanisms: intracardial (auto-) regulation, and extracardially initiated regulation.

The former is brought about by intrinsic properties of the myocardium, mainly its ability to respond to increased extension (greater end-diastolic

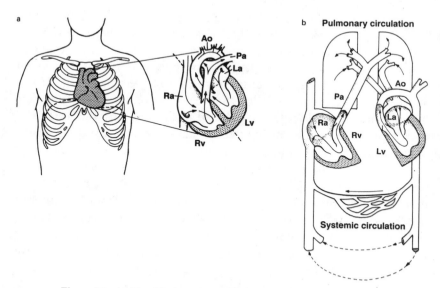

Figure 5.1   (a) Simplified anatomy of the opened human heart in a frontal view. (b) Schematic diagram of the connections of the two halves of the heart with the pulmonary and systemic circulations. Ra, right atrium; La, left atrium; Rv, right ventricle; Lv, left ventricle; Ao, aorta; Pa, pulmonary artery.

volume) with the development of a higher contractile force or of a higher stroke volume. The latter mechanism operates under the control of the endocrine and autonomic nervous systems. It enables the heart either to overcome a higher pressure or to eject a larger stroke volume without increased muscular extension, i.e., without increase in the end-diastolic volume. Under normal conditions the autoregulation is bought into play when changes in filling occur without a general increase in physical activity. This applies, for instance, to the coordination of the output of the two ventricles, to changes in the position of the body that affect venous return, to an acute increase in the blood volume (transfusion), etc.

The energy the heart requires for its mechanical work comes primarily from the oxidative decomposition of nutrients. In this regard cardiac and skeletal muscle differ fundamentally, for the latter can obtain a large part of the energy needed to meet short-term demands by anaerobic processes, and the "oxygen debt" that is built up can be repaid later. This is not possible with the heart, because it is exclusively dependent on oxidative energy supply. Although the weight of the heart of an adult is only about 0.5% of the total body weight, the O_2 demand of the heart is about 10% of the total resting O_2 consumption (about 25 ml of O_2/min). When the body is performing hard work, the O_2 consumption of the heart can rise to four times the resting level.

The coronary vessels, which supply the heart, are part of the systemic circulation, but they exhibit special features that are closely related to the way the heart operates. There are two coronary arteries, both arising from the base of the aortic root (Fig. 5.2). The right coronary artery supplies most of the right ventricle; the rest of the heart is supplied by the left coronary artery. Venous drainage is mainly through the coronary

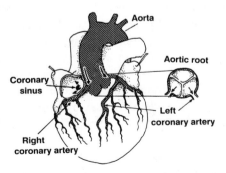

Figure 5.2 Human heart in a frontal view with the main coronary arteries originating from the aortic root. Venous return occurs via the coronary sinus to the right atrium.

sinus, which flows into the right atrium. At rest the total coronary blood flow amounts to about 250–300 ml/min; this is about 5% of the cardiac output. During normal resting activity the heart withdraws more oxygen from the blood than do the other organs. Of the 20 ml/dl of O_2 in the arterial blood, the heart extracts around 14 ml/dl. Therefore, when the load on the heart increases and more oxygen is required, it is essentially impossible to increase the rate of extraction. Increased O_2 requirement must be met primarily by increased blood flow, brought about by dilation of the vessels. The strongest stimulus to dilation of the coronary vessels is O_2 deficiency. A decrease in the oxygen content of the blood by only 5% leads to marked coronary vasodilation.

Because cardiac metabolism relies so heavily on oxidative reactions to provide energy, it is understandable that a sudden interruption of circulation (*ischemia*) results in extensive loss of function within a few minutes; contractions grow progressively weaker, a marked dilation develops, and after about 6–10 min the heart stops beating. Since coronary circulation is normally maintained by the pumping action of the heart itself, a strong diminution of the contractile force of the organ must also impair its blood supply, thus, in turn, weakening contraction still more. The same sequence of events takes place if cardiac contractile insufficiency is due to a heavy disturbance of the excitatory processes, i.e., ventricular fibrillation (see Secs. 5.8 and 5.10). When a circulatory breakdown affects the entire organism, the brain suffers irreversible damage after ischemia lasting only 8–10 min.

5.2   Origin and Spread of Excitation

Myocardial fibers, like nerve or skeletal muscle fibers, are excitable structures. The cell boundaries, which can be seen in the microscope as intercalated disk, offer no obstacle to the conduction of excitation. Since the musculature of the atria and ventricles forms a netlike structure, it behaves as a syncytium that means excitation arising anywhere in the atria or ventricles spreads out over all the unexcited fibers. This property provides the explanation of the all-or-none response of the heart; that is, when stimulated the heart either responds with excitation of all its fibers or gives no response. In a nerve or skeletal muscle, by contrast, each cell responds individually, so that only those fibers exposed to suprathreshold excitation discharge conducted impulses, while the others remain at rest.

Autorhythmicity and Geometry of Propagation

The rhythmic pulsation of the heart is maintained by excitatory signals generated within the heart itself. Under suitable conditions, therefore, a heart removed from the body will continue to beat at a constant frequency. This property is called *autorhythmicity*. Ordinarily, the spontaneous rhythmic triggering of excitation is performed exclusively by the specialized cells of the pacemaker and conducting system. The various elements in this system are diagrammed in Fig. 5.3.

Normally the heartbeat is initiated in the sino-atrial (SA) node, in the wall of the right atrium near the superior vena cava. When the body is at rest, the SA node drives the heart at a rate of about 70 impulses/min. From the SA node the excitation first spreads over the working myocardium of both atria. The only pathway available for conduction to the ventricles is indicated in Fig. 5.3. All the rest of the atrioventricular boundary consists of nonexcitable connective tissue. As the excitation propagates through the conducting system it is briefly delayed in the atrio-ventricular (AV) node. Propagation velocity is high (about 2 m/s) through the remainder of the system—the common bundle, the left and right bundle branches and their terminal network, the Purkinje fibers—so that the different ventricular regions are excited in rapid succession. From the Purkinje fibers, excitation spreads at a speed of about 1 m/s over the ventricular musculature.

Hierarchy of Pacemaker Activity and Artificial Pacemakers

The autorhythmicity of the heart is not entirely dependent on the operation of the SA node, since the other parts of the conduction system are also spontaneously excitable. But the intrinsic rhythm of these cells becomes considerably slower, the farther away from the SA node. Under normal conditions, therefore, these cells are triggered by the more rapid buildup of excitation in the higher centers, before they have a chance to trigger themselves. The SA node is the leading *primary pacemaker* of the heart, because it has the highest discharge rate.

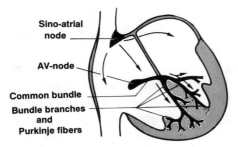

Figure 5.3 Diagram of the arrangement of the pacemaker and conducting system of the human heart as seen in frontal section.

If for any reason the SA node should fail to initiate the heartbeat, or if the excitation is not conducted to the atria (sino-atrial block), the AV node can substitute as a *secondary pacemaker* at a frequency of 40–60/min. If there should be a complete interruption of conduction from the atria to the ventricles (*complete heart block*), a *tertiary center* in the ventricular conducting system can take over as pacemaker for ventricular contraction. With respect to pacemaker activity, the SA node can be termed the *nomotopic* (in the normal place) center, and the remainder of the system the *heterotopic* (in an abnormal place) centers.

In the case of the above-mentioned complete heart block, atria and ventricles beat entirely independently of one another, the atria at the frequency of the SA node and the ventricles at the considerably lower frequency of a tertiary center (30–40/min). When there is a sudden onset of total heart block, several seconds can elapse before the ventricular automaticity "wakes up." In this pre-automatic pause an insufficient supply of blood to the brain may cause unconsciousness and convulsions (*Adams-Stokes syncope*). If the ventricular pacemakers fail altogether, the ventricular arrest leads to irreversible brain damage and eventually to death.

When conduction along the bundle branches is interrupted, the result is an incomplete heart block, as long as at least one branch or subdivision of a branch remains functional. In this case the excitation spreads out from the terminals of the intact conduction system and eventually covers the whole ventricular myocardium; the time required for complete excitation, of course, is considerably longer than normal. In the absence of autorhythmicity it is therefore possible to keep the blood in circulation by artificial electrical stimulation of the ventricles. Electrical stimulation can sometimes be continued for years. The stimuli are generated by subcutaneously implanted, battery-driven miniature pacemakers and conducted to the heart by wire electrodes.

5.3   Elementary Processes of Excitation and Contraction

The action potential of the cardiac muscle cells, like that of neurons or skeletal muscle fibers, begins with a rapid reversal of the membrane potential, from the resting potential (about -90 mV) to the initial peak (about $+30$ mV; see Fig. 5.4). This rapid phase of depolarization, lasting only 1–2 ms, is followed by a special feature of the myocardium, a prolonged plateau. This is terminated by repolarization to the resting potential. The action potential of the cardiac musculature lasts about 200–400 ms — more than 100 times as long as that of a skeletal muscle or nerve fiber (see Chapter 3). The functional consequences, as we shall see, are considerable (see Sec. 5.4).

Ionic Mechanisms of Excitation

The action potential is generated by a complicated interplay of membrane potential changes, changes in ionic conductivity, and ion currents. Fundamentals of the ionic theory of excitation have been discussed in detail in Chapter 3. Here we shall give only a short recapitulation, with reference to the specific peculiarities of cardiac muscle (for details see Fozzard and Arnsdorf, 1986; Noble, 1984). The resting potential of the myocardium is primarily a potassium potential, maintained by an electrogenic Na^+ pump. As in the neuron, the rapid upstroke phase of the action potential is brought about by a brief pronounced increase in Na^+ conductance, g_{Na}, which results in a massive Na^+ influx (see Fig. 5.4). This initial Na^+ influx, as in the neuron, is very rapidly inactivated. Hence, further mechanisms are required for the considerable delay in repolarization of the cardiac muscle tissue. These are (1) an increase in Ca^{2+} conductance (g_{Ca}) with delayed onset and slow decline, which causes a depolarizing influx of calcium (slow inward current) and (2) a decrease in K^+ conductance (g_K) with depolarization, which reduces the repolarizing K^+ outward current.

Repolarization of the myocardium results from a gradual decrease in g_{Ca} and an increase in g_K due to the more negative membrane potential. The decrease in g_{Ca} diminishes the slow inward current, and the increase in g_K enhances the K^+ outward current. When the membrane is at its resting potential, the depolarizing and repolarizing currents are in balance.

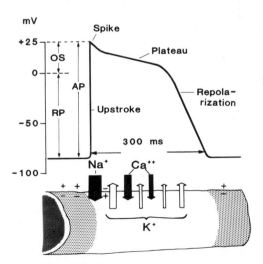

Figure 5.4   Top: General form of the action potential of a cardiac muscle cell: RP, resting potential; OS, overshoot; AP, amplitude of action potential. Bottom: Direction and magnitude of the ionic currents during the action potential. The magnitude is indicated by the thickness of the arrows.

Mechanism of Contraction

While the excitatory processes take place at the surface membrane of the cardiac muscle cell, the contractile machinery is located in the interior. The main constituents of this machinery are the "contractile proteins": actin (molecular weight 42 kD) and myosin (about 500 kD). The way in which they interact to bring about contraction is described by the sliding-filament theory (Huxley and Hanson, 1954). Actin and myosin form the thin and thick myofilaments of the myofibrils. These are contractile bundles about 1 μm in diameter, which are subdivided by the Z disks into compartments about 2.3 μm long, called *sarcomeres*. The structure of a sarcomere is illustrated in Fig. 5.5. There is a regular arrangement of the two filaments, with the myosin filaments forming the A band (isotropic band). On either side of the A bands are regions containing only thin filaments, which therefore appear light; these isotropic I bands extend to the Z lines.

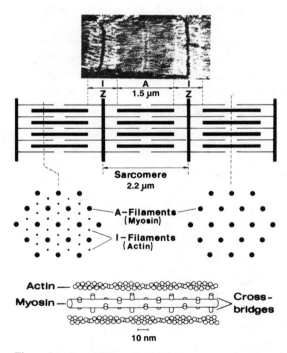

Figure 5.5   Top: Electron micrograph of mammalian cardiac muscle representing the functional unit of a myofibril. The relationship between this element and the whole cell can be derived from Fig. 5.6. Middle: Banded structure of the myofibrils; arrangement of the myosin and the actin filaments. Bottom: Actin filaments and myosin filaments with cross-bridges.

In the relaxed muscle the ends of the thick and thin filaments usually overlap only slightly. When the myofilaments contract, the thin actin filaments slide over the thick myosin filaments, moving between them toward the middle of the sarcomere. During the sliding process neither the myosin nor the actin filaments themselves shorten.

Excitation-Contraction Coupling

The transmission of an action potential from the excited cell membrane to the myofibrils in the depths of the cell requires several sequential processes in which calcium ions play a key role in that they initiate contraction. This occurs when the concentration of intracellular free calcium rises above about 10^{-7} mol/l. Two intracellular tubular systems are involved in this process: the transverse tubular system (*T system*) and the longitudinal system (*sarcoplasmic reticulum*, see Fig. 5.6). The T system is formed by invagination of the outer membrane and transmits the action potential to the interior of the cell. The longitudinal system represents an intracellular calcium store. This system is in close contact with the T system via so-called *lateral cisternae*. When the action potential is propagated along the surface and into the T system, it causes an influx of calcium ions. The accumulation of calcium near the lateral cisternae favors a further mobilization of calcium from the intracellular stores (*calcium-triggered calcium release*, see Fig. 5.6).

Upon repolarization of the membrane, the transmembrane influx of calcium ions ceases, and the calcium ions are removed by ATP-driven calcium pumps as well as by exchange mechanisms located in the longitudinal system and in the surface membrane (Fig. 5.6). Relaxation occurs when the myoplasmic concentration of activating calcium ions is reduced below about 10^{-7} mol/l. This reduction inhibits the interaction of actin and myosin cross-bridges, which then detach.

5.4   Stimulation, Propagation, and Refractoriness

Elementary Mechanisms

As in other excitable tissues, action potentials of cardiac cells are elicited by depolarization of the resting membrane to threshold (Fig. 5.7). Except for the sino-atrial and the atrioventricular nodal cells, all myocardial fibers contain fast Na^+ channels, and the activation of these channels by depolarization is responsible for the initiation of the action potential. Normally, the spontaneous depolarization of pacemaker cells initiates excitation in the circumscribed pacemaker area as soon as the threshold is attained (see Sec. 5.5). Excitation is then propagated due to depolarization of still-resting fibers by their coupling to closely adjacent excited ones. Artificial stimulation works by the same mechanism.

In order to stimulate the heart as a whole, it is sufficient to stimulate only a fraction of about 50 closely coupled cardiac cells. Hence, the stimulus intensity required to do this is about the same for small, isolated myocardial preparations as for the whole organ. If a single isolated cardiac cell in culture is stimulated through intracellular electrodes, a current intensity of about 1 nA (with a duration of 5 ms) is required to elicit an

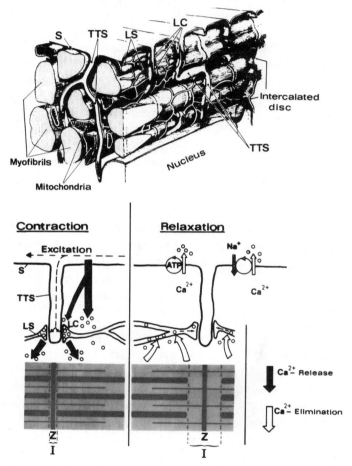

Figure 5.6   Top: Ultrastructure of a myocardial cell: S, sarcolemma; TTS, transverse tubular system; LS, longitudinal system (sarcoplasmic reticulum); LC, lateral cisternae. Bottom: Diagram of the interplay of excitation, calcium movement, and activation of the contractile apparatus. Events at the onset of contraction (left) and during relaxation (right).

action potential. If, on the other hand, a myocardial cell is in its normal connection within the tissue, suprathreshold intracellular stimulation, depending on the coupling resistance, requires more than 100 nA (Johna, 1989).

With extracellular stimulation, as it is applied in experiments on isolated cardiac tissue or by implanted pacemakers, threshold current strength for a given pulse duration depends strongly on the geometry of the electrodes and their position with respect to the cellular alignment in cardiac muscle. Under optimal conditions (electrodes diameter 0.7–1.0 mm; pulse duration 0.5 ms), pacemaker stimulation of the heart can be achieved with only 10–14 μA from an extracellular electrode (Irnich, 1973).

A sufficiently strong and long-lasting rectangular current pulse can stimulate twice: first, in the region of the cathode when the circuit is closed (*cathodal make*), and second, in the region of the anode when the circuit is opened (*anodal break*) (Fig. 5.7). This latter effect is in part due to an increase in recovery from inactivation of the sodium channels. In part it is brought about by a decrease in potassium conductance during hyper-polarization. This declines slowly when the stimulus is switched off, and thus leads to a transient depolarization to threshold (dotted line in Fig. 5.7). For more details about cardiac stimulation, see Chapter 6.

Conduction of the Action Potential

Conduction in heart muscle occurs by local circulating currents, in the same fashion that conduction is brought about in nerve (see Fig. 3.10). These currents have as their source the electromotive force (emf)

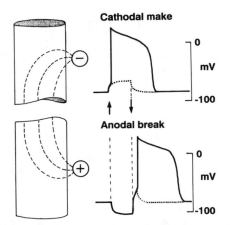

Figure 5.7   The two types of pulsed DC stimulation; changes in membrane potential during cathodal make and during anodal break. Action potentials are elicited when the membrane potential attains the threshold. Action potential duration about 300 ms.

of the sodium gradient across the membrane of the excited region. The current displaces the charge stored on the adjacent membrane, depolarizing the membrane toward its threshold for increasing sodium conductance. This process will take place more quickly as more current flows to the resting regions, and as the capacity that has to be discharged becomes smaller (Fozzard, 1979).

The structure and the functional properties of the ventricular conduction system allow rapid spread of excitation all over the ventricles, thus bringing about nearly synchronized contraction. This is especially pronounced in the hearts of large animals, where, as compared with the ordinary myocardium, the fibers of the conducting system exhibit considerably greater diameters with a correspondingly smaller internal resistance. As already mentioned, the intercalated disks offer no obstacle to the conduction of cardiac excitation. This can be attributed to special structures in the apposed portions of cardiac membrane, the so-called *gap junctions* or *nexuses*, which provide the structural correlate of the electrical coupling between cardiac cells.

Disturbances of propagation (slowing or block) are usually attributed to a reduction of the excitatory inward currents: the fast sodium inward current in the atrial and ventricular myocardium or the conducting system, and the slow calcium inward current in the AV node. The sodium inward current is reduced by depolarization due to inactivation of the voltage-dependent sodium channels. Depolarization itself may be due to various factors, such as loss of internal or increase of external potassium or decrease of potassium conductance. However, the fast sodium inward current can also be reduced in the absence of depolarization by, for example, drugs with local anesthetic effects, which are frequently used as antiarrhythmics. In the AV node, where fast sodium channels are absent, inhibition of the slow inward current by certain calcium antagonists (for instance, verapamil) can increase the normal AV delay.

If the excitatory inward currents are critically reduced, the ability of propagation will also depend on the geometry of the fibers: Normally, the wave of excitation will procede without difficulty in a direction that branches off from the main direction at an angle of more than 90°. Under less favorable conditions, block of conduction can preferably occur at such branchings (Spach et al., 1982). Moreover, block of conduction may also be due to an increase of the internal longitudinal resistance when conductance via the gap junctions becomes reduced. Such a change occurs when the intracellular concentration of free calcium rises above the maximal physiological concentration of about 10^{-5} mol/l. This can happen when calcium leaks into damaged cardiac cells. A similar effect is initiated by a rise of the intracellular concentration of hydrogen ions (acidosis) (Délèze, 1970; de Mello, 1972).

Definition and Mechanism of Refractoriness

When an action potential has been elicited in a cardiac muscle cell, this cell cannot be reexcited by the way of conduction or by stimulation until its membrane potential has repolarized to a certain level. The terms *absolute* and *relative refractory period* are used for phases of abolished and diminished responsiveness, respectively. Figure 5.8 shows how these are related to the action potential. During the absolute refractory period the cell is inexcitable, and during the subsequent relative refractory period excitability gradually recovers. Thus a new action potential can be elicited sooner with a stronger stimulus. Action potentials generated very early in the relative refractory period do not rise as sharply as normal action potentials, and have a lower amplitude and a shorter duration (Fig. 5.8).

The term *effective refractory period* is used to define the phase from the beginning of the excitatory cycle during which no propagated excitation can be elicited. This period lasts slightly longer than the absolute refractory period. When the threshold is not exactly determined, but instead stimuli of a constant suprathreshold diastolic strength (for instance, twice diastolic threshold) are used to determine the refractory period, its duration will comprise the absolute and part of the relative refractory period, during which the stimulus remains ineffective. This phase is usually called the *functional refractory period.*

The chief cause of refractory behavior is the inactivation of the fast Na^+ channels during prolonged depolarization. Not until the membrane

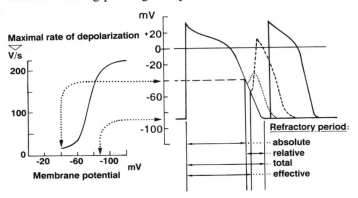

Figure 5.8   Left: Dependence of the maximal rate of depolarization of the action potential (as an indirect measure of the availability of the fast sodium channels) on the membrane potential prior to excitation. Right: Different types of refractoriness as related to the cardiac action potential.

has repolarized to about $-40\,mV$ do these channels begin to recover. The duration of the refractory period is therefore, as a rule, closely related to the duration of the action potential. When the action potential is shortened or lengthened, the refractory period changes accordingly. But drugs that act as local anesthetics, inhibiting the initial Na^+ influx or retarding its recovery after inactivation, can prolong the refractory period without affecting action potential duration. In the absence of drugs, a similar behavior is observed in the AV nodal cells, which, after termination of an action potential, remain nonexcitable for a short time. The refractory period of these cells is thus not only voltage- but also time-dependent.

Another phenomenon that might be considered in this context is the *supernormal period*. This appears as a transient decrease in threshold for stimulation below its diastolic level immediately following the relative refractory period. This phase coincides with the terminal phase of repolarization, when the membrane potential has not yet fully recovered. It can explain the response to electrical stimuli at this instant which is otherwise ineffective. The decrease in threshold (increase in excitability) during the supernormal period can be attributed to the fact that the distance between actual membrane potential and threshold potential is reduced because the membrane retains moderate depolarization. This is more pronounced when the time course of the terminal phase of repolarization is slow, and it disappears if repolarization occurs abruptly.

The prolonged refractory period protects the musculature of the heart from too-rapid reexcitation, which could impair its function as a pump. At the same time, it prevents recycling of excitation in the muscular network of the heart, which would interfere with its rhythmic activity. Because the refractory period of the excited myocardial cells is normally longer than the time taken for spread of excitation over the atria or ventricles, a wave of excitation originating at the SA node or a heterotopic center can propagate over the heart only once and must then die out, for it encounters refractory tissue everywhere. Reentry (defined in Sec. 5.9) thus does not normally occur.

An action potential triggered immediately following the relative refractory period of the preceding impulse is normal, as Fig. 5.8 shows, in upstroke rate and amplitude. Its duration, however, is distinctly less than that of the preceding action potential. In fact, there is a close relationship between the duration of an action potential and the interval that preceded it, and thus between duration and repetition rate. The main cause of this phenomenon is an increase in the potassium conductance, which outlasts the repolarization phase of the action potential and returns only gradually to the basal level. When the interval between action potentials is short, the increased K^+ conductance accelerates repolarization of the next action potential.

When cardiac muscle is exposed to 50- or 60-Hz alternating current, it will respond in a way illustrated by Fig. 5.9. The first depolarizing AC half-wave triggers an action potential, and then remains ineffective until the end of the corresponding (functional) refractory period, when a new action potential starts. For the reasons mentioned above, the duration of the second action potential is shorter, and so is its refractory period. Thus, 60-Hz AC, when applied to cardiac muscle, leads to a rhythmic response, but at a frequency increased in inverse proportion to the refractory period.

5.5   Regular and Ectopic Pacemakers

Elementary Events in Impulse Formation

The working myocardium of atria and ventricles is not automatically active: as outlined above, action potentials are generated by spread of excitation. In all cardiac muscle cells that are capable of autorhythmicity, depolarization toward the threshold occurs spontaneously. This elementary process of excitation can be observed directly by intracellular recording from a pacemaker cell. As shown in Fig. 5.10, the repolarization phase of such an action potential is followed—beginning at the *maximal diastolic potential*—by a slow depolarization which triggers a new action potential when the threshold is reached. The *slow diastolic depolarization* (pacemaker potential) is a local excitatory event, not propagated as the action potential is.

Actual and Potential Pacemakers

Normally only a few cells in the SA node are in fact responsible for timing the contraction of the heart (actual pacemakers). All the other

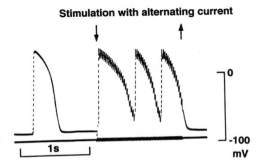

Figure 5.9   Electrical response of a single myocardial cell to stimulation with 50-Hz AC for 1.3 s. Microelectrode recording from isolated myocardium of rhesus monkey.

fibers in the specialized tissue are excited in the same way as the working musculature, by conducted activity. That is, these potential pacemakers are rapidly depolarized by currents from activated sites before their intrinsic slow diastolic depolarization reaches threshold. Comparison of the two processes, as illustrated in Fig. 5.10, shows how a potential pacemaker can assume the leading role when the actual pacemaker ceases to function. Because the slow diastolic depolarization of the potential pacemaker, by definition, takes longer to reach threshold, its discharge rate is lower. In the working myocardium there is no automatic depolarization; the upstroke of the action potential triggered by the imposed current rises sharply from the resting-potential baseline.

According to current opinion, the slow diastolic depolarizations of the SA node are mainly caused by the slow decline of the K^+ conductance

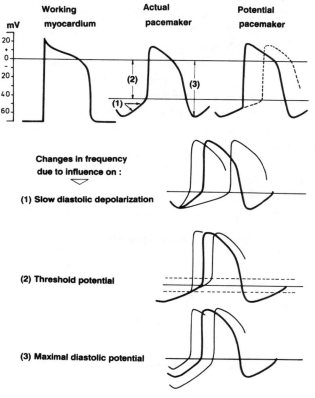

Figure 5.10   Top: Typical forms of action potentials in working myocardium as compared with actual and potential pacemakers. Bottom: Different mechanisms by which changes in frequency of pacemaker cells are brought about.

that had increased during repolarization. This requires a relatively high background Na^+ conductance to produce depolarization. In the ventricular conducting system the background Na^+ conductance is normally low. Therefore the membrane potential reaches relatively high levels just after the action potential, which permits extensive recovery of the rapid Na^+ system. The subsequent diastolic depolarizations involve a special ionic channel that does not operate in the SA node.

Ectopic Pacemakers

The capacity for spontaneous excitation is primitive rather than a highly specialized function of myocardial tissue. In the early embryonic stage, all the cells in the heart are spontaneously active. As fetal differentiation proceeds, the fibers of the prospective atrial and ventricular myocardium lose their autorhythmicity and develop a stable, high resting potential. But the stability of the resting potential can be lost under various conditions associated with partial depolarization of the membrane (cat-electrotonus, stretching, hypokalemia, Ba^{2+} ions) (see Fig. 5.11). Then the affected fibers can develop diastolic depolarizations like those of natural pacemaker cells, and in some circumstances can interfere with the rhythm of the heartbeat. On the other hand, depolarization due to elevated K^+ does not produce autorhythmicity, because a concomitant rise in K^+ conductance inhibits spontaneous activity. A center of autorhythmicity apart from the regular pacemaker tissue is called an *ectopic center* or *ectopic focus*.

5.6 Effects of Autonomic Nerves and of Changes in Electrolyte Composition

The cardioregulatory centers in the brain exert a direct influence on the activity of the heart, by way of sympathetic and parasympathetic nerves. This influence governs the rate of beat (chronotropic action), the systolic contractile force (inotropic action), and the velocity of

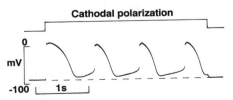

Figure 5.11 Induction of automatic activity in a cell of working myocardium (cat ventricle) by cathodal polarization.

atrioventricular conduction (dromotropic action). These actions of the autonomic nerves are mediated in the heart, as in all other organs, by chemical transmitters—acetylcholine in the parasympathetic system and noradrenaline in the sympathetic system.

Parasympathetic and Sympathetic Innervation

The parasympathetic nerves supplying the heart branch off from the vagus nerves on both sides in the cervical region. The fibers on the right side pass primarily to the right atrium and are concentrated at the SA node. The AV node is reached chiefly by cardiac fibers from the left vagus nerve. Accordingly, the predominant effect of stimulation of the right vagus is on heart rate, and that of left vagus stimulation is on atrioventricular conduction. The parasympathetic innervation of the ventricles is sparse; its main influence is indirect, by inhibition of the sympathetic action.

The sympathetic nerve supply, unlike the parasympathetic, is nearly uniformly distributed to all parts of the heart. The sympathetic cardiac nerves come from the lateral horns of the upper thoracic segments of the spinal cord, and make synaptic connections in the cervical and upper thoracic ganglia of the sympathetic trunk, in particular the stellate ganglion. The postganglionic fibers pass to the heart in several cardiac nerves. Sympathetic influences on the heart can also be exerted by catecholamines (mainly adrenaline) released from the adrenal medulla into the blood.

Chronotropy, Inotropy, and Dromotropy

Stimulation of the right vagus or direct application of acetylcholine to the SA node causes a decrease in heart rate (*negative chronotropy*); in the extreme case, cardiac arrest can result. Sympathetic stimulation or application of noradrenaline increases the heart rate (*positive chronotropy*). When vagus and sympathetic nerves are stimulated at the same time, the vagus action usually prevails, and is followed by a delayed sympathetic influence when stimulation is terminated (Fig. 5.12). Modification of the autorhythmic activity of the SA node by these autonomic inputs occurs primarily by way of a change in the time course of the slow diastolic depolarization. Under the influence of the vagus, diastolic depolarization is retarded, so that it takes longer to reach threshold (Fig. 5.10). In the extreme case, diastolic depolarization is eliminated and the membrane actually becomes hyperpolarized. The sympathetic fibers act to increase the rate of diastolic depolarization and thus shorten the time to threshold (see Table 5.1).

Because the positive chronotropic action of the sympathetic nerves extends to the entire conducting system of the heart, when a leading pacemaker center fails the sympathetic input can determine when and to what extent a subordinate center takes over as pacemaker. In the same way, however, an ectopic focus or rhythmicity can be stimulated to greater activity, so that the danger of arrhythmia increases.

Change in heart rate in itself has a considerable effect on the strength of myocardial contraction. In addition, the autonomic nerves to the heart act directly on mechanical force generation. The vagus acts to reduce the strength of contraction of the atrial myocardium (*negative inotropic acton*). Sympathetic activity increases the strength of contraction in both atrial and ventricular myocardium (*positive inotropic action*).

Change in heart rate in itself has a considerable effect on the strength of myocardial contraction. In addition, the autonomic nerves to the heart act directly on mechanical force generation. The vagus acts to reduce the strength of contraction of the atrial myocardium (*negative inotropic action*). Sympathetic activity increases the strength of contraction in both atrial and ventricular myocardium (*positive inotropic action*).

An influence of the autonomic nerves on the conduction of excitation can normally be demonstrated only in the region of the AV node. The sympathetic fibers accelerate atrioventricular conduction and thus shorten the interval between the atrial and ventricular contractions (*positive dromotropic action*). The vagus—especially on the left side—retards atrioventricular conduction and in the extreme case can produce a transient complete AV block (*negative dromotropic action*). These effects of the autonomic

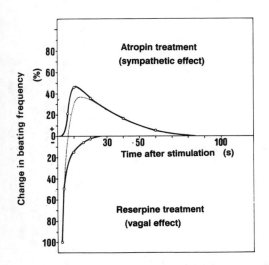

Figure 5.12 Changes in beating frequency of isolated right atrium (guinea pig) following AC (50-Hz) field stimulation (10 V/cm, 1 s). Dotted line: uninfluenced atria (controls). Vagal effects are derived from the difference between controls and atropine treatment, sympathetic effects from the difference between controls and reserpine treatment.

Table 5.1. *Main actions of several influences on the electrical and mechanical activity of the heart*

	SA node Pacemaker activity	AV node v_{max} and conduction velocity	Purkinje system v_{max} and conduction velocity	Purkinje system Pacemaker activity	Atrial and ventricular myocardium Action potential duration	Atrial and ventricular myocardium Force of contraction
Warming	+ +	+ +	+	+	– –	–
Cooling	– –	– –	–	–	+ +	+
Acidosis	–	0	0	–	+	–
Alkalosis	+	0	0	+	–	+
K_e increase	–	–	– –	– –	–	– –
K_e decrease	0	0	0	+ +	0	+
Sympathetic (noradrenaline)	+ +	+ +	0	+	+	+ +
Parasympathetic (acetylcholine)	– –	– –	0	(–)	In atrium – –	In atrium – –
O₂ deficiency	–	–	–	0	–	–

Note: $\dot{V}_{max}$ = maximal rate of depolarization of action potential. +, increase; + +, large increase; –, decrease; – –, large decrease; 0, no prominent effect.

transmitter substances are associated with a particular feature of the cells in the AV node. As discussed above, the fibers of the AV node closely resemble those of the SA node. Because there is no rapid inward Na^+ current, the upstroke is relatively slow and hence the conduction velocity is low. The vagus acts to decrease the rate of rise still further, whereas sympathetic activity increases it, with the corresponding effects on the velocity of atrioventricular conduction.

Vagal and Sympathetic Tone

The ventricles of most mammals, including humans, are influenced predominantly by the sympathetic system. By contrast, the atria can be shown to be subject to the continual antagonistic influence of both vagus and sympathetic nerves; this effect is most clearly evident in the activity of the SA node. It can be observed, for example, by intersecting or pharmacologically blocking one of the two sets of nerves; the action of the opponent then dominates. When the vagus input to the dog heart is removed, the rate of beating increases, from ca. 100/min at rest to 150/min or higher; when the sympathetic input is removed, it falls to 60/min or less. This maintained activity of the autonomic nerves is called *vagal* and *sympathetic tone*. Because the rate of the completely denervated heart (the autonomic rate) is distinctly higher than the normal resting rate, it can be assumed that under resting conditions vagal tone predominates over sympathetic tone.

Mechanism of Autonomic Transmitter Actions

The effects of vagal stimulation and of application of the parasympathetic transmitter, acetylcholine, are attributed to one fundamental action—an increase in the K^+ conductance of the excitable membrane. In general, such an influence is expressed in the tendency of the membrane potential to oppose depolarization. This tendency is evident in both the retardation of the slow diastolic depolarization in the SA node, described above, and in a shortening of the action potential of the atrial myocardium, which in turn weakens the contraction. The reduction of the rate of rise of the action potential in the AV node can also be explained on this basis, in that a stronger outward K^+ current counteracts the slow inward Ca^{2+} current. In the ventricular myocardium, by contrast, the sympathetic-antagonistic action dominates—that is, the main action is inhibition of noradrenaline release from the sympathetic nerve endings.

With respect to the mechanisms by which the sympathetic fibers (or their transmitters) act, there is convincing experimental evidence that they increase the slow inward Ca^{2+} current (increased Ca^{2+} conductance).

That is, the contractile force becomes greater (positive inotropic action) because this effect has intensified the excitation-contraction coupling. The positive dromotropic action on the AV node is also likely, in view of the above considerations, to be related to enhancement of the slow inward Ca^{2+} current. As yet there is no satisfactory explanation of the mechanism of the positive chronotropic sympathetic action. At the SA node, enhancement of the slow inward current is probably involved. In the case of the Purkinje fibers, however, an influence on a specific, hyperpolarization-activated pacemaker current is regarded as more likely.

The actions of autonomic transmitter substances are thought to involve binding of the transmitters to certain molecular configurations on the effector cell (the word "receptor" is used both for these subcellular structures and for sensory cells). The effects of noradrenaline and adrenaline on the heart, described above, are mediated by so-called *β receptors.* Sympathetic effects can be prevented by *β-receptor blockers* such as dichloroisoproterenol (DCI) and pronethalol. In the heart, as in other organs, the deadly nightshade poison atropine acts as an antagonist to the parasympathetic effects of acetylcholine.

In addition to the efferent autonomic supply, the innervation of the heart comprises a large number of afferent fibers, divided among the vagus and sympathetic nerves. Most of the vagus afferents are myelinated fibers originating in sensory receptors in the atria or the left ventricle.

Apart from the myelinated afferent fibers from specialized sensory receptors, the main group of fibers leaving the heart comes from dense subendocardial plexuses of nonmyelinated fibers with free endings; these fibers run in the sympathetic nerves. It is probably these fibers that mediate the severe, segmentally radiating pains experienced when circulation within the heart itself is impaired (angina pectoris, myocardial infarction).

Effects of the Ionic Environment and of Drugs

Of all the features of the extracellular solution that can affect the activity of the heart, the K^+ concentration is of the greatest practical importance. An increase in extracellular K^+ has two effects on the myocardium: (1) the resting potential is lowered because the gradient K_i^+/K_e^+ is less steep, and (2) the K^+ conductance of the excitable membrane is increased—as it is by acetylcholine in the atrial myocardium. Doubling of the K^+ concentration, from the normal 4 mmol/l to about 8 mmol/l, results in a slight depolarization accompanied by increased excitability (threshold for stimulation reduced) and increased conduction velocity. Moreover, this influence causes suppression of heterotopic centers of rhythmicity. A larger increase in K^+ (over 8 mmol/l) reduces excitability (threshold for stimulation increased) and reduces conduction velocity. When the extra-cellular K^+ concentration is lowered to less than 4 mmol/l, the stimulating influence on pacemaker activity in the ventricular conducting system dominates. The enhanced activity of heterotopic centers can lead to cardiac arrhythmias (Table 5.1).

The excitability-reducing action of large extracellular K^+ concentra-tions is turned to advantage during heart operations, to immobilize the heart briefly for the surgical procedures (cardioplegic solutions). While the heart is inactive, circulation is maintained by an extracorporeal pump (heart-lung machine). Impairment of cardiac function due to increased blood K^+ during extreme muscular effort or in pathological conditions can be largely compensated by sympathetic activity.

Under normal conditions, changes in extracellular Ca^{2+} are of limited interest for the function of the heart, because the neuromuscular excit-ability is much more sensitive to such influences. Under experimental conditions, changes in the extracellular Ca^{2+} concentration rapidly affect the force of cardiac contraction. Complete excitation-contraction un-coupling can be achieved by the experimental withdrawal of extracellular Ca^{2+}; under such conditions the action potential of the myocardium re-mains almost unchanged, but it is no longer accompanied by a mechanical response.

An effect similar to that of removing the extracellular Ca^{2+} can be achieved by means of Ca^{2+} antagonists (verapamil, nifedipine, diltiazem, etc.), which block Ca^{2+} influx during the action potential. On the other hand, the amplitude of contraction can be increased both by raising the extracellular Ca^{2+} concentration and by agents that enhance Ca^{2+} influx during the action potential (adrenaline, noradrenaline). In clinical practice the heart action is made more forceful by administering *cardiac glycosides* (digitalis, ouabain).

Table 5.1 summarizes the most important physical and chemical influences on excitation and contraction of the heart; only the dominant effects are considered.

5.7   Electrocardiogram

As excitation spreads over the heart, an electric field is produced that can be sensed on the surface of the body. The changes in magnitude and direction of this field in time are reflected in alterations of potential differences measurable between various sites on the body surface. The electrocardiogram (ECG) is a representation of such potential differences as a function of time. It is thus an indicator of cardiac excitation—not contraction!

Because the directly measured potentials amount in most cases to less than 1 mV, the commercially available ECG recorders incorporate electronic amplifiers. The amplifier inputs include capacitive coupling—highpass filters with a cutoff frequency near 0.1 Hz (a time constant of 2 s). Therefore, DC components and very slow changes of the potentials at the metal recording electrodes, which would be distracting, do not appear at the output. All electrocardiographs have a built-in means of monitoring amplitude in the form of a 1-mV calibration pulse set to cause a deflection of 1 cm.

ECG Form and Nomenclature; Relation to Cardiac Excitation

With electrodes attached to the right arm and left leg, the normal ECG looks like the curve shown in Fig. 5.13. There are both positive and negative deflections (waves), to which are assigned the letters P through T. By convention, within the QRS group, positive deflections are always designated as R and negative deflections as Q when they precede the R wave or as S when they follow it. By contrast, the P and T waves can be

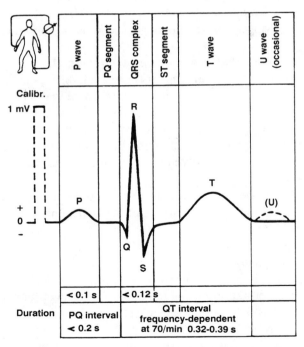

Figure 5.13   Normal form of the ECG with bipolar recording in the direction of the long axis of the heart. The times below the ECG curve are important limiting values for the duration of distinct parts of the curve.

either positive or negative. The distance between two waves is called a *segment*. An interval comprises both waves and segments (for instance, the PQ interval, see Fig. 5.13). The RR interval, between the peaks of two successive R waves, corresponds to the period of the beat cycle and is the reciprocal of beat rate (60/RR interval (s) = beats/min).

Before proceeding to analyze the sources of the ECG curve, the general significance of its elements may be considered: An atrial part and a ventricular part can be distinguished. The atrial part begins with the P wave, the expression of the spread of excitation over the atria. During the subsequent PQ segment the atria as a whole are excited. The dying out of excitation in the atria coincides with the first deflection in the ventricular part of the curve, which extends from the beginning of Q to the end of T. The QRS complex is the expression of the spread of excitation over both ventricles, and the T wave reflects recovery from excitation in the ventricles. The intervening ST segment is analogous to the PQ segment in the atrial part, indicating total excitation of the ventricular myocardium. Occasionally the T wave is followed by a so-called U wave; this probably corresponds to the dying out of excitation in the terminal branches of the conducting system.

The PQ interval is the time elapsed from the onset of atrial excitation to the onset of ventricular excitation, and is normally less than 0.2 s. A longer PQ interval indicates a disturbance in conduction in the region of the AV node or the bundle of His. When the QRS complex extends over more than 0.12 s, a disturbance of the spread of excitation over the ventricles is indicated. The overall duration of the QT interval depends on heart rate. When the heart rate increases from 40 to 180/min, for example, the QT duration falls from about 0.5 to 0.2 s.

Origin of ECG

As a wave of excitation passes over a cardiac muscle fiber, a potential gradient dV/dx is generated, the magnitude of which depends on the momentary phase of excitation. At the front of the wave there is a steep gradient corresponding to the amplitude of the action potential. During the repolarization phase, there appear much smaller gradients in the opposite direction. To a first approximation the excited myocardial fiber behaves in the physical sense as a *variable dipole*, the magnitude and direction of which are symbolized by an arrow (*vector*). By definition, the dipole vector points from minus to plus, that is, from the excited to the unexcited region; an excited site, as seen from the outside, is effectively electronegative as compared with an unexcited site. At every moment during the excitatory process, dipole fields across the surface of the heart sum to an integral vector. As this occurs, a large fraction of the vectors will neutralize one another, as observed from outside the system, because they exert equal effects in opposite directions.

When excitation spreads over the atria (P wave), the predominant direction of spread is from top to bottom; that is, most of the individual depolarization vectors point toward the tip of the heart and thus generate an integral vector pointing toward the apex. When the atria are excited as a whole, the potential differences disappear transiently, for all the atrial fibers are in the plateau phase of the action potential (PQ segment). Only when the excitation moves into the ventricular myocardium do demonstrable potential gradients reappear. Spread of excitation over the ventricles begins on the left side of the ventricular septum and generates an integral vector pointing toward the base of the heart (beginning of QRS). Shortly thereafter, spread toward the apex predominates (largest QRS vector). During this phase, excitation moves through the ventricular wall from inside to outside. Spread through the ventricles is completed with the excitation of a basal region of the right ventricle, at which time the integral vector points toward the right and up (end of QRS). While the excitation was spreading over the ventricles (QRS), it died out in the atria. When the ventricles are totally excited (ST segment), the potential differences disappear briefly, as they did during atrial excitation (PQ segment).

If repolarization of the ventricles took place in the same sequence as depolarization and at the same rate, the behavior of the integral vector during recovery would be expected to be approximately the opposite of that during the spread of excitation. This is not the case, for the following reasons. First, the process of repolarization is fundamentally slower than that of depolarization. Moreover, the rates of repolarization are not the same in the different parts of the ventricles. Repolarization occurs sooner at the apex than at the base, and sooner in the subepicardial than in the subendocardial layers of the ventricles. Thus, during the ventricular recovery phase (T wave), the direction of the integral vector hardly changes; it points to the left.

The different curve forms obtained with the arrangement of leads ordinarily used, on extremities and chest wall, are basically projections of the momentary integral vectors onto certain lead axes.

ECG Recording

A distinction is made between bipolar recordings and so-called unipolar recordings. In the former, two recording electrodes are placed at defined sites on the body surface and the potential difference between these electrodes is monitored. In the latter, only one recording electrode is placed at a defined site, and the potential is measured with respect to a reference electrode. This electrode can be thought of as positioned at the null point of the dipole, between positive and negative charge. In clinical practice, the following recording arrangements are the most commonly used today (Fig. 5.14).

Limb leads:
  Bipolar: standard Einthoven triangle (leads I, II, III)
  Unipolar: Goldberger's augmented limb leads (aVR, aVL, aVF)
Chest leads:
  Bipolar: so-called small chest triangle of Nehb (D, A, I), not shown in Fig. 5.14
  Unipolar: Wilson's precordial leads (V1–V6)

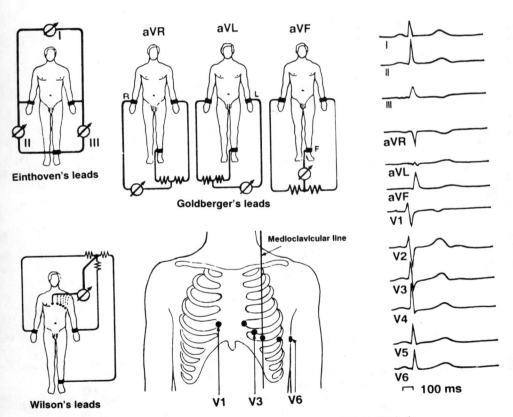

Figure 5.14 Arrangement of ECG leads in common use. Right: Typical curves recorded from a healthy subject.

Einthoven's Triangle. Because in bipolar recording from the limbs by the method of Einthoven the arms and legs act as extended electrodes, the actual recording sites are at the junction between limbs and trunk. These three points lie approximately on the corners of an equilateral triangle, and the sides of the triangle represent the lead axes. Figure 5.15 illustrates the way in which the relative amplitudes of the various ECG deflections in the three recordings are derived from the projection of the frontal plane vector onto the associated lead axes.

Types of QRS Axis Orientation. The direction of the largest integral vector (the chief vector) during the spread of excitation is called the *electrical*

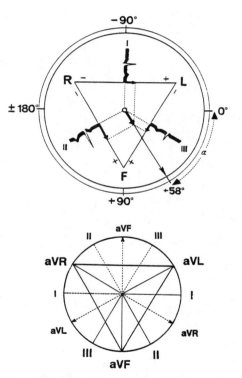

Figure 5.15   Top: Einthoven's triangle. The recording sites at the extremities are represented as the corners of an equilateral triangle, and the sites of the triangle correspond to the lead axes. Angle α of the QRS electrical axis indicated on the outer circle. The projection of an integral vector with angle α + 58° on the three axes is shown, and the magnitude of the various deflections in each axis is indicated by the customary curves. Bottom: Summary of axis orientations with the unipolar (Goldberger) and bipolar (Einthoven) limb leads.

axis of the heart. When the spread of excitation is normal, its direction in frontal projection agrees well with the anatomical long axis of the heart. Therefore limb recordings can be used to infer the orientation of the heart. The various categories are based on the angle α between the electrical axis and the horizontal. In the normal range (shown at the bottom in Fig. 5.15), the angle to the horizontal varies from 0° to 90°. Angles above the horizontal are given a negative sign. The general categories of QRS axis orientation are: normal range $(0° < \alpha < +90°)$; right axis deviation $(+90° < \alpha < +180°)$; left axis deviation $(-120° < \alpha < 0°)$.

For the construction of the electrical axis from the ECG by means of Einthoven's triangle (Fig. 5.15, top), two lead pairs suffice, for the third can be derived from the other two. At each instant during the excitatory cycle it holds that: deflection in II = deflection in I + deflection in III (downward deflections having negative sign).

In Goldberger's method, the voltage measured is that between one extremity—for example, the right arm (lead aVR)—and a reference electrode formed by voltage division between the two other limbs (see Fig. 5.14). With aVR recording, the lead axis on which the vector loop is projected is represented by the line bisecting the angle between I and II in the Einthoven triangle (Fig. 5.15). The axes for aVL and aVF are found in an analogous way.

Unipolar Precordial Leads. Whereas the limb leads just described are fundamentally related to the frontal projection of the vector, the unipolar precordial leads of Wilson provide information chiefly about the horizontal vector projection. A reference electrode is produced by joining the three limb leads, and an exploring electrode records from specific points on the chest at the level of the heart (see Fig. 5.14). A positive deflection is seen when the instantaneous vector, projected onto the appropriate axis, points toward the recording site. If it points in the opposite direction, the deflection is negative.

Use of the ECG in Diagnosis

The ECG is an extremely useful tool in cardiological practice, for it reveals changes in the excitatory process that cause or result from impairment of the heart's activity. From routine ECG recordings the physician can obtain information pertaining to the following.

Heart rate: Differentiation between the normal rate (60–90/min at rest), tachycardia (over 90/min), and bradycardia (below 60/min).

Origin of excitation: Decision whether the effective pacemaker is in the SA node or in the atria, in the AV node or in the right or left ventricle

Abnormal rhythms: Distinction among the various kinds and sources (sinus arrhythmia, supraventricular and ventricular ectopic beats, flutter and fibrillation)

Abnormal conduction: Differentiation on the basis of degree and localization, delay or blockage of conduction (sino-atrial block, AV block, right or left bundle-branch block, fascicular block, or combinations of these)

ORS axis orientation: Indication of anatomical position of the heart; pathological types can indicate additional changes in the process of excitation (unilateral hypertrophy, bundle-branch block, etc.)

Extracardial influences: Evidence of autonomic effects, metabolic and endocrine abnormalities, electrolyte changes, poisoning, drug action (digitalis), etc.

Primary cardiac impairment: Indication of inadequate coronary circulation, myocardial O_2 deficiency, inflammation, influences of general pathological states, traumas, innate or acquired cardiac malfunctions, etc.

Myocardial infarction (complete interruption of coronary circulation in a circumscribed area). Evidence regarding localization, extent, and progress

It should, however, be absolutely clear that departures from the normal ECG—except for a few typical modifications of rhythmicity or conduction—as a rule give only tentative indications that a pathological state may exist. Whether an ECG is to be regarded as pathological or not can often be decided only on the basis of the total clinical picture. In no case can one come to a final decision as to the cause of the observed deviations by examination of the ECG alone.

5.8   Abnormalities in Cardiac Rhythm as Reflected in the ECG

A few characteristic examples may indicate how disturbances of rhythmicity or conduction can be reflected in the ECG. The recordings, where not otherwise indicated, are from Einthoven's limb lead II (see Fig. 5.14).

We first consider an analytical diagram as it is frequently used to illustrate the origin and spread of excitation in the heart. The bar SA in Fig. 5.16 symbolizes the rhythmic discharge of the SA node. The successive stages in the spread of excitation are shown from top to bottom, with the absolute refractory periods of the atria (A) and ventricles (V) indicated along the abscissa by rectangles.

Figure 5.16a shows a normal ECG with the pacemaker in the SA node and the QRS complex preceded by a P wave of normal shape. Above the ECG trace, the process of excitation is diagrammed in the described way.

Rhythms Originating in the AV Junction

A source of rhythmicity in the AV junctional region (the AV node itself and the immediately adjacent conductile tissue) sends excitation back into the atria (including the SA node) as well as into the ventricles (Fig. 5.16b). Because excitation spreads through the atria in a direction opposite to normal, the P wave is negative. The QRS complex is unchanged, conduction occurring normally. Depending on the degree to which the retrograde atrial excitation is delayed with respect to the onset of ventricular excitation, the negative P wave can precede the QRS complex [Fig. 5.16b, (1)], disappear in it (2), or follow it (3). These variations are designated as upper, middle, and lower AV junctional rhythms.

Rhythms Originating in the Ventricles

Excitation arising at an ectopic focus in the ventricles spreads over various paths, depending on the source of the excitation (Fig. 5.16c). Because myocardial conduction is slower than conduction through the specialized system, the duration of spread through the myocardium is usually considerably extended. The differences in conduction path can cause pronounced deformation of the QRS complex.

Extrasystoles

Beats that fall outside the basic rhythm and temporarily change it are called extrasystoles. These may be *supraventricular* (SA node, atria, AV node) or *ventricular* in origin. In the simplest case an extrasystole can be interpolated halfway between two normal beats, and does not disturb the basic rhythm (Fig. 5.16d). Interpolated extrasystoles are rare, since the basic rhythm must be slow enough that the interval between excited phases is longer than an entire beat. When the basic heart rate is higher, a ventricular extrasystole is ordinarily followed by a so-called *compensatory pause*. As shown in Fig. 5.16e, the next regular excitation of the ventricles is prevented because they are still in the absolute refractory period of the extrasystole when the excitatory impulse from the SA node arrives. By the time the next impulse arrives the ventricles have recovered, so that the first postextrasystolic beat occurs in the normal rhythm. But with supraventricular extrasystoles or ventricular extrasystoles that penetrate back to the

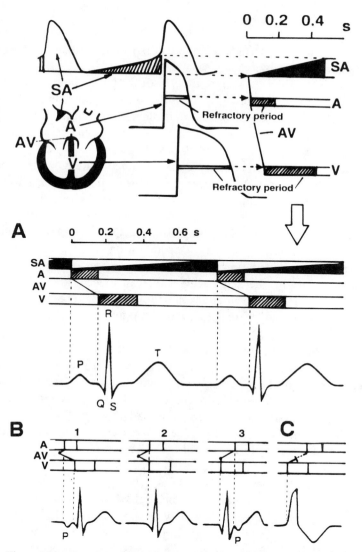

Figure 5.16. Top: Schematic drawing to illustrate the relationship between the spread of excitation in the heart and its representation by the analytical diagram. The rhythmic discharge of the SA node is symbolized in the bar SA of the diagram. The successive stages in the spread of excitation are shown from top to bottom, with the absolute refractory periods of the atria (A) and ventricles (V) indicated along the abscissa. AV summarizes the total atrio-ventricular conduction. Below: Diagrams of the normal time course of cardiac excitation (a) and of various disturbances (b–g). (b) Excitation generated at various parts of the AV junctional region. (c) Excitation originating in the ventricles spreads more slowly, and the QRS complex is severely deformed. (d) Interpolated ventricular extrasystoles of different origins. (e) Ventricular extrasystole with fully compensatory pause; S = normal SA interval. (f) Supraventricular extrasystole with incomplete compensatory pause. (g) Complete (third-degree) AV block.

SA node, the basic rhythm is shifted (Fig. 5.16*f*). The excitation conducted backward to the SA node interrupts the diastolic depolarization that has begun there, and a new cycle is initiated.

Atrioventricular Disturbances of Conduction

The ECG observed in cases of *complete AV block* is shown in Fig. 5.16*g*. As previously mentioned, the atria and ventricles beat inde-

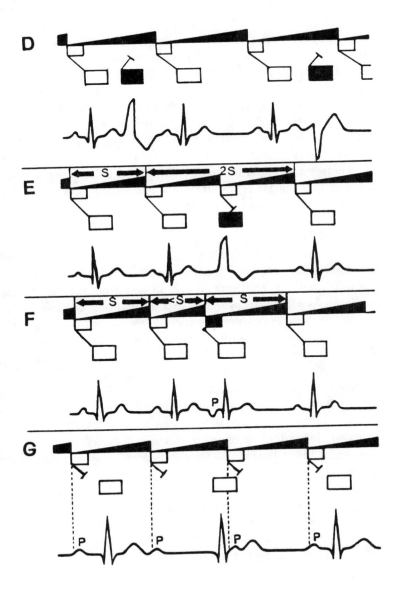

pendently of one another—the atria at the rate of the SA node, and the ventricles at the lower rate of a tertiary pacemaker. *Incomplete AV block* is characterized by interruption of conduction at intervals, so that (for example) every second or third beat initiated by the SA node is conducted to the ventricles (2:1 or 3:1 block, respectively).

Atrial Flutter and Fibrillation

Atrial flutter and fibrillation are arrhythmias resulting from an uncoordinated spread of excitation, so that some atrial regions contract at the same time as others are relaxing (functional fragmentation). Atrial flutter is reflected in the ECG by waves with a regular sawtooth shape and a frequency of 220–350/min, which take the place of the P wave (Fig. 5.17a). Because of temporary AV block due to the refractory period of the ventricular conducting system, normal QRS complexes appear at nearly regular intervals. In the ECG associated with atrial fibrillation (Fig. 5.17b), atrial activity appears as high-frequency (350–600/min) irregular intervals (absolute arrhythmia). There is a continuum of intermediate states between atrial flutter and fibrillation. In general, the hemodynamic effects of these atrial arrhythmias are only slight, and the patients are frequently quite unaware of the arrhythmia.

Ventricular Flutter and Fibrillation

When the ventricles are affected by the same sort of disturbance, the consequences are much more severe. Because the electrical activity is uncoordinated, the ventricles do not fill and expel the blood effectively. This happens with flutter as well as with fibrillation. Circulation is arrested and unconsciousness ensues; unless circulation is restored within minutes, death results. The ECG during ventricular flutter exhibits high-frequency, large-amplitude waves (Fig. 5.17c), whereas the fluctuations associated with ventricular fibrillation are very irregular, changing rapidly in frequency, shape, and amplitude (Fig. 5.17d). If flutter persists for several minutes, it changes more and more into fibrillation. Flutter and fibrillation can be set off by many kinds of heart damage: oxygen deficiency, coronary occlusion (infarction), cooling, overdoses of drugs or anesthetics, etc. Ventricular fibrillation is also the most common acute cause of death in electrical accidents.

5.9   Mechanism of Flutter and Fibrillation

There are two alternative concepts for the mechanisms underlying flutter and fibrillation. The first, originally proposed by Engelmann (1875) and later supported by Rothberger and Winterberg (1941), Scherf

(1947), and others, assumes *ectopic automaticity* as the main cause of fibrillation. The second concept, proposed by Mines (1914), Garrey (1914), and Lewis et al. (1920), postulates that *circulating excitation waves (reentry)* are responsible for the fibrillatory arrhythmia. The first case assumes that there are one or more ectopic foci that, by firing at high rates, overcome the regular formation and conduction of impulses in the heart. In the second case fibrillation is attributed to a primary disturbance in the spread of excitation, in which excitatory waves circulate throughout more or less extended regions of the heart.

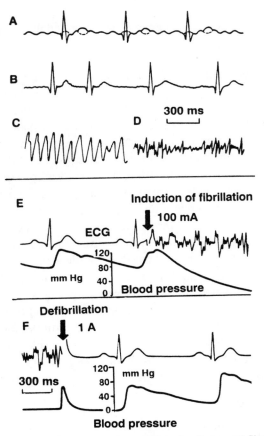

Figure 5.17 (*a–d*) ECG changes during flutter and fibrillation. (*a*) Atrial flutter with sawtooth-shaped flutter waves and 4:1 AV conduction. (*b*) Atrial fibrillation resulting in absolute arrhythmia of the ventricles. (*c*) Ventricular flutter. (*d*) Ventricular fibrillation. Time scale in F valid for all curves. (*e, f*) Induction and termination of ventricular fibrillation by electric current.

It cannot be ruled out at the moment that either of the two mechanisms might operate under certain conditions. However, most of the phenomena that accompany the induction of fibrillation by any means can be explained convincingly only by the hypothesis of reentry. Direct evidence in favor of this hypothesis was presented by Allessie et al. (1973) for isolated rabbit atria, and by Janse et al. (1980) for the left ventricles of isolated perfused porcine and canine hearts after coronary occlusion. These authors used multielectrode recordings from the outer or inner surface of the heart to follow the spread of the excitation wave.

The following sections present an analysis of the mechanism of reentry and its implications for the vulnerability of the heart to ventricular fibrillation. It is shown that a nonhomogeneous state of excitability creates electrically unstable conditions in the heart, which make it susceptible to fibrillation.

Conditions for Reentry; Anatomical and Functional Pathways

A basic requirement for reentry is the existence of interconnected pathways along which excitation can proceed continuously. Since the days when Mines and Garrey developed the concept of reentry, many efforts have been made to identify anatomical structures in the heart that might represent reentry pathways. Rosenblueth and Garcia Ramos (1947) were able to show experimental evidence for such structures in the atrial wall surrounding the ostia of the great veins. A second pathway for reentry was shown by Wit et al. (1972) in meshes formed by Purkinje fibers (see Fig. 5.18), the excitability of which had been depressed. Moreover, Allessie et al. (1977) demonstrated reentry in a flat sheet of the atrial wall without macroscopically visible circuit pathways, showing that the development of reentry does not strictly depend on any preexisting ringlike anatomical structure, but may occur when the electrophysiological properties of myocardial tissue create a functional state that offers comparable conditions. Thus, a refractory zone may simulate a nonexcitable anatomical obstacle around which reentry can occur. In any case, reentry requires an excitation wave with a refractory zone shorter than the given interconnected pathway.

Length of the Excitation Wave

In a myocardial strand assumed to be infinite in length, excitation would proceed in a wavelike fashion. The distance between the excited and recovered parts is about 0.3 m, as can be calculated by multiplying the conduction velocity (about 1.0 m/s) by the duration of the myocardial action potential (about 0.3 s). It is clear that such a large excitation wave cannot bring about reentry in a human heart, because it is longer than any conceivable cardiac pathway. Thus, in order to allow reentry to continue, the excitation wave must be shortened, either by decreasing the velocity of conduction, shortening the refractory period, or both. To what

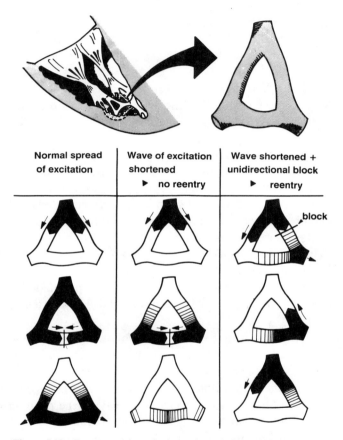

Figure 5.18   Representation of a loop-shaped piece of the ventricular conducting system (top) to illustrate the normal spread of excitation and the development of reentry in this structure. Absolute refractory zones of the excitation waves are represented in black, relative refractory zones are dashed. Left: The wave of excitation travels along the conducting system from top to bottom. Where the wavefronts meet, they extinguish each other because none of them can proceed into the absolute refractory zone of the other one. The same events will occur if the waves become shortened (middle). Right: Reentry is possible only if shortening of the waves of excitation is combined with transient block of conduction in one direction. This may occur when the heart is stimulated during its early relative refractory period.

extent these alterations are necessary will depend on the relationship between the size of the possible pathways and the length of the excitation wave.

The consistent relationship between heart size and probability of fibrillation was an early finding of Garrey (1914), who first emphasized that there must be a critical myocardial mass (or size) for reentry to occur. Furthermore, this relationship explains the well-known fact that sustained fibrillation is more likely in large hearts, whereas small hearts tend to defibrillate spontaneously.

Mechanisms of Abbreviated Refractory Periods

Several mechanisms can lead to an abbreviation of the cardiac excitatory process: (1) interval-dependent shortening of the cardiac action potential with increasing frequency of excitations, (2) shortening due to metabolic disturbance caused by ischemia, and (3) shortening due to the electrotonic interaction between excited (depolarized) tissue proximal to a conduction block and unexcited tissue beyond the block.

1. An action potential initiated soon after a preceding one is found to be considerably shorter. The shorter the interval between the two action potentials, the shorter the duration of the second one (*interval-duration relation*). This influence depends on the rate of recovery from inactivation of the depolarizing ionic currents and on the rate of inactivation of the repolarizing currents (see Chapter 3). If an action potential is initiated before these processes are complete, its duration will be shorter than usual (Carmeliet, 1977).

2. In mild ischemia or in the early phase following coronary occlusion, there is a marked decrease in action potential duration accompanied by shortening of the refractory period (Elharrar et al., 1977). In more severe ischemia, the action potential duration decreases further but the refractory period tends to lengthen, thus displaying the phenomenon of *postrepolarization refractoriness* (Lazzara et al., 1975).

3. If an action potential is propagated in excitable tissue, its shape depends not only on the active membrane properties of the excited tissue, but also on the passive membrane properties of the tissue still unexcited. During the fast upstroke of the action potential, the active membrane area supplies current to the neighbouring resting area, thus initiating excitation. This local circulating current, which depolarizes the resting membrane, simultaneously tends to repolarize the active membrane and to shorten the action potential. If propagation fails at a site of a conduction block, it can hasten the repolarization of the active membrane area proximal to the block. In this way, the duration of the action potential

proximal to the site of a conduction block can be shortened by about 50% (Sasyniuk and Mendez, 1971).

Mechanisms of Slow Conduction

Very low conduction velocities of about 0.1 m/s or less can be obtained experimentally in myocardial preparations perfused with solutions containing both potassium in high concentrations and epinephrine (Cranefield et al., 1971). In such a medium the fast inward current is inactivated by the potassium-induced depolarization, and simultaneously the slow inward current is enhanced by epinephrine. The result is an action potential, known as a *slow response*, which is slowly propagated with a conduction velocity of about 0.02 to 0.08 m/s. Under these conditions, it is possible to induce reentry within a loop of Purkinje fibers shorter than 15 mm.

Another possible mechanism of conduction delay, which extends the experimental evidence for reentry, occurs as a result of the electrotonic spread of excitation across a segment of inexcitable tissue (Jalife and Moe 1981). An inexcitable segment—for instance, an ischemic zone—may be electrically well coupled with surrounding tissue but be unable to generate an all-or-nothing response. Thus, the spread of excitation across such tissue will be determined by the local circuit current that causes a depolarization in the excitable tissue beyond the inexcitable zone. A depolarization of sufficient amplitude may elicit an action potential in the distal tissue, although with marked delay. Such an electrotonically mediated conduction delay can result in apparent conduction velocities of less than 0.01 m/s, and thus may allow reentry along circuit pathways only about 2–3 mm in length (Antzelevitch and Moe, 1981). Moreover, if the conduction delay across an inexcitable gap outlasts the refractory period of the tissue proximal to the gap, the excitation of the tissue beyond the gap can in turn reexcite the proximal tissue by electronic transmission. This phenomenon is called *reflected reentry*.

More recently, investigations of the conditions that favor reentry have drawn attention to changes in conduction velocity that depend on the direction of propagation as related to the fiber orientation. It has been observed that propagation of the excitation wave is slower in the direction transverse to the long cell axis than in the direction of this axis.

Measurements of the axial resistance exhibit considerably higher values in the direction transverse to the long cell axis than in the longitudinal direction (Clerc, 1976; Spach et al., 1981). A possible explanation for the increased transverse axial resistance may be given by the fact that a substantial part of the resistance is located in the cell-to-cell junctions.

Because the current in the transverse direction must pass through a junction every cell width, there are more cell junctions per unit length in the transverse direction and hence the effective axial resistance will be greater.

Unidirectional Block and One-Way Conduction

The implicit assumption in considering an abbreviated excitation wave as a condition favoring reentry is that the wave propagates in the loop-shaped pathway in only one direction. However, this condition does not necessarily result from the abbreviation of the excitation wave. If an excitation wave enters a loop-shaped pathway (a preexisting, anatomical ring-like structure, as illustrated schematically in Fig. 5.18) it will branch off and travel in both directions around the loop. The two newly arising excitation wave fronts will then encounter and extinguish each other somewhere opposite to the entrance to the loop, because neither of them is able to propagate into the refractory zone of the other one. Therefore, in order to accomplish reentry, the excitation wave must be forced to travel in only one direction; that is, conduction must be blocked in one branch of the loop (Fig. 5.18, right). Furthermore, the blocked area must permit conduction in the opposite direction of the excitation wave arriving via the unblocked branch of the loop. If the effective refractory period of the area proximal to the block is shorter than the conduction time along the pathway, it will be able to reenter the loop. Thus, the occurrence of unidirectional conduction is an essential condition for reentry.

Probably the most important cause of unidirectional conduction block in cardiac tissue is given by nonuniformity in the recovery of excitability leading to spatial variations in refractoriness. As illustrated in Fig. 5.18 (right), the two pathways of the loop differ in their refractoriness. An impulse entering the system is therefore blocked in the right limb because of the still-existing refractory state of this region, whereas it can travel freely in the left limb where excitability has already recovered. After some delay (due to the travel time), the latter impulse is able to reenter the right limb, where the refractory period has meanwhile expired. Experimental evidence for the significance of nonuniformity in excitability was shown by Allessie et al. (1976), who found that reentrant tachycardias could be induced in a sheet of rabbit atrial tissue only if the excitation wave was elicited at a region of markedly inhomogeneous refractoriness.

Unidirectional conduction can also be associated with electronically mediated propagation across a functional obstacle. Successful electrotonic transmission across an inexcitable gap in Purkinje fiber preparations is dependent on the mass of excited tissue proximal to the gap. If this tissue mass is too small to provide enough current for the excitation of a larger mass of tissue beyond the gap, conduction will fail. However, conduction in the opposite direction can succeed because the larger tissue mass is able to exert a sufficient depolarizing influence (Jalife and Moe, 1976).

Nonuniformity of active or passive cell properties leading to asymmetry in conduction are inherent features of cardiac tissue. Although not apparent under normal conditions where the safety factor of propagation is high, this nonuniformity can become obvious by the occurrence of unidirectional block if propagation is critically impaired, for instance, during propagation of a premature beat.

5.10   Vulnerable Period: Threshold for Fibrillation

It is well known that atrial or ventricular fibrillation can be induced by electrical stimulation coinciding with late atrial or ventricular systole—the so-called vulnerable period. This term was introduced by Wiggers and Wégria (1939) to express the fact that the susceptibility of the heart to ventricular fibrillation is more or less confined to a certain phase of the cardiac cycle.

In the ventricles the nonuniformity in recovery of excitability is maximal preceding the apex of the T wave of the ECG (Fig. 5.19). At this time, electrical stimulation elicits an excitation wave that encounters some regions fully recovered, others partially recovered, and others still absolutely refractory. Propagation of an electrically induced wavefront can thereby be initiated preferentially in certain directions, thus setting the stage for multiple reentry—the electrophysiological basis of ventricular fibrillation. The normal heart is susceptible to ventricular fibrillation only during this phase; at other times stimulation is ineffective or results only in a single extrasystole.

In addition to the induction of fibrillation by electrical stimulation, a spontaneously generated extrasystole that occurs during a highly nonuniform phase of recovery may also induce fibrillation. This is the basis of the so-called *R-on-T phenomenon* in the ECG, which is considered a preliminary symptom of imminent fibrillation. It is also important to remember that under certain conditions the duration of the vulnerable period may span a longer time should marked nonuniformity in recovery persist. This is the case after a premature beat that exaggerates nonuniformity of recovery (Wégria et al., 1941). Furthermore, the vulnerable period may extend even beyond the T wave if the expiration of the refractory period of the myocardium lags behind the repolarization. This has been demonstrated during ischemia, where postrepolarization refractoriness occurs (Williams et al., 1974). Under such conditions, late stimulation or a late coupled premature beat may induce tachycardia or ventricular fibrillation (El-Sherif et al., 1975). In recent years it has become customary to use the term "vulnerability" in a more general sense to describe the degree to which the heart is susceptible to fibrillation, including the expansion or shift of the susceptible phase within the excitatory cycle. Those changes are reflected in the strength of the stimuli required to induce the arrhythmias, i.e., in the threshold for fibrillation. The greater the susceptibility of the heart to fibrillation, the smaller the intensity of the stimulus required to induce the arrhythmia, and this is equivalent to a decrease of the fibrillation threshold.

The Threshold for Fibrillation
The fibrillation threshold is defined as the minimal current intensity required to produce this arrhythmia. This minimal intensity of

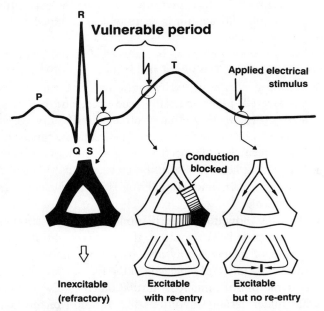

Figure 5.19   Diagram to explain the vulnerable period of the ventricles. The triangle-shaped figures below the ECG curve symbolize (as in Fig. 5.18) the branched network of the conducting system or the myocardium. In the vulnerable period the conduction pathway is still partially refractory, so the wave of excitation generated by stimulation can propagate in only one direction and induce reentry in the same way as illustrated in Fig. 5.18. The ventricles would still be nonexcitable if they were stimulated earlier; at a later time reentry is no longer possible because normal spread of excitation again takes place.

current can be considered as an index of the extra amount of nonuniformity in excitability that must be added via the applied current to the inherent nonuniformity of the heart to achieve the fibrillatory condition. This extra nonuniformity is brought about in part by the direct influence of the current itself, which favors depolarization below the cathode and repolarization below the anode (see Chapter 4). If the inherent non-uniformity of excitability of the heart is low, the required increase in nonuniformity brought about by the applied current must be large, and thus the fibrillation threshold will be high. On the other hand, if the inherent nonuniformity is more pronounced, for instance, because of ischemia, then little additional extra nonuniformity will be required and the fibrillation threshold will be low. Thus, the diseased heart may be more susceptible to fibrillation by electric shock.

Relation to the Threshold for Stimulation

Because the determination of the fibrillation threshold is based on electrical stimulation, the absolute value of the fibrillation threshold must also depend on factors that determine the stimulation threshold. These factors include such purely physical influences as the size and position of the electrodes, the current waveform, the duration of current flow, and the transition resistance. Only if these circumstances can be kept constant can relative changes in the fibrillation threshold be attributed to changes in the vulnerability of the heart. In Fig. 5.20, mean values for both thresholds are compared as the stimulus duration is increased from 0.5 ms to several milliseconds (Younossi et al., 1973). The values of the stimulation threshold reflect the *strength-duration curve* for the isolated whole heart. The curve drawn from the values for the fibrillation threshold shows an essentially similar shape, but it is shifted to a higher level by a factor of 10 to 20. Similar findings have been obtained by in-situ measurements from different species (Brooks et al., 1955).

The different levels of the two thresholds can be explained by the different conditions under which a cardiac response is elicited. In order to attain the stimulation threshold, a local density of current just high enough to stimulate a number of cells (about 50) with fully recovered excitability is required. By contrast, to produce fibrillation the cardiac response must be elicited during the early phase of recovery, and that requires a higher current intensity. Furthermore, to elicit premature responses at different sites of the heart, the current must spread throughout a certain mass of tissue, thus increasing the probability of multiple reentry. For the same reason, the two thresholds respond in an opposite way to changes in the electrode size (Fig. 5.20). Enlargement of the electrodes leads not only to a more extensive extrapolar current spread, but also to a decrease in the current density. The former effect is obviously responsible for the decrease in the fibrillation threshold; the latter increases the threshold for stimulation.

Effects of Alternating and Direct Current

The fibrillatory efficacy of a current can no longer be predicted from its stimulatory efficacy when the current covers more than one cardiac cycle. This observation may be derived from a comparison of the fibrillation and stimulation thresholds determined with alternating current (AC) and with a direct-current pulse (DC) at a constant stimulus duration of 1 s (Fig. 5.21). The threshold for stimulation with DC is clearly below the value obtained with AC. This result can be explained by the higher slope of the rising phase of the DC pulses compared with the sinusoidal AC waves. By contrast, the fibrillation threshold for DC is considerably higher than for AC (Hohnloser et al., 1982).

It is clear that the explanation for this result must be sought in causes other than the stimulatory efficacy of the two kinds of current. These causes can be derived from Fig. 5.9. The single fiber responded to

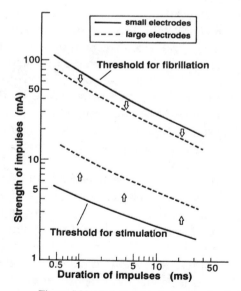

Figure 5.20   Relationship between stimulation threshold and fibrillation threshold in the isolated guinea pig heart ($n = 14$) for rectangular pulses (duration 0.5–50 ms). Note opposite effect of changes in size of stimulation electrodes on the two thresholds. Large electrodes were about 0.7 cm^2, small electrodes 0.05 cm^2.

50-Hz AC with a series of action potentials at a frequency that depended
on the duration of its refractory period and on the current strength. When
the isolated heart is stimulated under comparable conditions, it shows a
series of extrasystoles. A DC pulse exerts its stimulating effect only
during the make-or-break phase of the current (Fig. 5.7). However, under
the influence of long-lasting suprathreshold DC, the single fiber shows
an additional effect, namely, the development of spontaneous activity
(Fig. 5.11). The myocardial fiber then behaves like a pacemaker cell, exhibit-
ing diastolic depolarizations and rhythmic firing. The frequency of these
spontaneous action potentials, however, is much lower than during AC
flow of comparable strength. Similarly, the frequency of extrasystoles of
the whole heart is lower during DC flow than during AC, even at a higher
current intensity. It seems that the differences in the fibrillation thresholds
for AC and DC can be related to the different numbers of repetitive
extrasystoles that preceded the onset of fibrillation (see also Sect. 6.5).

Significance of Single and Repetitive Extrasystoles

Wégria et al. (1941) were the first to discuss a decrease of the
fibrillation threshold after a premature response. They deduced this idea
from the fact that the vulnerable period of a premature beat is prolonged
and extends nearly to the end of the T wave, thus indicating a further

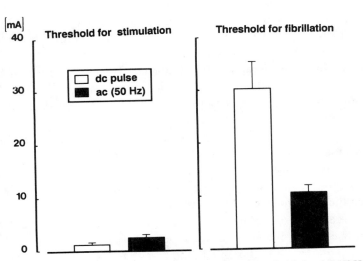

Figure 5.21   Right: Comparison of fibrillation thresholds for AC (50 Hz) and
DC at a constant duration of 1 s. Mean values ± SD from guinea pig heart
($n = 5$). Left: Corresponding thresholds for stimulation. Size of electrodes was
0.05 cm² in all experiments.

state of nonuniform excitability. Han and Moe (1964) confirmed this conclusion by means of multiple-electrode recordings, which showed that the dispersion of the refractory period of nearby regions of the dog heart increased after a premature beat and thereby enhanced the intrinsic vulnerability; they measured a 35% decrease of the fibrillation threshold after a premature response (Han et al., 1966). This result was confirmed by Hauf et al. (1977), who found the decrease of the fibrillation threshold for a premature response to be 34% in the guinea pig heart.

A series of extrasystoles reduces the fibrillation threshold more so than does a single extrasystole (Roy et al., 1977). The dependence of the fibrillation threshold on the number of the repetitive extrasystoles that precede the onset of the arrhythmia explains why the fibrillation threshold decreases with increasing stimulus duration, i.e., with an increasing number of prefibrillating extrasystoles. After the induction of a certain number of extrasystoles closely succeeding one another, the nonuniformity of excitability is at its maximum and cannot be increased further by inducing a greater number of extrasystoles. Hence, with increasing duration of AC flow, the fibrillation threshold reaches a minimum current plateau that is usually not much higher than the stimulation threshold (see Fig. 6.9).

5.11   Electrical Defibrillation and First Aid

The first observations on electrical defibrillation date from about the same time as the induction of fibrillation by electric current was first described. Nevertheless, electrical defibrillation was not performed in humans before 1947 (Beck et al., 1947). Defibrillation requires short pulses of comparatively high intensity, applied through a large area. In case of success, fibrillation is terminated instantaneously, and normal cardiac activity is restored, usually without any sign of injury.

Defibrillation can be explained by synchronized stimulation of all excitable regions of the fibrillating heart by the electric shock. In this way, the heart as a whole is converted into an absolutely refractory state which prevents circus movement from going on. This state is comparable with the absolute refractory period at the end of the normal spread of excitation (Fig. 5.19). When interrupted in this way, reentry does not reappear, provided there are no premature excitations.

The large area of application of the defibrillation pulse and its comparatively high intensity of up to several amperes are required to achieve a suprathreshold stimulus strength in all regions of the fibrillating heart where excitable gaps are present at the instant of the shock. Since the excitable gaps perform circus movement, too, defibrillation can also be effective if comparatively weak shocks are applied as a series of pulses in a circumscribed area via an intracardiac pacemaker electrode. Under these conditions, defibrillation occurs through the abolition of the excitable gaps, not instantaneously, but successively when they pass the region of stimulation.

Without doubt, electrical defibrillation is the most effective measure presently available against ventricular fibrillation. However, its application will lead to full success only when the circulatory arrest due to fibrillation does not exceed several minutes. Otherwise irreversible damage of the brain must be expected. When the circulatory arrest lasts only a few seconds, unconsciousness may ensue as the first still-reversible disturbance of the central nervous function. After about 2 min spontaneous respiration stops, and, increasing with time, the heart itself becomes more and more depressed, because its blood supply is interrupted, too.

Thus, if for any reason electrical defibrillation cannot be performed immediately, first aid measures have to be undertaken to maintain a minimum circulation and ventilation. This is possible by means of external *cardiac massage* combined with *artificial respiration*. The rhythmically exerted compression of the heart between the breastbone and the vertebral column leads to a unidirectional blood flow because backward flow is impeded by the cardiac values. Artificial mouth-to-mouth-respiration can be performed alternating with cardiac massage. This is effective because the oxygen content of the expiratory air is about 60% of that in the inspiratory air. Cardiac massage and mouth-to-mouth respiration are life-saving measures often indispensable to bridge the time from the incidence of ventricular fibrillation to the availability of electrical defibrillation— provided, however, that they are initiated as early as possible and performed continuously, perhaps for hours.

6

Cardiac Sensitivity to Electrical
Stimulation

6.1   Introduction

Analyses of electric shock have, in the past, typically treated
consumer power exposures (50- and 60-Hz alternating currents, and direct
currents). However, there are many contexts where the electrical stimulus
waveshape or repetition pattern may be much more complex. Complex
current exposures may be accidental, may be an incidental by-product of
a medical procedure, or may be the intended waveform of a medically
therapeutic or prosthetic device. We need to understand the potential
human reactions to complex current stimuli in order to determine
acceptable operational standards for such devices.

It is important to understand the relationships between cardiac excitability and stimulus features such as duration, frequency, repetition
pattern, waveshape, electrode configuration, current density, and current
pathway. Much of our knowledge of these effects is derived from studies
directed toward therapeutic uses of cardiac stimulation, such as cardiac
pacing and defibrillation. The knowledge gained from such studies can
be applied to electrical safety concerns, but such data are limited in their
applicability: it is impossible to test all conceivable conditions of stimulus
waveforms and methods of application. For this purpose, the development
of a generic computational model is advocated, as discussed in Sec. 6.10.

We can better evaluate cardiac hazards if we understand the principles
of cardiac electrophysiology presented in Chapter 5. The reviews in
Chapters 3 and 4, concerning principles of excitable membranes, will also
be referred to in this chapter.

6.2   Threshold Sensitivity with Respect to Cardiac Cycle

The sensitivity of the heart to externally applied electrical
stimulation depends critically on the timing of the stimulus with respect

to the cardiac cycle. Figure 6.1 (adapted from Jones and Geddes, 1977, and Geddes, 1985) shows the threshold sensitivity for both excitation and fibrillation as a function of the initiation time of brief (0.5–5 ms) rectangular current stimuli. The data were obtained in experiments with 120 dogs, with stimulation of the myocardium via a helical bipolar electrode. The waveform drawn below the figure illustrates the action potential (AP) that might be measured across the membrane of a single myocardial cell. The lower waveform shows a corresponding ECG that might be simultaneously measured with cutaneous electrodes. The horizontal axis of the figure applies to the initiation time of the stimulus with respect to an ongoing cardiac cycle, measured as a fraction of the QT interval of the ECG waveform. The QT interval is equivalent to the AP duration, as suggested by the drawings below the graph.

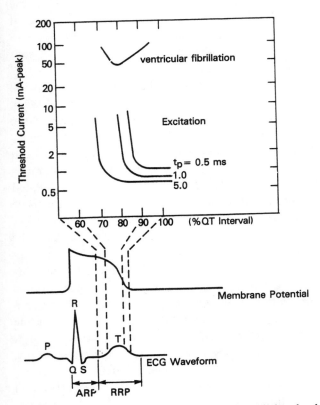

Figure 6.1   Excitation and ventricular fibrillation thresholds: upper, stimulation thresholds; middle, membrane potential; lower, ECG waveform. (Adapted from Geddes, 1985.)

The period labeled "ARP" is the *absolute refractory period* for the excitable cardiac tissue, during which time the cellular membrane is nearly fully depolarized. During this period it is impossible to initiate a new AP process, no matter how strong the stimulus. The period labeled "RRP" is the *relative refractory period*—this is the repolarization phase of the AP process, during which the membrane may be reexcited by a sufficiently strong stimulus. Following the relative refractory period is the *diastolic period*, during which the tissue is most sensitive to excitation. The refractory behavior of cardiac tissue is largely a result of the *deactivation variable* of the cellular membrane during the AP process (refer to Sec. 3.2).

The excitation curves shown in Fig. 6.1 reflect the refractory condition of the excitable cardiac membrane, showing a threshold that declines sharply during the RRP to a minimum plateau that is sustained until the next cardiac cycle. Ventricular fibrillation (VF), however, can be induced only in a narrow window during the repolarization phase of the heart—this is the so-called vulnerable period of the heart. The temporal point in the heart cycle when an extrasystolic excitation is most likely to cause VF displays a bell-shaped distribution. The parameters of that distribution (mean $\pm$ S.D.) were found to be 85% ($\pm$9%) of the heart's QT duration as revealed in experiments with sheep (Ferris et al., 1936), and 72% ($\pm$9%) in experiments with dogs (Jones and Geddes, 1977). Table 6.1 lists for the Jones and Geddes experiments the ratio of the minimum fibrillation current during the vulnerable period to the minimum excitation current during the diastolic period. The indicated ratio will depend on the conditions of stimulation, including the stimulus waveform, the electrode size, and the locus of excitation. The tabulated values apply to stimulation of dog hearts via a bipolar helical electrode sutured to the myocardium.

Only a few cardiac cells need to be excited to produce an extrasystolic contraction (Geddes, 1985), whereas ventricular fibrillation typically requires several conditions, namely: (1) excitation, (2) during the repolarization phase, (3) involving a critical mass of the heart. According to one theory, ventricular, fibrillation is due to the initiation of excitation at a site that is partially refractory as the normal AP propagates over the surface of the heart [see Chapter 5, and Antoni (1979)]. Reexcitation at that stage can result in an extraneous AP that fails to extinguish naturally, but rather circulates in an uncoordinated fashion that fails to produce pumping. This AP circulation is called "reentry" in this schema. It should not be assumed that excitation outside the vulnerable period is inconsequential to cardiac safety considerations. As pointed in Sec. 6.4, extra-

Table 6.1. *Ratio of VF to excitation thresholds (excitation by monophasic current pulse)*

Current duration (ms)	Ratio I_F/I_{Ex}
0.5	145 ± 55
1	100 ± 27
2	112 ± 37
3	87 ± 25
4	81 ± 12
5	63 ± 37
10	$52 \pm ?$

Notes: I_F determined during vulnerable period; I_{Ex} determined during diastolic period. Experiments with dogs ($n = 120$).
Source: Data from Jones and Geddes (1977). Stimulation via helical bipolar electrode sutured to myocardium.

systolic excitation can significantly increase the susceptibility of the heart to fibrillation by a subsequent excitatory stimulus.

The heart is more sensitive during the diastolic interval to stimulation at the site of a cathodal electrode than at the anode. Neurons are also more sensitive to cathodal stimulation because of the spatial concentration of depolarization in the vicinity of a cathode, as opposed to the less intense and more diffuse depolarization near an anode (see Sec. 4.4). Nevertheless, cardiac tissue can be preferentially excited with an anodal stimulus during the relative refractory period (Brooks et al., 1955; Cranfied et al., 1957). Figure 6.2 (from Antoni, 1979) illustrates anodal and cathodal thresholds. An anodal pulse applied during the relative refractory period repolarizes the heart to a level that depends on the strength of the stimulus. If the repolarization is sufficient, the membrane may spontaneously revert to a depolarized state if the anodal stimulus is terminated within the AP time frame—the result can be the initiation of an extraneous AP (Hoffman and Cranfield, 1960). Cranfield and colleagues (1957) also demonstrated that, in the relative refractory period, anodal break may cause excitation at some minimum threshold, but may fail to produce excitation above a higher stimulus value. This phenomenon produces a "dead zone" between the anodal and cathodal thresholds where excitation does not occur.

When the stimulus exceeds the VF threshold by a sufficient amount, a fibrillating heart may be defibrillated. This results from wide-scale excitation that places the heart in a homogeneous state. As a result, there exists a threshold window within which fibrillation is possible. Below the lower limit of the window, VF is not possible because the stimulus is too feeble to cause excitation; above an upper limit, VF is avoided by wide-scale excitation. Consequently, exposure to high voltage or high current can result in massive injuries, without death from VF. This observation is discussed more thoroughly in Sec. 10.2.

6.3   Strength-Duration Relations for Unidirectional Currents

An experimental relationship between the threshold intensity of a rectangular current stimulus and its duration defines a *strength-duration* (S-D) relationship. We have seen in Chapter 4 that experimental S-D data concerning cardiac excitation are frequently fitted with the mathematical expression that Weiss (1901) originally derived for nerve excitation [Eq. (4.15)] or the expression derived by Lapicque [Eq. (4.17)]. The Weiss formulation is sometimes referred to as a "parabolic" relation, and that of Lapicque as an "exponential" relation. As shown in Chapter 4, the parameter τ_e in either the Weiss or Lapicque formulation may be

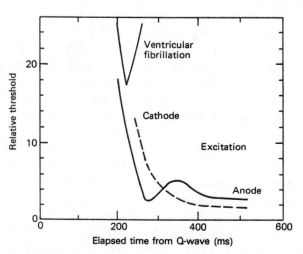

Figure 6.2   Excitation and ventricular fibrillation thresholds. [Adapted from Antoni (1979). Reprinted by permission of VCH Publishers, Inc., 220 East 23rd St., New York, N.Y. 10010 from: *Progress in Pharmacology*, (Stuttgart, Gustav Fisher Verlag), vol. 2/4, pp. 5–12, fig. 3 (page 6).]

determined by the ratio of minimum charge to minimum current:

$$\tau_e = \frac{Q_0}{I_0} \tag{6.1}$$

Equation (6.1) represents a simple and practical relationship for determining the parameter τ_e in both the parabolic and exponential relationships. Figure 4.3 plots normalized S-D curves from the parabolic and exponential formulations, along with normalized thresholds from a myelinated nerve model. The horizontal axis is the normalized threshold current I_t/I_0. The three models closely agree, with only minor differences in thresholds for normalized durations between 0.1 and 10.

Although the Weiss and Lapicque formulas were originally applied to neural excitation thresholds, these curves have been used to fit empirical data for cardiac excitation.[†] For example, the exponential relationship has provided an excellent fit to ventricular excitation thresholds in dog and turtle hearts (Pearce et al., 1982), as well as excitation thresholds of sheep cardiac Purkinje fibers (Dominguez and Fozzard, 1970). Equally good fits have been obtained with the parabolic relation for excitation thresholds of dog hearts (Jones and Geddes, 1977). Comparison of experimental S-D curves for cardiac excitation in dogs reveals that both formulations give a good fit to experimental data, with no clear preference between the two (Mouchawar et al., 1989). Table 6.2 summarizes data derived from a variety of sources listed in the final column of the table. The table is organized by experimental end points consisting of excitation, fibrillation, and defibrillation. The column labeled "τ_e" lists empirical time constants, which in most cases were determined by the ratio Q_0/I_0 [see Eq. (6.1)], or as specified in the cited work.

The τ_e values listed in Table 6.2 have a geometric mean of approximately 3.0 ms—a value that is roughly a factor of 10 above S-D time constants typically observed for neural excitation (see Sec. 4.5 and Table 7.1). Some general observations are pertinent to the information presented in Table 6.2.

Electrode Size. Irnich (1980) reports that τ_e increases monotonically with electrode size, as illustrated in Fig. 6.3. The Irnich data can be fitted by a relationship of the form $\tau_e = K A^{0.56}$, where A is the electrode area and K is a scaling constant. The direction of this relationship follows that reported for neural excitation (Pfeiffer, 1968). Smyth et al.'s data (1976) correspond

---

[†]As used here, excitation refers to the production of an extrasystolic contraction—usually during the diastolic period of a previous beat.

Table 6.2. *Strength-duration data for cardiac stimulation by square-wave current waveforms*

Item	τ_e (ms)	Species	Preparation	Stimulus locus	Electrode	Notes	Reference
A. Thresholds of excitation							
1	3.8	Guinea pig	Isolated tissue	Papillary muscle	Small		Weirich et al. (1985)
2	1.75	Rabbit	Isolated heart	Left ventricle	1 cm^2		Roy et al. (1985)
3	2.86 ±0.99	Dog	In vivo	Trans-chest	8.1 cm diam.		Voorhes et al. (1983)
4	2.14 ±0.97	Dog	In vivo	Ventricle	Small		Pearce et al. (1982)
5	7.31 ±2.22	Turtle	In vivo	Ventricle	Small		Pearce et al. (1982)
6	0.68 to 3.94	*	*	*	$4{-}94 \text{ mm}^2$	(a)	Irnich (1980)
7	0.206	Dog	In vivo	Myocardium	See note	(b)	Jones and Geddes (1977)
8	1.1	Human	In vivo	*	12 mm^2		Fozzard and Schoenberg (1972)
9a	2.1	Dog	in vivo	Ventricle	Loop-and-catheter cathode	(c)	Thalen et al. (1975)
9b	0.73	Dog	In vivo	Ventricle	Loop-and-catheter anode	(c)	Thalen et al. (1975)
10	3.75 ±0.50	Sheep	Long (>8 mm) Purkinje fiber	Near end	Pointlike	(d)	Fozzard and Schoenberg (1972)
11	29.3	Sheep	Short Purkinje fiber	Near end	Pointlike	(d)	Fozzard and Schoenberg (1972)

12	2.6 ±0.88	Sheep	Purkinje fiber	Near end	Pipette	(d)	Dominguez and Fozzard (1970)
13a	2.3	Human	In vivo	*	0.32 cm^2	(e)	Smyth et al. (1976)
13b	1.7	Human	In vivo	*	0.18 cm^2	(e)	Smyth et al. (1976)
14	1.49	Dog	In vivo	Myocardium	7.9 mm^2		Mouchawar et al. (1989)
B. Thresholds of fibrillation							
15	7.7	Guinea pig	Isolated heart	Whole heart	*		Weirich et al. (1985)
16	(~36.0)	Guinea pig	Isolated heart	*	*	(f)	Hohnloser et al. (1982)
17	1.70	Dog	In vivo	Myocardium	See note	(b)	Jones and Geddes (1977)
C. Thresholds of defibrillation							
18	3.90	Dog	Isolated heart	Whole heart	Uniform field	(g)	Geddes et al. (1985)
19	2.76	Dog	Isolated heart	Right ventricle	Catheter	(g)	Wessale et al. (1980)
20	4.01	Dog	In vivo	Trans-chest		(g)	Bourland et al. (1978)
21	2.98	Pony	In vivo	Trans-chest		(g)	Bourland et al. (1978)
22	6.50	Dog	In vivo	Ventricle	*		Koning et al. (1975)

*Information sketchy or not available.

(a) τ_e monotonically increasing with electrode size.

(b) Electrode pair (1-cm separation) wrapped around 2-mm-OD tube and sutured to myocardium.

(c) Average of loop and catheter electrode time constants.

(d) τ_e strongly dependent on length of excised fiber, intracellular electrode.

(e) Data from 21-month electrode implant.

(f) Fibrillation induced by anodal break during vulnerable period of extrasystole; stimulus duration 30 to 100 ms. Time constant not indicative of single excitation (see text).

(g) Trapezoidal pulses with slopes not exceeding 15%.

closely to the Irnich relationship. For instance, for an electrode area of either 32 or 8 mm², the values of τ_e indicated by Smyth et al.'s are 2.3 or 1.7 ms, respectively. These points correspond very closely to the Irnich curve shown in Fig. 6.3. Thalen et al. (1975), however, reported that the S-D time constant for stimulation of dog hearts remained approximately constant over a range of electrode sizes that resulted in 3:1 differences in the excitation threshold.

Theoretical considerations (Jack et al., 1983, pp. 275, 418) predict that the S-D time constant should increase along with increases in the size of the area stimulated. With a small electrode, charge is applied rapidly to the membrane in a concentrated area, but more slowly to distant parts of the membrane. Consequently, with a small electrode, the membrane becomes more efficiently depolarized, and also acquires a local voltage gradient. These factors contribute to lower excitation thresholds and to shorter S-D time constants. On the other hand, if a larger area of membrane is excited, the time required for depolarization will be increased, and, because the depolarization is more diffuse, larger stimulus intensities will be required for excitation. Jack et al. (1983, p. 418) reported that the size of the stimulated area affects the time constants much more in muscle than in nerve tissue.

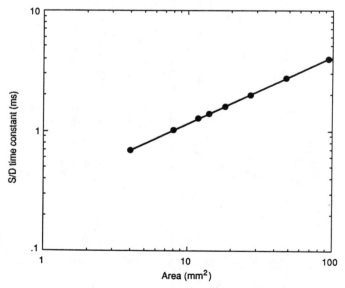

Figure 6.3   Relationship between S-D time constant and electrode area for cardiac excitation; data from Irnich (1980).

Polarity. With cathodal stimulation, the site of depolarization is concentrated near the electrode; and with anodal stimulation, it is spread more gradually over distant areas [see Sec. 4.4 and Reilly et al. (1985)]. As explained above, both excitation thresholds and S-D time constants are expected to be reduced as the area of stimulation is reduced. Consequently, it is expected that cathodal stimulation would result in shorter time constants and lower excitation thresholds than anadol stimulation. Thalen et al. (1975) found that cathodal excitation thresholds were indeed smaller than anodal thresholds by a factor of 2.6, in agreement with theoretical expectations; however, the observed S-D time constants with anodal stimulation were smaller than with cathodal time constants by a factor of 2.5, a result not in conformance with the above arguments. Further work is needed to reconcile these experimental observations with theoretical expectations.

Size of Tissue Preparation. When the stimulating current is introduced through an intracellular microelectrode, the values of τ_e for small cardiac tissue samples depend critically on the size of the preparation. In the experiments of Fozzard and Schoenberg (1972), unusually large (29.3-ms) values of τ_e were seen with small preparations. However, long samples (> 8 mm) of Purkinje fibers yielded values of τ_e that appeared typical of whole-heart preparations (3.75 ms). The direction of this effect follows theoretical expectations for intracellular excitation (Jack et al., 1983). See Chapter 4 for additional discussion of relationship between S-D time constants and the size of the stimulated tissue.

Fibrillation by Prolonged Currents. The S-D curves discussed above apply to single cardiac responses, in which case we expect the curve to follow the Weiss or Lapicque formulas [Eqs. (4.15) and (4.17)], so that the S-D time constant may be determined by Eq. (6.1), or other curve-fitting techniques. However, when the duration of the stimulating current is prolonged beyond the S-D time constant, the fibrillation threshold may continue to drop due to the effects of multiple excitations (see Sec. 6.4). If we attempt to fit such data to the Weiss or Lapicque formula, we may obtain time constants that appear to exceed that of the fundamental excitation process. The data presented by Hohnloser et al. (1982), for example, display an unusually long value of τ_e (item 16 of Table 6.2). The S-D curve for these experiments declined steadily with stimulus duration, out to 60 ms, after which an increase in threshold was seen. According to the authors, the fact that fibrillation could be initiated either by cathodal make or anodal break accounted for the S-D relationship that was observed. The timing of these events with respect to the cardiac cycle is of major importance, according to the authors. Knickerbocker (1973) also observed fibrillation thresholds that declined steadily with prolonged stimulation out to 2 s—the maximum period of stimulation in his experiment (refer to Sec. 6.8 and Fig. 6.11).

Consistency of Data. There is a wide spread of reported τ_e values, with a geometric mean of about 3.0 ms. The thresholds of fibrillation listed in Table 6.2 demonstrate a particularly wide variance in comparison with the other categories, although it is difficult to judge the extent to which this is simply a consequence of relatively few data points. In general, it is more difficult to establish minimum fibrillation thresholds than excitation thresholds, because the former depend more critically on the on and off times of the stimulus current with respect to the cardiac cycle, as well as on the number of prior extrasystolic excitations (refer to Sec. 6.5).

6.4   Strength-Frequency Relationships for Sinusoidal Stimulation

Before considering how oscillatory currents produce cardiac excitation, it is helpful to make comparisons with corresponding properties of neural excitation. Figure 4.15 illustrates S-D curves from a neuroelectric model for three types of stimuli: a monophasic rectangular waveform, a symmetric biphasic rectangular waveform, and a single cycle of a sine wave. Compared with the S-D curve for the monophasic pulse, the thresholds for the sinusoidal stimuli rise at durations that are less than the S-D time constant. This upturn occurs because the current reversal of the sinusoidal cycle tends to reverse an action potential that was started by the initial phase of the stimulus. In order to compensate for this reversal, the biphasic wave requires a higher current to achieve threshold. This cancellation effect is most effective as the phase duration t_p is reduced. For long t_p, thresholds depend more on the rate of rise and peak value of the current. It should not be construed that thresholds for all oscillatory stimuli are necessarily elevated above monophasic stimuli of the same phase duration. Indeed, if the sinusoidal cycle is repeated as an oscillatory waveform, thresholds may decrease with successive oscillations, as noted for neural stimulation in Fig. 4.18. For prolonged oscillatory stimuli, the excitation threshold may eventually reach a value below that for a single monophasic pulse.

In Chapter 4 we presented a functional relationship for strength-frequency (S-F) curves obtained from monophasic S-D curves by considering that a sinusoidal half-cycle corresponds to a single monophasic pulse. This simplification would, however, fail to account for high- and low-frequency upturns in S-F curves relative to S-D curves. A better empirical fit to S-F data from the myelinated nerve model was described as $I_t = I_0 K_H K_L$, where I_t is the threshold current, I_0 is the minimum threshold current, and K_H and K_L are high- and low-frequency terms defined by Eqs. 4.36 and 4.37, respectively. These empirical expressions involve high- and low-frequency transition parameters f_e and f_0, respectively. While the equations were derived from a neuroelectric model, they also provide a reasonably good fit to cardiac stimulation data if the equation parameters are chosen correctly. The equations are, however, empirical, with only rough heuristic connections to any underlying physical law.

Table 6.3 includes f_e values determined from S-D curves published by several sources listed in the right-hand column. The table is organized by separate categories for beat rate effects, excitation, and fibrillation. The tabulated data are for continuous cycle stimuli except for items 2 and 4, which apply to single-cycle stimulus waveforms. The data listed under the heading f_e were determined by a rough curve fit of the experimental data to Eq. (4.36). Excluding item 1, the geometric mean of the f_e values in Table 6.3 is 115 Hz. Item 1 was excluded from this average because it was felt that the mechanism affecting the beating rate in this experiment is fundamentally different from the other two categories by virtue of the unusually small value of f_e, and the sharply defined frequency sensitivity of the threshold current. The value $1/(2f_e) = 4.3$ ms may be compared with the S-D time constant $\tau_e = 3$ ms derived from the data in Table 6.2. This comparison appears adequately consistent considering the great diversity of experimental methods represented in Tables 6.2 and 6.3.

Figure 6.4 plots a number of experimental threshold data points taken from S-D curves published by several of the sources listed in Table 6.3. The plotted data have been normalized by the minimum threshold current in the various experiments. Figure 6.4 also shows a plot of the empirical formulas defined by Eqs. (4.35)–(4.37), in which the following relationships have been used:

Cardiac excitation parameters:
$f_e = 115$ Hz
$f_0 = 10$ Hz
$a = 1.45$ for single-cycle stimulus
$\quad\quad 0.9$ for continuous-cycle stimulus
$b = 0.8$

The value of f_e has been taken from the geometric mean of the data listed in Table 6.3. The other values have been taken from the myelinated nerve model for neural excitation (refer to Chapter 4). Of the data plotted in

Table 6.3. *Strength-frequency data for cardiac stimulation by sinusoidal current waveforms*

Item	f_e (Hz)	Species	Preparation	Stimulus locus	Electrode	Notes	References
A. Thresholds of beat rate change							
1	3	Frog	Isolated heart	Whole heart	Uniform field	(a)	Kloss and Carstenson (1982)
B. Thresholds of excitation							
2	278	Guinea pig	Isolated tissue	Papillary muscle	AgAgCl	(b)	Weirich et al. (1985)
3	50	Guinea pig	Isolated heart	Aorta/apex	AgAgCl	(c)	Weirich et al. (1983)
C. Thresholds of fibrillation							
4	250	Guinea pig	Isolated heart	Whole heart	AgAgCl	(b)	Weirich et al. (1985)
5	72	Guinea pig	Isolated heart	Aorta/apex	AgAgCl	(c)	Weirich et al. (1983)
6	79	Dog	In vivo	Ventricle	3.1 cm^2		Kugelberg (1976)
7	148	Human	In vivo	*	4-mm diam.		Kugelberg (1976)
8	100	Dog	In vivo	Various	Various		Geddes et al. (1971)
9	100	Dog	In vivo	Trans-chest	8 × 10 cm	(d)	Geddes and Baker (1971)
10	120	Dog	In vivo	Trans-chest	$\frac{1}{2}$-in. wire	(e)	Geddes and Baker (1971)

*Information sketchy or not available.
(a) Thresholds for changing beat rate by 30%.
(b) Single-cycle stimulus.
(c) Current duration = 1 s.
(d) Current flow transverse to long axis of animal.
(e) Current flow longitudinal to long axis of animal.

Fig. 6.4, that of Weirich et al. (1985) apply to single-cycle stimuli; the other data apply to continuous-cycle stimulation. The empirical curves plotted in Fig. 6.4 provide a reasonably good fit to published experimental data when seen in a combined context. Better fits to the experimental data could be obtained if the parameters f_e and f_0 were determined separately for each individual experiment. The low-frequency term, K_L, should have an upper limit, K_{DC}, that accounts for the finite threshold observed with DC stimulation. Fibrillation thresholds for DC currents have been reported by several authors. The data of Ferris et al. (1936) indicate K_{DC} values of about 5.2; Biegelmeier (1987) reported values of K_{DC} in the range of 3 to 4 for durations of several seconds, and a value of about 1.0 for durations of a fraction of a second.

6.5   Duration Sensitivity for Oscillatory Stimuli

Cardiac sensitivity to oscillating currents is affected by a complex of factors not present with single DC pulses, including the following. (1) A current reversal can reverse an excitation process begun on an initial stimulus phase. (2) The excitation threshold resulting from a prolonged AC current steadily drops, and eventually may become lower than that

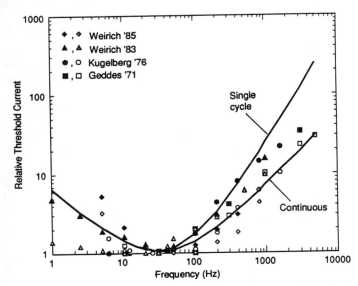

Figure 6.4   Strength-frequency data for cardiac stimulation by sinusoidal currents. Points are examples from published experimental data, solid curves represent analytic expression for single-cycle (upper) and continuous-cycle (lower) stimuli. Two examples are provided for each source, as indicated by open and filled symbols.

of a single DC pulse. (3) An AC current may promote a train of extra-systolic excitations, which successively make the heart more susceptible to ventricular fibrillation. The first two of these effects was demonstrated in Chapter 4 for the neural membrane—Figure 4.18 shows a threshold that oscillates with successive current phases, but nevertheless steadily declines on the average.

Excitation Sensitivity

The cancellation effect of an oscillatory stimulus has also been observed for cardiac stimulation. Figure 6.5 (adapted from Green et al., 1985) shows the ventricular fibrillation threshold versus the duration of a 50-Hz sine wave initiated at its peak. A threshold minimum occurs at approximately the point of current reversal (5 ms). Although the current dip is small in comparison with the variance of the measurements, it was reported to be consistently reproduced in several experiments. Further-more, when the stimulus was initiated at the zero crossing of the current waveform, the minimum was shifted to 10 ms. The effects of current reversal on the fibrillation threshold shown in Fig. 6.5 is relatively weak. But, by analogy with neural stimulation, a stronger effect might be seen if the stimulus frequency were to be increased. Indeed, biphasic current reversal of a 1-kHz waveform has been shown to require current levels that are significantly elevated relative to a monophasic waveform (refer to discussion concerning Fig. 6.17).

Stimulation by an alternating current involves a nonlinear integration of the current that crosses the excitable membrane. The membrane con-

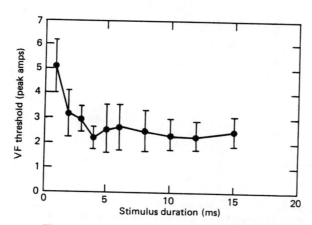

Figure 6.5   Ventricular fibrillation threshold versus duration of 50-Hz sinusoidal stimulus initiated at peak. Electrodes right fore limb to left hind limb of dogs ($n = 12$). Means and S.D. shown (from Green et al., 1985).

ductance appears to be approximately constant if the membrane depolarization voltage is well below the excitation threshold. However, as the depolarization voltage approaches the value needed for excitation, the conductance for membrane current influx and efflux becomes asymmetric. As a result, membrane depolarization caused by one phase of cyclic stimulus may be only partially offset by a current reversal on a succeeding half-cycle. This asymmetric process can result in a buildup of the membrane depolarization on successive cycles of an oscillatory stimulus. This process of membrane depolarization was demonstrated in the cardiac system with stimulation of the papillary muscle of rhesus monkeys by 50-Hz alternating currents (Antoni et al., 1969). In that study, AC depolarization effects were observed experimentally, and were also verified with computer simulation of the electrical response of Purkinje fibers.

Fibrillation Sensitivity

Another important consideration with oscillatory stimuli is the potential for producing a sequence of excitations—an effect that can significantly enhance the biological response that might otherwise result from a single excitation (see Sect. 5.10). In Fig. 4.19 we have seen that a train of action potentials on a nerve fiber can be produced in synchrony with the sinusoidal stimulus, where the rate of AP generation depends on the frequency and magnitude of the stimulus current. Oscillatory stimuli can also produce multiple excitations in the cardiac system. This phenomenon has important safety consequences, because the fibrillation threshold falls steadily with increasing number of extrasystolic excitations. Figure 6.6 illustrates this effect with experimental data from in-vivo excitation of dog hearts (from Sugimoto et al., 1967). In this figure, the vertical axis gives the VF threshold to a 60-Hz current; the horizontal axis gives the number of prior ventricular responses produced by a conditioning 60-Hz current with variable duration. The duration of current for six responses was about 1 s. Over the range of one to six ventricular responses, the average VF threshold drops by a factor of 30. When the conditioning current was changed to a train of unidirectional rectangular pulses, and the VF stimulus was a single DC pulse, a response curve nearly identical to that shown in Fig. 6.6 was obtained. Clearly, the ability of a stimulus to produce multiple cardiac excitations significantly affects its potency for causing fibrillation.

Table 6.4 compares excitation and ventricular fibrillation thresholds for in-vivo stimulation of dog hearts. The VF threshold exceeds the excitation threshold by a multiple of 150 when the stimulating current is a single DC pulse; with 60-Hz stimulation of 5 s duration, the multiple is only 2.4.

Table 6.4. *Comparison of AC and DC excitation and fibrillation thresholds—experiments with dogs*

	DC ($t = 10$ ms) (mA-peak)	AC (50 Hz, $t = 5$ s) (mA-peak)
Ex. thresholds	0.10 ± 0.05	0.11 ± 0.03
VF thresholds	15 ± 4.3	0.26 ± 0.08
Ratio VF/Ex.	150	2.4

Notes: VF stimuli applied during vulnerable period to in-vivo dog hearts ($n = 14$); 25-gauge bipolar needle electrodes spaced 5 mm apart, inserted 2 mm into epicardial surface. $\pm$ values indicate standard deviation.
Source: Data from Sugimoto et al. (1967).

Nevertheless, the 60-Hz excitation threshold is nearly the same as that for a single DC pulse. Table 6.5 summarizes AC and DC thresholds from the data of Hohnloser et al. (1982). With DC stimulation of 1 s duration, the ratio of the VF threshold to the excitation threshold is 23:1; with AC stimulation, the ratio is only 4.2:1.

Both alternating and direct currents can disrupt the normal rhythm of the heart and promote multiple excitations. In Fig. 3.17, for example, we have seen that theoretical models of ventricular myocardial fibers respond

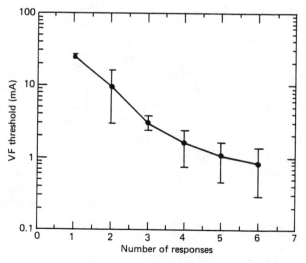

Figure 6.6   Ventricular fibrillation thresholds as a function of number of excitations from 60-Hz stimulation. Experiments with dogs ($n = 6$); mean $\pm$ standard deviation plotted. Needle electrodes inserted into epicardial surface. Data from Sugimoto et al. (1967).

Table 6.5. *Comparison of AC and DC excitation and fibrillation thresholds—experiments with guinea pig hearts*

	DC ($t = 1$ s) (mA-peak)	AC (50 Hz, $t = 1$ s) (mA-peak)
Ex. thresholds	1.3 ± 0.6	2.6 ± 0.8
VF thresholds	30 ± 8.5	11 ± 2.2
Ratio VF/Ex.	23.1	4.2

Notes: Isolated perfused guinea pig hearts ($n = 3$); ring-shaped electrode attached to apex of heart.
Source: Data from Hohnloser et al. (1982).

to steady outward membrane current with a series of extraneous ventricular responses. The greater fibrillation efficacy of 60 Hz relative to a DC pulse arises because the AC stimulus is a more efficient promoter of multiple ventricular responses. The lowering of the fibrillation threshold by an extrasystolic excitation as compared with a normally conducted beat arises because the extra beats create a higher degree of non-homogeneity within the heart, making it more susceptible to fibrillation (see Chapter 5). A burst of closely spaced extrasystoles especially enhances this effect (Hohnloser et al., 1982). The maximum achievable rate of extrasystolic excitation is enhanced by the fact that the AP duration of an extrasystole can be significantly shorter than the normal AP duration. This phenomenon was demonstrated by simulation of Purkinje fiber response: Fig. 3.14, for example, shows that repetitive stimulation of the Purkinje fiber results in AP durations that are about one-half of that for a normal cardiac beat.

Figure 6.7 illustrates the relationship between current magnitude and the number of extrasystolic excitations induced by a 3-s burst of 50-Hz alternating (solid dots) or direct (open dots) current introduced into isolated guinea pig hearts (Hohnloser et al., 1982). The terminal values of each curve show the fibrillation threshold. It is evident that both AC and DC stimulation are capable of producing multiple excitations, but the AC stimulus is much more effective.

Figure 6.8, from Weirich et al. (1983), demonstrates the relationship between the extrasystolic rate and the fibrillation threshold for sinusoidal currents of 1 s duration with frequency ranging from 1 to 1000 Hz. The upper solid line represents the fibrillation threshold; the lower solid line represents the threshold for inducing an excitation during the diastolic interval of the heart; the broken lines represent current levels at which the extrasystolic rate is constant. The thresholds in Fig. 6.8 have a minimum around 30 Hz. Above that frequency, fibrillation and diastolic thresholds follow constant excitation rates of 11 and 8 per second, respectively. Below 30 Hz, fibrillation and diastolic thresholds also rise, but are associated with decreasing excitation rates. The upturn in thresholds above 30 Hz may be explained by the fact that as the AC frequency is increased, the alternate phases of alternating current have an increasing tendency for mutual cancellation. This explanation is strengthened by the observation that thresholds do not rise when the sinusoidal stimulus is replaced by a train of 1-ms unidirectional pulses (Weirich et al., 1983). Figure 6.8 also shows a low-frequency upturn that is associated with decreasing excitation rate as the frequency is lowered. Although a low-frequency upturn is expected for an excitation curve (see Fig. 6.4), it is not easy to explain the association of VF thresholds with decreasing excitation rates because it becomes less probable that an excitation will coincide with the vulnerable period of the heart.

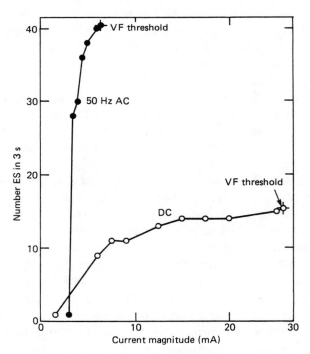

Figure 6.7   Relationship between current strength and number of extrasystoles for DC and 50-Hz AC stimulation of 3-s duration. Terminal values represent ventricular fibrillation thresholds. Perfused guinea pig hearts; ring electrode at apex of heart. (From Hohnloser et al., 1982.)

The Z Relationship for Fibrillation by AC Currents

A variety of experiments have demonstrated that when VF thresholds are plotted versus duration of an AC stimulus, the result is a sigmoidal-shaped curve in which an asymptotic maximum is achieved at short durations (a few cycles of stimulation), and an asymptotic minimum is reached at long durations (several seconds). This so-called Z relationship was evident in the early experiments of Ferris et al. (1936), and was later substantiated by others.

Figure 6.9 illustrates the Z relationship for fibrillation by AC stimulation as determined by Roy et al. (1977). In these experiments, stimulation was via catheter electrodes of various areas, introduced into the right ventricle of dog hearts. The authors obtained an empirical fit of thresholds with the following equation:

$$I_T = I_0 \left[1 - \exp(-K_n N^b) \right]^{-1} \qquad \text{(for } N \geqslant 3) \qquad (6.2)$$

where I_0 is the minimum current corresponding to the longest duration (1000 cycles of 60 Hz); K_n and b are empirical constants given in Table 6.6. The value of K_n listed for an area of 500 mm² was obtained by this writer using logarithmic extrapolation of the K_n values at the three areas tested ($A = 0.22, 14,$ and 90 mm²). This extrapolated value represents a rough estimate of the threshold dependence for cardiac excitation over a large

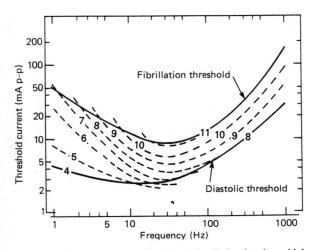

Figure 6.8   Threshold cardiac stimulation by sinusoidal current of 1-s duration. Ring electrode at heart apex. Isolated, perfused guinea pig hearts ($n = 4$). Upper solid curve shows VF threshold, lower solid curve shows excitation threshold during diastolic interval, broken curves show contours of constant estrasystole number. Mean thresholds shown. (From Weirich et al., 1983.)

Table 6.6. *Empirical constants for Eq. (6.2)*

Area (mm²)	K_n	b	I_{max}/I_{min}
(500)	(3.7×10^{-3})	(1.5)	(50)
90	1.8×10^{-3}	1.5	90
14	8.3×10^{-4}	1.5	200
0.22	1.1×10^{-4}	2.0	1000

Notes: Empirical constants for $A = 90$, 14, and 0.22 mm² were determined by Roy et al. (1977). Values for $A = 500$ mm² have been extrapolated from Roy et al.'s data. Values listed for I_{max}/I_{min} represent ratio of upper and lower plateaus of fibrillation thresholds.

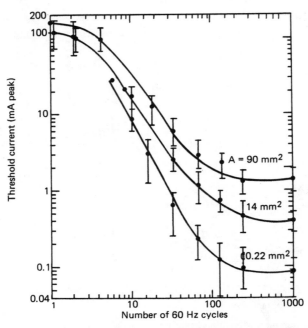

Figure 6.9   Relationship between VF threshold and number of 60-Hz cycles. Separate curves apply to area of catheter electrodes inserted into right ventricle. Mean and S.D. values shown for 50 dogs. (From Roy et al., 1977, © 1977 IEEE.)

area. Equation (6.2) provides an excellent fit to Roy et al.'s experimental data, provided that $N \geqslant 3$; for $N \leqslant 3$, measured thresholds reach an asymptotic maximum, whereas Eq. (6.2) continues to rise, thereby overstating the experimental thresholds. An inference drawn from these experiments is that the threshold dependence on electrode area is a strong function of the duration of the AC stimulus. At 1000 cycles of 60-Hz stimulation ($t = 17$ s), there is an 18:1 ratio of thresholds between the largest and smallest electrodes tested (90 and 0.22 mm^2); with 1 cycle of current, the ratio of thresholds is perhaps 2:1. The final column of Table 6.6 lists the ratio of the upper and lower plateaus of the Z curves, based on the data from Roy et al. The upper plateau has been determined from Eq. (6.2) at $N = 3$. It can be seen that the plateau ratio diminishes as the area is increased. The value listed for $A = 500$ mm^2 is an estimate of the value that would apply to large-area contracts or possibly to whole-heart stimulation.

Figure 6.10 illustrates VF thresholds in guinea pigs versus the duration of AC stimulation having frequencies of 5, 50, and 200 Hz. The curves for 50 and 200 Hz differ by a nearly constant multiple that reflects the frequency sensitivity of the excitation process (refer to Sec. 6.4). The curve at 5 Hz, in contrast, converges to the 50-Hz thresholds at short durations,

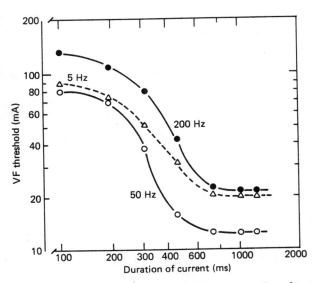

Figure 6.10 Fibrillation threshold versus duration of current for three frequencies. Data from isolated perfused guinea pig hearts. (Adapted from Antoni, 1985.)

and to the 200-Hz thresholds at long durations. The indicated VF thresholds depend in a complex way on the individual extrasystolic excitations, their rate, and their timing with respect to the cardiac cycle. At 1 s stimulation, the difference in thresholds for the 50- and 200-Hz curves closely correspond to the differential sensitivity of diasystolic stimulation (see Fig. 6.8). The 5-Hz VF threshold is, however, elevated relative to that at 50 Hz because of its lower rate of extrasystolic production. As the duration of stimulation is shortened, the extrasystolic production rate for the three curves becomes less disparate. Consequently, for short stimulus durations, VF thresholds more closely reflect diastolic excitation sensitivity at all three stimulation frequencies.

Figure 6.11 illustrates VF thresholds for 20-Hz AC and DC currents applied to dogs (left foreleg to right hind leg). Both AC and DC stimuli result in steadily declining thresholds for increasing stimulus duration, although the rate of decline in AC thresholds is much greater. As explained above, the differential rate of decline arises because the extrasystolic rate associated with the AC waveform is greater than that for the DC waveform.

Considering the mechanisms discussed above, we can represent the duration sensitivity of 50/60-Hz fibrillating currents as shown in Fig. 6.12. It is assumed that short-duration stimuli coincide with the vulnerable period of the cardiac cycle. The thresholds in regions (1) and (2) follow a strength-duration curve for pulsatile stimuli—the assumed S-D time

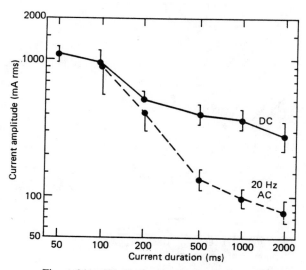

Figure 6.11   Fibrillation thresholds versus duration for DC and 20-Hz AC currents; in-vivo stimulation of dogs ($n = 10$–25). Error bars show 95% confidence limits. (From Knickerbocker, 1973, © 1973 IEEE.)

constant is 3 ms, in accordance with the analysis in Sec. 6.3. The lower curve in region (1) is drawn under the assumption that the stimulus current initiation time results in a unidirectional (monophonic) pulse, in which case the threshold converges to a t^{-1} relationship. If the initiation time is near a current reversal such that the current pulse is biphasic, the threshold will be elevated as indicated by the upper curve in region (1). For a balanced biphasic waveform, the relationship will be approximately $t^{-1.5}$, as suggested by the discussion in Sec. 6.4 and by analogy with neural membrane properties (see Chapter 4). For intermediate cases of biphasic asymmetry, the threshold would fall in between the two curves. In region (2), the stimulus duration is longer than the S-D time constant (indicated as t_0 in Fig. 6.12), and the threshold curve converges to a constant-current relationship. In region (3), the stimulus consists of more than one AC cycle, leading to multiple extrasystoles. The VF threshold drops with each successive excitation, reaching a minimum plateau in region (4).

The ratio, R, of the plateau thresholds in regions (2) and (4), and the transition duration t_1 and t_2, vary with the experimental conditions, including the tested species. Table 6.7 summarizes empirical values of R, t_1, and t_2 applying to 50/60-Hz exposure to the whole heart, or via a large-area cardiac contact electrode.

In applying animal data to humans, Biegelmeier and Lee (1980) argue that differences in heart beat rate between humans and animals should be taken into account, such that the ratio R and upper transition duration

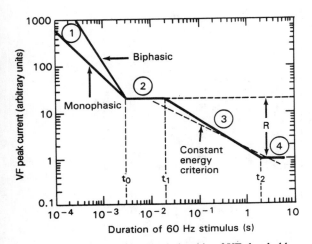

Figure 6.12   General relationship of VF threshold versus 60-Hz stimulus duration. Numbered portions of the curve represet regions of differing sensitivity (see text). Short-duration stimuli are assumed to coincide with vulnerable period.

Table 6.7. *Z-curve parameters for 50/60-Hz currents*

Species	Current path	Threshold ratio, R	Transition time (s)		Reference	Notes
			t_1	t_2		
A. Measured relationships						
Sheep	Right fore/left hind limb	10	0.27	1.4	Ferris et al. (1936)	(a)
Pig		10	0.03	4.0	Jacobsen et al. (1975)	
Dog	ventricle (5 cm² elect.)	50	0.05	2.0	Roy et al. (1977)	(a)
Dog	Fore/hind limb	25	0.02	2.0	Biegelmeier and Lee (1980)	(b)
Guinea pig	*	6	0.20	0.80	Antoni (1985)	
B. Postulated relationships for humans						
Human	Hand/foot	20	0.20	5.0	Biegelmeier and Lee (1980)	(c)
Human	Hand/foot	20	0.10	2.0	Biegelmeier (1987)	(c)

*Information sketchy or not available

(a) Area relationship extrapolated to 5 cm² (see text).

(b) Derived from data of Kouwenhoven et al. (1959), Scott et al. (1973), and Kiselev (1963).

(c) Postulated from animal data; adjustment for best rate; 50% probability curves.

t_2 would be increased in a human relative to an animal with a much faster beat rate. This idea is bolstered by the data in Table 6.7 for guinea pigs—the transition time t_2 and the ratio R are both quite short with respect to the data for larger animals having longer cardiac cycles. Biegelmeier proposes human VF thresholds as indicated in the lower section of Table 6.7. The bottom entry represents the recommendations of the International Electrotechnical Commission (IEC) for human safety criteria. The most recent IEC proposed standards (Beigelmeier, 1987) take into account the fact that the transition between regions (2) and (3) occurs with an S-shaped curve, rather than the straight-line segments represented in Fig. 6.12.

The extent to which Fig. 6.12 would apply to frequencies other than 50/60 Hz has not been established, although its general features are expected to be applicable to a much wider frequency range. The data of Antoni (1985) (Fig. 6.10) indicate that over the time interval of 0.1 to 1 s, the Z curve for 200 Hz is shifted above the curve for 50 Hz by an approximately constant multiplicative factor, consistent with the frequency sensitivity discussed in Sec. 6.4. If the duration of stimulation were to be reduced below 0.1 s to a fraction of a 200-Hz cycle, we would expect the 50-Hz and 200-Hz curves to converge to the same curve indicated in regions (1) and (2).

6.6   Energy Criteria and Impulse Currents

Engineers in North America have traditionally evaluated safe current thresholds for 60-Hz exposure using the so-called *electrocution equation* of Dalziel (1960, 1968):

$$I_f = Kt^{-1/2} \tag{6.3}$$

Where I_f is the presumed minimum fibrillation threshold, t is the duration of exposure, and K is a scaling constant that depends on body weight and the assumed percentile rank of sensitivity. The intended limits on t are from 8.3 ms to 5 s. Equation (6.3) is fundamentally an energy criterion in which I^2t is a constant. Values of K for various body weights and percentile ranks are given in Table 6.8, based on Dalziel's criteria (see Fig. 6.15). The values listed under the category "MNF" were said to apply to a "maximum nonfibrillating" current. The 50 percentile rank was also represented as an *average* in some of Dalziel's writings. Dalziel concluded that VF thresholds are linearly related to body weight, as discussed in Sec. 6.7. Chapter 11 discusses the electrocution equation further, and compares it with other safety criteria.

Dalziel based the time relationship in Eq. (6.3) on his analysis of the

Table 6.8. *Values of K for electrocution equation*

Body weight (kg)	I_F (mA) for percentile rank		
	50%	0.5%	MNF[a]
20	177	78	61
50	368	185	116
70	496	260	156

[a]MNF indicates "maximum nonfibrillating current."
Source: Based on data from Dalziel (1968).

50/60-Hz fibrillation data from dogs developed by Ferris et al. (1936) and Kiselev (1963). Figure 6.13, for example, shows one of Dalziel's curve fits to Ferris's and Kiselev's combined data involving 45 dogs. The open points represent the minimum fibrillation thresholds obtained at each tested duration. The closed points represent Dalziel's estimates of the 0.5 percentile rank values, obtained by extrapolating the statistical distribution of VF thresholds pertaining to each tested value of t. The solid line shows Dalziel's energy fit to the extrapolated points; the broken line has been added by this writer to indicate a Z-curve interpretation

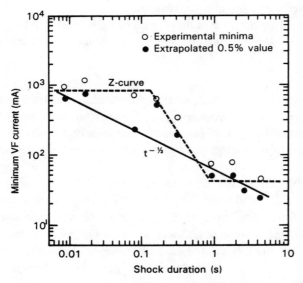

Figure 6.13   Minimum VF thresholds obtained in experiments with 45 dogs using data of Ferris and Kiselev. (Adapted from Dalziel, 1968, © 1968 IEEE.)

of the same data. Except for one extrapolated point at 0.083 s, the minimum experimental points and the extrapolated point both appear to fit a Z curve.

It was not unreasonable for Dalziel to have postulated a constant energy fit to the empirical data available to him. For one thing, the $t^{-1/2}$ relationship does provide a reasonable fit to the data when viewed over the range of durations from 0.08 to 5 s, as indicated in Figs. 6.12 and 6.13. Also, an energy criterion has a certain intuitive appeal to many engineers. Furthermore, the role of extrasystolic excitations in lowering VF thresholds was not widely known at the time that Dalziel published his work.

Although the electrocution equation may not correspond strictly to theoretical expectations, it does give a reasonable fit to the VF threshold/ time relationship for 60-Hz currents with durations exceeding 1 cycle. A difficulty, however, is that some analysts have attempted to ascribe an energy criterion to a wide variety of waveforms, including short-duration impulse currents. This view has probably been reinforced by other work of Dalziel (1953), in which he suggests an energy criterion for short-duration impulse currents.

It should be clear from preceding discussions that the excitation or fibrillation capability of a stimulus is not predicted by its energy content. We have seen in Sec. 6.3 that short-duration unidirectional currents instead produce a well-defined strength-duration curve that is unrelated to constant energy. For durations well below the strength-duration time constant, the unidirectional stimulus threshold converges to a constant charge criterion (see Sec. 6.3). If the transient is biphasic, the stimulus becomes less effective (see Sec. 6.4).

Short-duration charge thresholds for electrodes placed on the heart have been reported for a variety of open-heart experiments and procedures. For cutaneous electrodes, direct experimental data on transient thresholds are lacking, and we need to draw inferences from data applying to longer stimulus durations. Referring to Fig. 6.12, the charge threshold for a monophasic current may be estimated as

$$Q_0 = I_L R t_0 \tag{6.4}$$

where Q_0 is the minimum charge threshold for short durations ($t < 3$ ms), and I_L is the peak current threshold for a long-duration ($t = 3$ s) stimulus; R and t_0 are defined in Fig. 6.12. As an example, consider the median threshold for hand-to-foot exposure of an individual weighing 20 kg. From Table 6.10 below, Dalziel's peak fibrillation threshold is $0.102\sqrt{2} = 0.144$ A. If we use $R = 20$ and $t_0 = 3$ ms, the mean fibrillation charge threshold from Eq. (6.4) is calculated as 8.6 mC. That value compares favorably with the minimum fibrillating charge of 5mC observed experimentally by Pelaska (1963) using capacitor discharges across the thorax of dogs. Cardiac arrhythmias were observed by Pelaska with capacitor discharges as low as 0.25 mC. It is probable that the later value does not represent a minimum threshold for arrhythmias, because stimuli below 0.25 mC were apparently not tested by Pelaska.

When a capacitor discharge is lead through an inductance, a decaying oscillatory stimulus is produced, with a net charge displacement still given by $Q = CV$. The excitation potency of an oscillatory discharge can, under some circumstances, be greater than a purely capacitive discharge, since the amount of charge initially displaced can exceed the stored charge on the capacitor (Reilly and Larkin, 1985b). Indeed, fibrillation thresholds have been observed to be lower when inductance is added to a capacitive discharge circuit (Dalziel, 1953), although this is not always the case (Pelaska, 1965). Neural excitation thresholds, expressed in terms of net capacitor charge, have also been observed to be lower when inductance is added to the discharge circuit (Reilly and Larkin, 1985b). In order to account for the overall effect of added inductance, we would have to consider the initial displaced charge, and the frequency and decay time constant of the oscillatory current waveform. The interrelationship among these factors has not presently been defined.

6.7   Body-Size Scaling

Much of our empirical knowledge of electrical hazards has been derived from animal experimentation. In interperting such data for human applications, we must make some judgements concerning scaling from animal data. Differences in physiology, anatomy, and size may all have a bearing on the scaling method. Body size is one variable that has been studied systematically.

Experimental Body-Size Data

Much of our knowledge of electrical safety is derived from animal experimentation. Early experiments, for example, made use of the dog and rat for application to human safety (Kouwenhoven and Langworthy, 1931; Hooker et al., 1932; Kouwenhoven, 1949). Several studies have suggested that average thresholds increase with the size of the experimental animal. In early studies, Ferris and colleagues (1936) plotted the VF thresholds for several species of animal (guinea pig, rabbit, cat, sheep, pig, calf, and dog), and postulated a linear relationship between body weight and the size of the animal, as shown in Fig. 6.14. The thresholds were also correlated with the weight of the heart itself, although the relationship appeared to be a nonlinear one. Dalziel later used Ferris's data, along with Kiselev's (1963) data on dogs to obtain a linear regression fit:

$$I_f = 3.68\,W + 28.5 \tag{6.5}$$

where W is the body weight (kg), and I_f is the average VF threshold (mA-rms) applied between fore and hind limbs for a duration of 3 s. The correlation coefficient for Eq. (6.5) was reported as $r = 0.74$. Figure 6.15 plots Dalziel's regression line, along with the average thresholds and weights of the larger animals tested by Ferris and Kiselev; the middle line shows Dalziel's 0.5 percentile rank, which he obtained by extrapolating the statistical distributions of the VF thresholds. Dalziel proposed a non-fibrillating current safety criterion for a hypothetical 50-kg person as

$$I_f = 116t^{-1/2} \qquad (\text{mA-rms}) \tag{6.6}$$

In this formula, the time relationship follows Eq. (6.3), and the body-weight relationship follows the lower curve in Fig. 6.15. The minimum non-fibrillating criterion of Dalziel has been widely used for safety analysis in North America (see Chapter 11 for additional details).

The body-weight relationship was examined more recently by Geddes

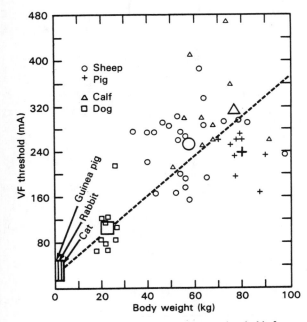

Figure 6.14   Ventricular fibrillation thresholds for several species of animal, plotted against body weight. Large symbols indicate average values. Current path, left fore limb–right hind limb; duration, 3 s. (From Ferris et al., 1936, © 1936 AIEE.)

Table 6.9. *Body-weight regression parameters*

Electrodes	K	a	r
Fore/fore	69.4	0.533	0.887
Left fore/right hind	29.7	0.510	0.929
Left fore/left hind	33.6	0.437	0.749

Notes: Five species of animal ($n = 104$); 60-Hz
current; applicable durations, 2–5 s.
Source: Data from Geddes et al. (1973).

and colleagues (1973). Using data from 104 animals of several species
(rabbits, puppies, one monkey, dogs, goats, and ponies), regression fits
were obtained as a power law of the form:

$$I_f = KW^a \tag{6.7}$$

Table 6.9 lists Geddes' regression coefficients for several electrode arrangements. Although these coefficients were obtained with a stimulus duration

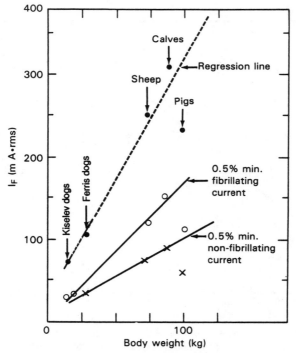

Figure 6.15   Relationship between fibrillating current and body weight for
3-s shocks of 60 Hz (from Dalziel, 1968, © 1968 IEEE).

Table 6.10. *Median fibrillation thresholds (mA) for 3-s shocks using regression formulas of Dalziel and Geddes*

W (kg)	Geddes			Dalziel
	F/F	r.F./l.H.	l.F./l.H.	F/H
20	343	137	124	102
50	558	218	185	213
70	668	259	215	286
100	808	311	251	397

Notes: Column headings indicate current pathway: F, H indicate fore, hind limb; r, l indicate right, left.
Source: Data are based on regression formulas of Dalziel (1968) and Geddes et al. (1973).

of 5 s, it would not be unreasonable to apply the same values over the range 2–5 s, since little threshold variation was observed over that range. Variations in the coefficient K indicate a sensitivity to the current pathway, which is discussed in some detail in Sec. 6.10.

Table 6.9 indicates that the VF threshold is nearly proportional to the square root of body weight, rather than the linear relationship given by Eq. (6.7). Nevertheless, the two regression formulas are in reasonable agreement over a fairly wide range of body weights, as seen in Table 6.10, especially for the front-to-hind limb thresholds. For small human body weights (20 kg), the Dalziel formula gives a more conservative estimate of VF thresholds; at large body weights (100 kg), the Geddes relationship is more conservative.

A body-weight relationship is not universally accepted for the VF threshold in human safety applications. Points of view on the subject were presented in a specialists' panel meeting (Geddes and Antoni, 1985). In that discussion, Prof. Biegelmeier argued effectively against a body-weight formulation, pointing out that when the VF thresholds of several species are plotted against body weight (as in Fig. 6.14), one indeed observes an apparent body-weight relationship. But when the thresholds of a single species (sheep, pigs, dogs) are plotted against body weight, the statistical correlation largely disappears. If we examine the data points for individual species in Fig. 6.14, for example, the indicated regression line does not appear to be justified. Rather, the regression line may represent an interspecies dependency that is dominated by factors other than body weight per se.

As a consequence of the arguments stated above, the IEC does not include a body-weight law in its safety criteria. Rather, the IEC bases its criteria on experimental data from dogs (Biegelmeier, 1986, 1987). This procedure is considered by the IEC to be conservative in view of the demonstratedly lower VF thresholds of dogs as compared with humans. Especially pertinent are experimental thresholds obtained for both dogs and humans using the same arrangements of electrodes placed in the heart, or on the myocardium (Green et al., 1972; Rafferty et al., 1975a, 1975b). Considering several end points (ECG irregularities, pump failure, ventricular fibrillation), median thresholds for humans were observed to be a factor of 2 or more above corresponding thresholds for dogs.

A number of researchers have considered the dog heart to provide a reasonably conservative model for human safety applications. In the experiments of Green and Rafferty cited above, the smallest current having an observable effect—that of heart rhythm disturbance—occurred at 60 µA in dogs, and at 80 µA in humans. Consequently, the authors recommended for open-heart surgery a maximum 50-Hz leakage current of 60 µA for a wire placed in the right ventricle. That value, however, would not represent an absolute lower limit on the VF threshold for open-heart exposure. Roy and colleagues (1976) found the minimum VF current threshold to be inversely related to the size of the contact electrode—a minimum value of 18 µA-rms was observed for right-ventricle exposure of dog hearts to 60-Hz currents via a contact electrode having an area of 0.224 mm². The threshold voltage of the contact electrode relative to the surrounding tissue was, however, relatively insensitive to electrode area— the minimum voltage necessary to produce VF was found to be in the vicinity of 100 mV-rms, almost without regard to electrode size.

Scaling Arguments

As shown in Sec. 4.3, the electric field aligned with the long axis of an excitable cell is responsible for the depolarizing force on the cell's membrane. Current density is, however, a more commonly cited variable in analyses of electric shock. For a given quantity of current injected through the limbs, current density in the heart will generally diminish as the size of the animal is increased. The direction of this effect corresponds to the body-weight formulas mentioned above. On the other hand, fibrillation requires not just excitation at some locus, but a critical mass of excited tissue. One might argue that in the larger heart, the mass of the heart would tend to counterbalance its reduced current density. If the fibrillation threshold were related to the fraction of body current intercepted by the heart, as suggested by Bridges (1985), then we would conclude that the fibrillation threshold ought to be about the same for animals of nearly similar geometries, but of different sizes and weights.

Considering the reentry theory for electrically introduced fibrillation (see Chapter 5), one would except that, below some critical size, a small heart ought to be more difficult to fibrillate than a large heart. This could occur if the reentry paths were too small to simultaneouly support states that were excited, partially refractory, and resting. Experimental evidence indeed shows that the heart of a small animal tends to revert spontaneously from a fibrillated to a normal state (Zipes, 1975). This observation, however, says little about the *threshold* of fibrillation.

Gross anatomical features can complicate the scaling of data from animal human. Most experimental data have been derived from quadrupeds and applied to a bipedal human. The geometry of the heart and thorax is considerably different between a quadruped and a human, and it is likely that considerations other than body weight are important to the scaling question (Lee, 1966). A further complication concerns the position of the animal during experimentation, because the heart shifts significantly within the thorax, depending on the position of the animal (Bridges, 1985). Such shifts could cause the heart to alter its position with respect to a region of high current density when current is introduced through the limbs.

Although there are uncertainties associated with body-weight scaling, it is clear that interspecies differences do exist, and that smaller animals generally have lower VF thresholds than larger animals. Less clear is the appropriate scaling method within a single species, especially humans. It will require further experimental work to resolve remaining uncertainties.

6.8   Statistical Distribution of Thresholds

Much of the data presented in previous sections treat excitation thresholds in terms of mean or medians. When analyzing hazardous currents, we would like to know the statistical distribution of reaction thresholds. A variety of experiments on both human and animals indicate the log-normal distribution for fibrillating currents applied between the front and hind limbs. An example is shown in Fig. 6.16 for fibrillating currents in dogs, using combined data from the experiments of Ferris and Kiselev. The log-normal distribution plotted as a straight line in the figure provides a much better fit than the linear-normal distribution suggested by early investigators (Dalziel, 1968).

Table 6.11 summarizes distribution parameters extracted from the data of several investigators who applied currents across the limbs of experimental animals. The experimental data were provided in terms

Table 6.11. *Experimental statistical distribution parameters for fibrillating currents (front/hind limb paths)*

Species	Current type	Duration (s)	Normalized I_F(mA)[a]		Data source[b]
			10%	90%	
Sheep	60 Hz	0.03–3.0	0.59 ± 0.16	1.90 ± 0.76	Ferris et al. (1936)
Dogs	60 Hz	0.008–5.0	0.65 ± 0.10	1.84 ± 0.57	Kouwenhoven et al. (1959)
Dogs	50 Hz	3.0	0.67	1.57	Kiselev (1963)
Dogs	50 Hz; pulsed[c]	$1.5t_c$	0.70 ± 0.08	1.48 ± 0.17	Jacobsen et al. (1975)
Dogs	20 Hz	0.1–2.0	0.64 ± 0.06	1.59 ± 0.18	Knickerbocker (1973)
Dogs	DC	0.05–2.0	0.71 ± 0.07	1.49 ± 0.12	Knickerbocker (1973)
Overall data ($n = 29$)			0.66 ± 0.10	1.69 ± 0.47	

[a] I_F values are normalized by median.
[b] Data of Ferris, Kouwenhoven, and Kiselev are summarized by Dalziel (1968).
[c] Duration was 150% of cardiac period.

of VF thresholds for current durations ranging from 8 ms to 5 s, and with several different waveform types. The thresholds listed in Table 6.11 give the observed 10 and 90 percentile ranks, normalized by the median value. The ± values indicate the standard deviation of the normalized percentile ranks across the various experimental distributions. In all, 29 distributions were represented in the cited reports. There was no recognizable sensitivity in the normalized distributions with respect to the duration of current, the waveform type, or the tested species. In view of this invariance, it is reasonable to represent an overall statistic as listed at the bottom of the table.

Table 6.12 provides a summary of log-normal rank values corresponding to a distribution having a 10 percentile rank of 0.66, consistent with

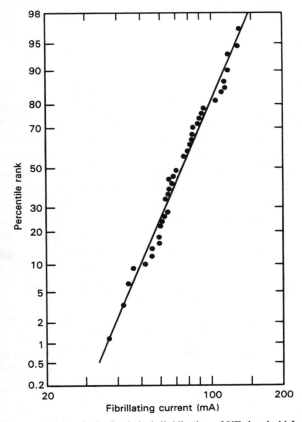

Figure 6.16   Statistical distribution of VF threshold for dogs; 3-s shocks applied front-to-hind limb. Combined data from Ferris et al. (1936) and Kiselev (1963), as summarized by Dalziel (1968).

Table 6.12. *Log-normal percentiles for fibrillating currents (front/hind limb paths)*

Percentile rank (%)	Normalized $I_F{}^a$
99	2.14
95	1.67
90	1.51
75	1.24
50	1.00
25	0.80
10	0.66
5	0.60
1	0.47
0.5	0.43

[a]I_F has been normalized by median value.

the combined experiments of Table 6.11. The data are listed to the 0.5 percentile rank—a value frequently used in safety analysis. The data probably do not support percentile rank projections below that value. As shown in the table, the log-normal projection indicates that the 0.5 percentile rank lies below the median by the multiplicative factor 0.43. This value is not substantially different from the projections of Dalziel, who used a linear-normal model to fit experimental data below the 50 percentile rank.

Compared with the animal data discussed above, a much broader distribution has been reported for human patients undergoing open-heart surgery for value replacement. Watson and colleagues (1973) measured VF thresholds in 56 patients exposed to 50-Hz alternating current for several electrode arrangements, including 2.5-cm^2 disk electrodes placed on the epicardium, needle electrodes inserted into the myocardium, and bipolar pacing electrodes placed on the apex of the right ventricle. Figure 6.17 illustrates the distribution of fibrillating current. The original data of Watson et al. have been replotted on coordinates on which a straight line would conform to a lognormal distribution. The ratio of the 10 percentile rank to the median VF current for Watson et al.'s data averaged 0.42 for the three-electrode arrangements; the 90 percentile rank was above the median value by an average multiple of 2.54. These values indicate a significantly broader statistical spread than with the animal data summarized in Table 6.11. It is difficult to determine whether the greater spread in human open-heart measurements is a consequence of variations in the placement of the electrodes with respect to the most sensitive excitable regions of the heart, a reflection of the pathological state of the patient sample, or due to some other mechanism not yet identified.

6.9   Combined AC and DC Stimuli

We have seen from the discussion of Secs. 6.3–6.5 that cardiac sensitivity to monophasic DC pulses can be significantly different than sensitivity to alternating current. It follows that excitation and fibrillation thresholds for AC currents ought to be modified when a DC component is included. The effects of DC bias are illustrated in Fig. 6.18 for a sinusoidal stimulus at 1 kHz, with a duration of either 1 or 10 cycles (Roy et al., 1985). For these data, stimuli were provided to isolated rabbit hearts with a 1-cm electrode attached to the left ventricle and an indifferent electrode at a distant location. The vertical axis in Fig. 6.18 indicates the excitation threshold for cardiac pacing; the horizontal axis indicates the degree of DC bias as a percentage of the peak of the AC component. At the extremes of the horizontal axis, stimulus current was either fully cathodal or fully anodal; in between are intermediate cases of DC bias, with a symmetric biphasic wave represented at the center. Thresholds for symmetric sine waves are significantly above the purely anodal or purely cathodal

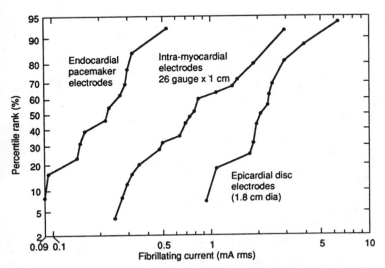

Figure 6.17   Distribution of 50-Hz fibrillating current in patients undergoing surgery for valve replacement. Straight lines on this format represent log-normal distributions. (Data from Watson and Wright, 1973.)

waveforms, reflecting the cancellation effects of current reversal (see Secs. 6.4–6.5). Thresholds for the monophasic currents are considerably below those for the balanced biphasic current. The cathodal threshold is lower than the anodal threshold by factors of 1.8 and 1.4 for the 1-cycle and 10-cycle stimuli, respectively.

Figure 6.19 illustrates cardiac excitation thresholds for various monophasic waveforms applying to the previously described experiments of Roy et al. In this case, the DC bias was such that the sinusoidal current was unidirectional. Stimuli consisted of biased sine waves with frequencies of 0.5, 1, 10, 50, 500 and 1000 kHz, and with durations ranging from 0.2 to 10 ms, as indicated on the horizontal axis of the figure. An additional stimulus waveform consisted of a rectangular pulse of the indicated duration. Thresholds are expressed on the vertical axis in units of transferred charge. The brackets shown on the curves encompass the mean excitation thresholds over the entire set of stimulus waveforms. When measured in charge units, thresholds among the various waveforms all followed the strength-duration curve of a rectangular monophasic pulse. Cathodal thresholds in Fig. 6.19 are consistently lower than anodal thresholds by an average ratio of 1.8 ± 0.2. This polarity sensitivity difference agrees with other cardiac sensitivity data reported by Thalen et al. (1975) and by Cranfield et al. (1975). Similar polarity differences have also been seen

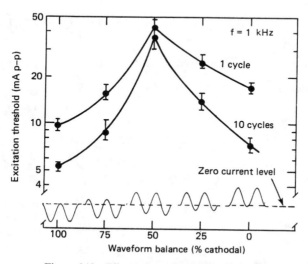

Figure 6.18   Effects of waveform bias on stimulation by 1-kHz sine wave. Isolated rabbit hearts, 1-cm^2 active electrode on left ventricle. Horizontal axis indicates degree of DC offset. Brackets indicate $\pm$ S.D. (Adapted from Roy et al., 1985.)

with peripheral nerve stimulation, where the polarity difference can be explained by the spatial distribution of polarizing currents nearby an active electrode (Sec. 4.4).

Effects of waveform bias illustrated in Figs. 6.18 and 6.19 apply to excitation thresholds with waveforms of limited duration (up to 10 ms). It is much more complex to determine VF thresholds for biased waveforms, particularly when the stimulus duration is sufficiently prolonged that multiple excitations may be produced. The experiments of Knickerbocker (1973) are valuable in assessing VF thresholds for mixed waveforms of prolonged duration. These experiments were carried out on anesthetized dogs, with electrodes attached from the left foreleg to the right hind leg. The stimuli consisted of DC or 20-Hz AC currents administered either singly or in combination; durations ranged from 200 to 2000 ms. Figure 6.20 gives a rough summary of the mean VF thresholds. The mix of AC and DC components is plotted on Cartesian coordinates in Fig. 6.20, with the points on the axes corresponding to pure DC (vertical axis) or pure AC (horizontal axis). The single-component thresholds correspond to the data of Fig. 6.11.

A biased AC waveform will be purely monophasic when the DC component is at least $\sqrt{2}I_{AC}$, where I_{AC} is the rms AC current. This relationship is represented by the diagonal line in Fig. 6.20, above which the waveforms are purely monophasic and below which they are biphasic with varying

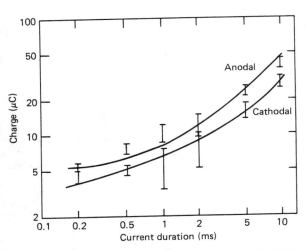

Figure 6.19 Strength-duration curves for stimulation by rectangular pulses and DC biased sine waves of frequency 0.5, 1, 10, 50, 500, and 1000 kHz. Rabbit hearts ($n = 5$); 1-cm^2 electrode at left ventricle. Brackets indicate range of mean thresholds for all waveforms. (Adapted from Roy et al., 1985.)

degrees of asymmetry. At all the durations studied, the pure AC thresholds were lower than the pure DC thresholds. The threshold contours for the mixed waveforms form patterns that depend not only on the excitatory power of the stimuli, but also on the associated excitation rate.

6.10   Electrodes and Current Density
Electrode Area

For a given amount of current into an electrode, average current density beneath the electrode increases inversely with the electrode area. At the same time, the smaller electrode results in a reduced amount of tissue that may be brought to excitation. A unified theory does not exist to predict the effect of electrode area on VF thresholds, although there is considerable empirical data on this issue.

Figure 6.21 summarizes VF thresholds for 60-Hz currents delivered with electrodes in contact with the heart (adapted from Roy, 1980). The individual points are average fibrillation thresholds taken from a variety of published data applying to 60-Hz stimuli of at least 2 s duration. Consid-

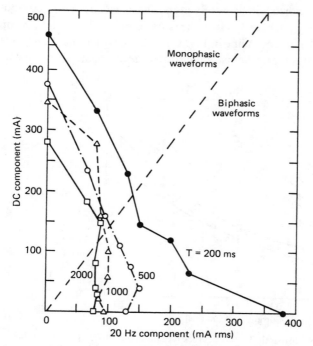

Figure 6.20   VF threshold contours for combined 20-Hz AC and DC currents of prolonged durations. Limb-to-limb electrodes with dogs. (Data from Knickerbocker, 1973.)

ering the diversity of experimental conditions and tested species, the consistency of the data is remarkable. The vertical axis of Fig. 6.21 expresses the ratio of threshold current to electrode contact area. Threshold current density determined this way drops as the electrode area is increased to about $5\,cm^2$, after which very little further drop is encountered. The minimum value plotted in the figure is about $0.5\,mA/cm^2$, a value that is supported by data from both human and dog heart experiments (Starmer and Whalen, 1973). The data shown in the figure represent experimental *averages*: minimum current density thresholds were typically about a factor of 2 below the averages for the various experiments evaluated by Roy. Using the log-normal model of Table 6.12, it is estimated that the "minimum" values observed by Roy might represent 1 percentile values of a statistical distribution.

Roy et al. (1986) suggest the following electrode-size relationship for VF thresholds based on experiments with dogs in which a disk electrode was applied to the epicardial surface of the right ventricle:

$$I_f = K d^c$$
$$
\begin{aligned}
c &= 0 && \text{for } d < 2\,\text{mm} \\
c &= 1.5 && \text{for } 4 \leqslant d \leqslant 15.9\,\text{mm} \\
c &= 2.0 && \text{for } d > 15.9\,\text{mm}
\end{aligned}
\tag{6.8}
$$

where I_f is the VF threshold, d is the electrode diameter area, and K is a scaling constant. Roy's relationship states that, below a lower critical size (2-mm diam. = 0.03-cm^2 area), the VF threshold current is directly proportional to electrode area (square of diameter). In between those critical sizes, current thresholds are proportional to the 1.5 power of electrode diameter (0.75 power of area). Analogous relationships were reported

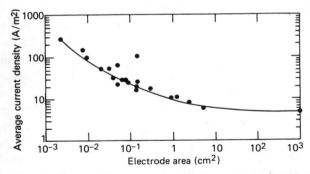

Figure 6.21   Average current density for VF thresholds with 60-Hz stimulating current of at least 2-s duration. Data points apply to a variety of different experiments. (Adapted from Roy, 1980.)

for excitation thresholds (Lindermans and Van der Gon, 1978), in which threshold current was found to be independent of electrode area for diameters below 0.4 mm, and proportional to the 1.5 power of diameter between 0.8 mm and 1.8 cm. Lindermans and colleagues justified their experimental results with theoretical calculations of current density beneath a contact electrode and the assumption of a *liminal length* of 0.3 mm for myocardial excitation, i.e., a minimum size of depolarized tissue necessary to initiate excitation. That liminal length is consistent with the value suggested in studies of individual Purkinje fibers (Fozzard and Schoenberg, 1972).

The relationships between electrode area and threshold current density will be different for excitation and fibrillation thresholds. In tests with isolated guinea pig hearts stimulated by DC pulses, Weirich and colleagues (1985) showed that thresholds of fibrillation expressed in units of current are lowered if the electrode size is increased, but the opposite effect is seen with thresholds of excitation. According to the authors, current thresholds of excitation go up with electrode size because a minimum current density is required for stimulation. To attain fibrillation, however, it is not sufficient merely to stimulate a few cells—rather, a critical mass of heart muscle must be excited to cause the degree of nonhomogeneity that is required for fibrillation.

The observations of Weirich et al. are supported by Table 6.13, which lists current and current density thresholds for excitation and ventricular fibrillation of dog hearts by 2-ms DC pulses (from Chen et al. 1975). When expressed in units of current magnitude, thresholds of excitation drop steadily as the electrode size is made smaller, whereas fibrillation thresholds do not differ greatly over the range of areas tested, and actually rise at the smallest tested area. The trends seen in Table 6.13 demonstrate that, as far as excitation is concerned, the heart is more easily excited by a given amount of current into a small electrode. With fibrillation, on the other hand, there exists a trade-off between the excitation threshold and the critical mass of excited tissue required to support fibrillation (Antoni, 1979). Table 6.14 gives additional data applying to the pacing of human hearts by 2-ms pulses (from Furman et al., 1967). The current densities listed here may be compared with the data for dog hearts given in Table 6.13. At an electrode area of 0.28 cm^2, the thresholds in the experiments with dog hearts are given as 1.8 mA/cm^2, and in the experiments with human hearts, as 3.0 mA/cm^2. Considering the different experimental conditions and tested species, these data are reasonably consistent. The data of Tables 6.13 and 6.14 provide further support for the proposition that experimental thresholds obtained from dogs provide conservative estimates for human applications.

Table 6.13. *Current density thresholds for stimulation of dog hearts by DC pulses (t = 2 ms)*

Electrode area (cm²)	VF threshold (mA/cm²)	(mA)	Ex. threshold (mA/cm²)	(mA)	VF/Ex. ratio
0.005	5000	25.0	14.0	0.07	357
0.08	275	22.0	2.0	0.16	140
0.28	104	29.1	1.8	0.50	58

Notes: Pulse duration = 2 ms. In vivo stimulation of dog hearts ($n = 20$). Active electrode implanted into myocardium.
Source: Data from Chen et al. (1975).

Table 6.14. *Current density thresholds for stimulation of human hearts by DC pulses (t = 2 ms)*

Electrode area (cm²)	Ex. Threshold (mA/cm²)	(mA)
0.12	3.27	0.39
0.28	3.00	0.84
0.49	3.14	1.54
0.87	2.30	2.00

Notes: In vivo stimulation of human hearts. Active catheter electrode inserted into right ventricular apex.
Source: Data from Furman et al. (1967).

Current Density

The foregoing discussion has dealt with current density thresholds inferred from experiments in which an electrode was held in contact with cardiac tissue. Current may also be introduced less directly, such as with electrodes applied to the body at locations distant from the heart, or with current induced by exposure to time-varying magnetic fields. In these cases, current may be applied more or less uniformly to the whole heart, rather than in a concentrated locus beneath a contact electrode.

Stimulation thresholds for whole-heart exposure are traditionally evaluated in terms of current density. Nevertheless, the electric field within the biological medium may be a more relevant parameter, as suggested by Starmer and Whalen (1973), who observed that VF thresholds are better correlated with spatial potential differences than with current density. This observation is consistent with theoretical and experimental studies with nerve fibers (Sec. 4.4), which show that the relevant excitation parameter is the electric field longitudinal to the fiber, rather than the current density per se. The electric field (E), and the current density (J), are simply related by the conductivity (σ) of the medium by $J = E\sigma$. Although current density is often cited as an experimental parameter, conductivity is not always given in published descriptions of experimental data. For that reason, the electric field may be difficult to ascertain.

Current density is most commonly determined by dividing stimulus current by the area of the electrode, as with the data in Fig. 6.21, and Tables 6.13 and 6.14. Current density determined this way will correctly give the *average* value beneath the electrode, but may understate the maximum density because current tends to concentrate at the edges of a contact electrode, as demonstrated in theoretical (Caruso et al., 1979) and experimental (Lane and Zebo 1967) studies (see Fig. 2.9).

Considering the possibility of nonuniform current distribution at the interface between an electrode and biological tissue, the values plotted on the vertical axis of Fig. 6.21 probably provide conservative estimates of the maximum current density. With this caveat in mind, the current density of $0.5\,\text{mA/cm}^2$ shown for the largest electrodes in Fig. 6.21 may be used as an estimate of the average current density threshold for VF by whole-heart exposure to prolonged ($>2\,\text{s}$) 60-Hz currents, and a value of $0.25\,\text{mA/cm}^2$ for a more conservative estimate of a 1 percentile experimental value. Table 6.4 indicates that excitation thresholds are about 40% of the 60-Hz fibrillation value for prolonged ($t > 2\,\text{s}$) 60-Hz and also for monophasic pulse stimulation ($t \approx 10\,\text{ms}$). Applying this factor to the VF current density results in an estimated average threshold of $0.2\,\text{mA/cm}^2$ for excitation by either prolonged 60-Hz or single DC pulses, and a 1 percentile threshold of $0.1\,\text{mA/cm}^2$. That minimum value is quite close to the value of $0.12\,\text{mA/cm}^2$ calculated as a minimum threshold for excitation of peripheral nerves (Table 4.2). The current densities cited above have been obtained in experiments with healthy animals. It is not known whether these values are representative of the human heart when it is in a pathological state (see Sec. 6.8).

Locus and Direction of Stimulus Current

Thresholds of fibrillation will vary with the location of electrodes, whether they are contacting the heart or are applied externally. Table 6.15 (adapted from Roy et al., 1987) lists VF thresholds for a 1-cm² electrode applied to five different regions of a pig's heart. An indifferent flexible

Table 6.15. *VF thresholds for 60-Hz
currents applied through a 1-cm² active
electrode (data listed in order of increasing
average threshold)*

Electrode position	n	VF threshold (ma)		
		min	avg	rms
1	16	0.66	0.98	0.23
2	14	1.06	1.42	0.52
3	12	1.19	1.82	0.42
4	15	1.09	1.83	0.51
5	14	1.35	2.17	0.71

Source: Data from Roy et al. (1987).

electrode (15 × 56 cm) was wrapped around the back and sides of the animal. Seven animals were tested using a 60-Hz stimulus of 5 s duration. The data in the table are listed in the order of increasing average current density; the stimulated regions are identified in Fig. 6.22. The thresholds span a range of 2.2:1 from the least to the most sensitive area. The most

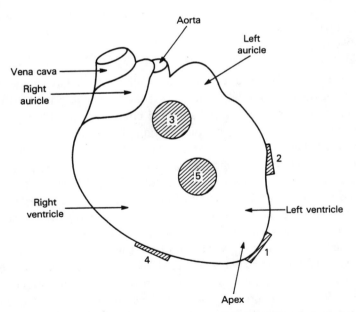

Figure 6.22   Diagram of pig's heart showing position of 1-cm² electrodes numbered in order of increasing average VF thresholds. (From Roy et al., 1987.)

Table 6.16. *Relative VF sensitivity with respect to pathway of 50/60-Hz current (large area, t ⩾ 3 s)*

Electrode positions	Relative I_F			
A. Hand-to-hand paths				
Left hand–right foot (left fore–right hind)		1.0		1.0
Left hand–left foot (left fore–left hind)		1.13		1.0
Right hand–left foot (right fore–left hind)	1.0			1.25
B. Hand-to-hand paths				
Right hand–left hand (right fore–left fore)	1.56	2.3		2.5
C. Foot-to-foot paths				
Left foot–right foot (left hind–right hind)	>50			
D. Misc. hand paths				
Right hand–back (right fore–back)				3.33
Left hand–back (left fore–back)				1.43
Right hand–chest (right fore–chest)				0.77
Left hand–chest (left fore–chest)				0.67
Hand–seat				1.43
E. Other paths				
Left chest–right chest	1.04		1.4	
Left foot–head	1.20			
Chest–back			1.0	
References	(1)	(2)	(3)	(4)

Sources: (1) Ferris et al. (1936); (2) Geddes et al. (1973); (3) Roy et al. (1986); (4) IEC (1982).

sensitive area is the apex; regions of the right ventricle have higher thresholds than those on the left ventricle. At the heart apex, the average and minimum thresholds listed in Table 6.15 compare favorably with the VF thresholds noted in the previous section if electrode area is properly considered. The authors pointed out that their results are consistent with the concept that a critical mass of cardiac muscle is required for sustained fibrillation—the apex, having the greatest muscle thickness, is associated with the lowest threshold. Excitation via external electrodes would be expected to have the lowest thresholds if the electrodes were oriented so as to provide the greatest exposure to the apex of the heart.

Table 6.16 indicates the variation of VF thresholds with respect to the pathway of 50/60-Hz currents delivered for long duration ($t > 3$ s) through electrodes placed at various locations on the body. The tabulated data, derived from the references listed at the bottom of the table, indicate relative thresholds with a single column. In column 1, for example, the value $I_f = 1.56$ for a hand-to-hand path indicates a finding that 56% more current was needed to induce VF than with a right hand-to-foot path having $I_f = 1.0$. A listing of 1.0 in one column does not necessarily indicate the same VF threshold as a listing of 1.0 in another column. The I_f data in the first three columns apply to experiments with animals (Ferris et al., 1936; Geddes et al., 1973; Roy et al., 1986); the values in the last column are criteria recommended in Europe by the IEC for evaluation of electrical hazards (IEC, 1982).

The relative current thresholds listed in the table are a result of several factors, including the following:

1. *Current density.* The current density in the vicinity of the heart will vary with the placement of external electrodes. In experiments with dogs, it has been reported that 7% of the applied current flows through the heart with a forelimb-to-hind limb path, whereas only 3% is intercepted by the heart for current applied across the forelimbs (Kouwenhoven et al., 1932).

2. *Excitable fiber orientation.* We have seen in Chapter 4 that current aligned with the long axis of an excitable fiber is much more effective in producing excitation than current flow in a transverse direction.

3. *Exposure of sensitive region.* As suggested by Table 6.15, we may expect a 2:1 difference in sensitivity for stimulation to different regions of the heart. Variation in the placement of external electrodes may provide more or less exposure to sensitive region of the heart.

4. *Anisotropic impedance properties.* The impedance of muscle tissue is lowest when the current is aligned with the direction of the muscle fibers (Chilbert et al., 1983). The conductivity parallel and prependicular to the fibers of ventricular muscle demonstrate a ratio of about 2:1 (Ruch et al., 1963). Accordingly, the direction of current would affect its distribution within the heart, and the mass of excited tissue.

5. *Polarity of DC stimulation.* We have seen in Chapter 4 that an excitable fiber will be depolarized most effectively at end structures that are oriented toward the cathode of a current source.

While the factors mentioned above may all play a role in the heart's sensitivity to electrode placement and the current path, it is not possible to separate out their relative importance on the basis of available data. However, a number of generalizations can be made, as noted below.

Longitudinal versus Transverse Current. Current flow along the long axis of the body (longitudinal flow) is associated with significantly lower thresholds than with current flowing in an orthogonal (transverse) direction. Hand-to-foot pathways are observed to have significantly lower thresholds than hand-to-hand paths—the ratio of thresholds for the two orientations is about 2:1. Table 6.10 lists average thresholds for longitudinal paths, attributed to subjects of different body weights.

Foot-to-Foot Paths. Available data indicate that current applied from one foot to another is very unlikely to result in VF. Ferris et al. (1936) found that current applied from foot to foot failed to produce VF at a current level that was 50 times greater than the VF threshold for hand-to-foot contacts. The relatively high thresholds for foot-to-foot contact can probably be explained largely on the basis of current density arguments: relatively little current reaches the heart when it is introduced through the legs or hind limbs.

Chest Electrodes. When one of the electrodes is placed on the chest, the VF threshold depends critically on its precise placement with respect to anatomical features of the heart. The lowest threshold for a chest electrode is obtained when it is placed directly over the apex of the heart (Geddes et al., 1973)—a finding consistent with open-heart tests (Roy et al., 1987). With large-area (300 cm²) electrodes on the chest, VF thresholds have been found to be 1.4 times lower when the current path was from the chest to the back of the animal than with a transverse (side-to-side) orientation (Roy et al., 1986). Thresholds of stimulation with a large-area chest electrode were measured typically between 40 and 70 mA with a minimum value of 20 mA in tests of humans (Zoll et al., 1985). 60-Hz fibrillation thresholds for 200-mm² chest electrodes on dogs averaged 68 mA (Roy et al., 1986).

DC Stimulation. As with AC stimulation, a longitudinal orientation of DC is associated with significantly lower thresholds than with transverse current flow. Fatal accidents in Europe involving DC exposure have been reported to occur only with a longitudinal current path (Biegelmeier, 1987). Additionally, the polarity of the DC electrodes significantly affects VF thresholds, with positive foot electrodes being the most hazardous. In early experiments (Ferris et al., 1936), it was found that the VF threshold was about 36% lower when the feet were at a positive potential relative to a hand electrode as compared with the opposite polarity. Biegelmeier (1987) reports that VF thresholds with positive foot and negative hand electrodes are about one-half the thresholds pertaining to the opposite direction of current flow.

6.11   Computational Models of Cardiac Excitation

We have seen that cardiac stimulation thresholds depend strongly on features of the stimulus waveform, including its timing with respect to the cardiac cycle. Experimental data have provided quantitative information on these relationships for specific waveforms, including rectangular and sinusoidal waveshapes. It would be useful to have a computational model that could be used to explore sensitivity of cardiac excitation to a broader class of stimulus conditions. Such a computational model would have to include the dynamic electrical response characteristics of the excitable tissue comprising the heart.

Models for the dynamic electrical properties of nerve tissue have been used successfully to study a variety of problems related to neural excitation by externally applied electric currents, as noted in Chapter 4. For example, a computational model of a myelinated nerve has been used to determine excitation thresholds for complex current waveforms, and has been applied to the determination of safety standards for currents induced by time-varying magnetic fields (Chapter 9). It is likely that excitation of cardiac tissue by externally applied currents could also be studied with appropriate computational models. Chapter 3 describes a model of the electrical properties of the excitable membrane, as developed originally by Hodgkin and Huxley, and later extended to the cardiac membrane. The models allow one to study the development of excitation in a single cardiac cell from a knowledge of the current crossing the cell membrane. Single-cell models have been used to study excitation sensitivities; Antoni and colleagues (1970), for example, used a single-cell adaptation of Nobel's (1962) model to explain experimental observations of polarization effects with 50-Hz sinusoidal stimulation. Despite their demonstrated utility, single-cell models have inherent limitations for electrical stimulation studies.

To use a single-cell model for electrical stimulation studies, we must be able to define the current crossing the membrane. However, this is not always known: When current is injected in the medium with an external electrode, the waveform of current crossing the membrane is not necessarily the same as that emanating from the electrode (McNeal, 1976). Single-cell representations are not readily adaptable to extracellular electrode placement, where the distribution of stimulus current along the membrane is an important factor. An additional limitation of single-celled models is that excitation must be determined on the basis of membrane depolarization rather than AP propagation. In neural excitation studies, these difficulties have been overcome with a representation of an extended axon, in which nonlinear membrane elements are interconnected in a one-dimensional array (see Sec. 4.4). A similar approach might be used for the cardiac system.

Ideally, an extended cardiac model would incorporate nonlinear, single-cell elements into a three-dimensional array. However, even two-dimensional descriptions, such as that studied by Joyner et al. (1975), involve formidable computational requirements if the appropriate non-linear equations are to be included. In order to make a computational model more tractable, Sharp and Joyner (1980) developed a numerical method involving a one-dimensional array based on cable equations; the individual elements in this array used nonlinear conductances defined by either the MNT or the BR model (refer to Sec. 3.3). The authors were able to demonstrate AP conduction velocities for the array that were consistent with experimental data. The effects of variations in coupling resistance between individual cells were also demonstrated. Henriquez and Plonsey (1987) also used a one-dimensional model to study the effects of junction resistance on AP development and propagation.

A computational model based on excitable cardiac membrane proper-ties would greatly facilitate further exploration of cardiac sensitivity to complex waveforms, such as brief biphasic currents, or repetitive pulse sequences. Such a model could likely be used to explore excitation thresh-olds. The development of a fibrillation model, however, would require complexities far beyond those needed to describe excitation, including a description of a spatial propagation of the excitation process. While spatial propagation properties have been explored in a number of models (Barta et al., 1987; Joyner et al., 1975; Eifler and Plonsey, 1975), a comprehensive model for electrical safety applications has yet to be demonstrated.

7

Sensory Responses to Electrical Stimulation

7.1   Introduction

Sensory sensitivity to electrical stimulation depends on a host of factors generally falling into categories associated with the stimulus waveform, its method of delivery, and subjective variables. In most situations involving electrical safety or acceptability, currents are applied to the body via cutaneous electrodes. There are also practical applications where electric currents may be applied subcutaneously, or induced internally via external electromagnetic fields. While the emphasis in this chapter is on electrocutaneous stimulation, many of the principles discussed here may be applied to other modes of stimulation. The reader is directed to Chapter 9 for additional discussion of peripheral nerve stimulation by time-varying magnetic field effects, or by induced shock within intense electric field environments.

7.2   Mechanisms of Electrical Transduction

Currents of a fraction of a microampere can be detected when the finger is gently drawn across a surface charged with small AC potentials (Grimnes, 1983b). Such levels are roughly 100 times less than commonly tested electrical thresholds. Detection of such small currents results from electromechanical forces arising from electrostatic compression across the *stratum corneum* (the outermost layer of dead skin cells). As analyzed by Grimnes, the electrostatic force K is

$$K = \frac{A\varepsilon v^2}{2d^2} \tag{7.1}$$

where A is the contact area, ε is the dielectric constant of the corneum, d is its thickness, and v is the instantaneous voltage. The compression of the corneum would not normally be sensed. But when the skin is moved along the charged surface, there is a vibratory frictional force on the finger that is maximized on each half-cycle of the alternating voltage. This vibrational force stimulates mechanoreceptors, and is responsible for the detection of microampere currents. Grimnes estimates that the minimum voltage contributing to a detectable vibration is in the vicinity of 1.5 V at 50 Hz.

The detection of microampere currents through mechanical vibration is, for most purposes, of passing interest, although it may be important for a researcher to know about it when designing perception tests. Of greater significance is the mode of detection when the current level is raised to roughly 0.1 mA or above. At that point, perception can be initiated by the electrical excitation of neural structures, according to the mechanisms discussed in Chapters 3 and 4.

The question of exactly what is excited with electrocutaneous stimulation, and the specific site of initiation, has not been studied systematically. At the lowest levels of stimulation, it is likely that peripheral structures are involved, since these are closest to the surface electrode. Among fiber classes, the larger-diameter myelinated fibers have the lowest electrical thresholds. A variety of circumstantial evidence presented in this chapter points to the involvement of one or another class of mechanoreceptor. The precise site of electrical stimulation is unknown; whether stimulation is initiated at the axon proper, at the site of the generator potential of sensory receptors, or along free nerve endings has not been demonstrated. Some evidence, however, exists, as noted in Chapter 4, that the site of initiation is near neural end structures, including receptors or free nerve endings. Electrocutaneous perception is a local phenomenon; subjects typically report sensation occurring locally at the electrode site, rather than remotely as might be supposed if the excitation occurred on the axons of deeper-lying nerves. It is only when the current is raised substantially above the perception level that distributed sensations are felt.

If the current is raised sufficiently above the threshold of perception, excitation of unmyelinated nociceptors becomes possible. Because of their higher electrical thresholds and generally deeper sites, these structures are not likely to be involved at perception threshold levels. At still higher current levels (some tens of milliamperes for long-duration stimuli), thermal detection due to tissue heating becomes possible. Neuroelectric thresholds may exceed thermal thresholds if the waveform of the electric current is inefficient for electrical stimulation, such as with sinusoidal currents of very high frequency ($> 10^5$ Hz), or pulsed currents of very short duration. But for most cases of practical interest, electrical thresholds are well below thermal thresholds.

The electrical stimulus is nonspecific as to the receptor class that might
be stimulated. Nevertheless, there is some degree of selectivity on the basis
of electrical properties—thresholds tend to be lower as the neuron is
closer to the corneum, as the fiber diameter is increased, and as its
orientation is aligned with the internally generated electric field (i.e.,
direction of current flow). At suprathreshold levels, we would expect an
array of different fiber types to be stimulated, including deeper-lying
structures such as small-diameter nociceptors and motor neurons.

One might question whether the electrical stimulus would feel like some
natural modality corresponding to the receptor type that is being excited.
In the author's own research, subjects most often selected punctate pressure
descriptors from a list of 10 choices for capacitive discharge stimulation
applied cutaneously to the forearm, as shown in Fig. 7.1 (Reilly and Larkin,
1985a). Similar results have been obtained for stimulation by 300-ms
square-wave pulses: At levels just above perception, subjects preferred
descriptors of distributed pressure; but as the stimulus level was increased,
they favored punctate pressure descriptors (Tashiro and Higashiyama,
1981).

Electrical stimulation by pulse trains evokes sensations closer to natural
stimuli when the nerve is stimulated directly by a percutaneous microelectrode. In the experiments of Vallbo and colleagues (1984), a microelectrode was inserted above the elbow into the median nerve, which

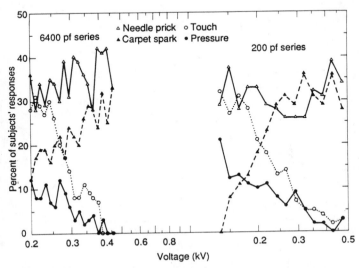

Figure 7.1   Qualitative ratings by subjects receiving suprathreshold
capacitive discharges. (From Reilly and Larkin, 1984.)

has a receptive field in the hand. When stimulated by a train of pulses, subjects reported localized sensations in the hand akin to steady pressure, transient pressure, or vibration, depending on the particular fiber being stimulated and the electrical stimulation parameters. The authors reported that stimulation of a single nerve fiber could elicit perceptible sensations for some classes of fibers.

Some understanding of the neural pathways for electrical stimulation can be inferred from experiments in which A or C fibers are selectively blocked in human subjects (Torebjörk and Hallin, 1973). When mainly A-fiber response is present, weak electric shocks are felt as tactile sensations, and strong shocks are perceived as a short, sharp blow without prolonged pain. When A-fiber response is blocked, there is an impaired discrimination of weak stimuli, and strong stimuli evoke painful sensations due to C-fiber activity.

The electrocutaneous stimulus is a very unnatural one in terms of its spatial distribution of stimulation, lack of receptor specificity, abnormal neural recruitment properties, abnormal temporal patterns of excited action potentials, and concurrent recruitment of sensory and motor neurons. These factors all contribute to a sensory quality that is unlike any for which nature has equipped us. These unnatural properties are

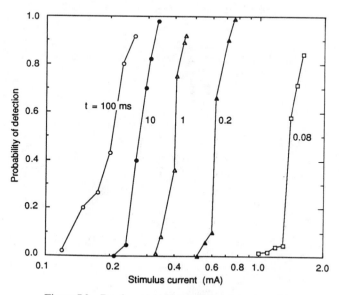

Figure 7.2 Psychometric functions for detection of square-wave pulses applied cutaneously; 0.5-cm^2 saline-treated electrode on forearm. (Data from Rollman, 1974.)

likely to contribute to the highly aversive quality that most people attribute to electrical stimulation that may not greatly exceed their perception threshold.

The relatively small dynamic range of electrical stimulation is reflected in extremely steep *psychometric functions*, i.e., the functional relationship between the stimulus level and the probability of detection. Figure 7.2, for example, illustrates psychometric functions applying to a single individual detecting cutaneously applied square-wave pulses of various durations (Rollman, 1974). The coefficient of variation (standard deviation to mean ratio) measured by Rollman was typically 0.08, a value that is significantly smaller than that observed for other sensory modalities. Because of the steep psychometric functions, a change of stimulus magnitude of only 20% typically elevates the detection probability from 10% to 90% for each curve shown in Fig. 7.2, even though the absolute median thresholds vary by more than 10 to 1. Figure 7.2 also demonstrates a significant waveform sensitivity in which thresholds are inversely related to the duration of stimulation. This phenomenon is described by strength-duration (S-D) curves, discussed further in the succeeding section.

7.3   Perception of Transient Monophasic Currents

Strength-Duration Relationships

Sensory sensitivity to electrical stimulation varies greatly with the applied stimulus, including its waveshape, duration, repetition pattern, polarity, and whether the stimulus is monophasic or biphasic. The relationship between the intensity of a monophasic current and the threshold of reaction has been the most systematically studied aspect of waveform dependency. The S-D relationships for neural excitation have a long history of exploration, beginning with the early studies of Weiss (1901) and Lapicque (1907). In Chapter 4, the early formulations are described in terms of an exponential [Eq. (4.15)] or parabolic [Eq. (4.17)] relationship. Although their mathematical expressions are different, the two formulas express similar results (Fig. 4.3).

Experimental S-D curves uniformly demonstrate that perception thresholds converge to a minimum current (I_0) as the current duration is made very long, and to a minimum charge value (Q_0) as it is made very short. The minimum current has traditionally been called the *rheobasic current*. Whether the exponential or parabolic formulations are used, the S-D curve may be described in terms of two parameters: the equivalent S-D time constant (τ_e), and the minimum (rheobasic) threshold. The time constant is a measure of the temporal responsiveness of the excitable system. In both exponential and parabolic formulations, the S-D time constant may be obtained by the ratio Q_0/I_0 [Eq. (4.20)].

The characteristic S-D relationship applies to both perception and suprathreshold values. Figure 7.3 illustrates neurosensory and neuromotor

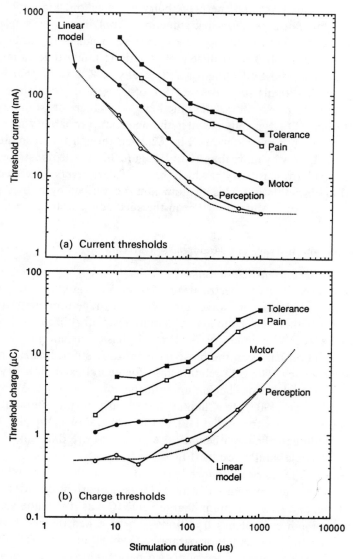

Figure 7.3   Strength-duration curves for sensory and motor reactions to square-wave pulses; forearm stimulation, 4-cm² electrode. (Data from Alon et al., 1983.)

S-D curves from the experiments of Alon and colleagues (1983). The upper set of curves represents thresholds in terms of the peak current of a monophasic square-wave stimulus. The lower set is given in terms of stimulus charge (product of amplitude and duration). The broken line indicates the exponential expressions [Eqs. (4.6) and (4.9)], adjusted such that the values Q_0 and I_0 correspond to the experimental perception threshold values. With this adjustment, the experimental and analytic curves agree remarkably well.

A variety research on sensory stimulation has been reported in terms of S-D relationships, as summarized in Table 7.1. The column headed τ_e refers to the S-D time constant, and the columns headed I_0 and Q_0 to the minimum current and charge thresholds. In most cases, the tabulated value of the S-D time constant has been determined by $\tau_e = Q_0/I_0$. In a few cases, where the minimum thresholds were not available, the S-D time constant was determined by $\tau_e = $ chronaxy$/0.693$ [see Eq. (4.19b)], or by a simple curve fit of Eq. (4.6). The table entries are subdivided according to categories of human perception, animal or nerve preparations, and a theoretical model for myelinated nerve. The "human" category includes both neurosensory and neuromuscular categories. The values of Q_0 and I_0 depend on a number of factors, such as electrode size, the locus of stimulation, the reaction threshold being tested, and whether the current is supplied transcutaneously or subcutaneously. These factors are discussed in detail in later sections of this chapter. The ratio Q_0/I_0 is much less sensitive to these factors.

Empirical time constants for neuromuscular reactions do not differ significantly from neurosensory values, although the absolute threshold for motor responses is typically higher. Although motor neurons typically have larger diameters than sensory neurons, and hence would be expected to have lower electrical thresholds, the fact that they are situated in deeper strata below the skin accounts for their higher thresholds. When the muscle is denervated, motor stimulation takes place via direct excitation of muscle fibers rather than motor neurons. In this case, motor thresholds rise significantly, and the time constants increase dramatically (Bauwens, 1971; Harris, 1971). Motor responses are discussed in greater detail in Chapter 8. The fact that the parameters of the S-D curves vary with conditions of neuropathy (Friedli and Meyer, 1984) suggests a potential diagnostic tool.

Although sensory time constants encompass a range similar to that noted for neural stimulation (Parts B and C of Table 7.1), there is a greater representation toward the larger values with sensory data. The geometric mean of τ_e values given in Table 7.1 for human perception is about 270 μs;

Table 7.1. *Strength-duration data for peripheral nerve stimulation*

Item	Subject	N	Stimulus locus	Electrode	Reaction	τ_e (μs)	I_o (mA)	Q_o (μC)	Reference
A. Human subjects									
1.	Human	6	Forearm	4-cm² carbon sponge	Perception	138	3.5	0.48	Alon et al. (1983)
2.	Human	6			Motor	127	8.5	1.08	Alon et al. (1983)
3.	Human	6			Pain	70	24.5	1.72	Alon et al. (1983)
4.	Human	6			Tolerance	142	33.8	4.80	Alon et al. (1983)
5.	Human	4	*	2-mm diam.	Pain	760	0.5	0.38	Notermans (1966)
6.	Human	3	Finger (cap. discharge)	1.1-mm diam.	Perception	200–900	—	0.12	Larkin and Reilly (1984)
7.	Human	*	Ulnar nerve, forearm	0.5-cm² elect. paste	Perception	480	0.25	0.12	Rollman (1975)
8.	Human	3	Finger pad	0.4 cm²	Perception	217	0.69	0.15	Hahn (1958)
9.	Human	*	Forehead; forearm	0.1 mm²	Perception	189	1.8	0.34	Girvin et al. (1982)
10.	Human	6	Ulnar nerve	*	AP	<260	2.0	<0.52	Heckmann (1972)
11.	Human	6	Ulnar nerve	*	Muscle twitch.	275	3.2	0.88	Heckmann (1972)
12.	Human	*	Denervated muscle	*	Muscle twitch.	8333	6.0	50	Heckmann (1972)
13.	Human	*	Enervated muscle	*	Muscle twitch.	30–700	*	*	Harris (1971)
14.	Human	*	Denervated muscle	*	Muscle twitch.	>4300	*	*	Harris (1971)
15.	Human	2	Forearm	3.9-cm² elect. paste	Muscle twitch	140	9.9	1.39	Crago et al. (1974)
16.	Human	12	Foot	8 mm dia.	Perception	295	1.1	0.32	Friedli & Meyer (1984)
17.	Human	8	Finger	8 mm dia.	Perception	383	0.5	0.19	Friedli & Meyer (1984)
B. Animal; in-vitro nerve preparation									
18.	Rat, cat	12	Tibialis ant. musc.	10mm, intra. musc.	Muscle twitch	85	5.2	0.44	Crago et al. (1974)
19.	Rat, cat	12	Peroneal nerve	*	AP	55	0.38	0.021	Crago et al. (1974)
20.	Rabbit	2	Tibial nerve (4–13 μm diam.)	0.1-mm diam., in vivo (subcutaneous)	AP	80–100	—	—	Reilly et al. (1985)
21.	Cat	*	A_β fiber	In vitro	AP	30	—	—	Li and Bak (1976)
22.	Cat	*	A_δ fiber	In vitro	AP	650	—	—	Li and Bak (1976)
23.	Toad	*	Myelinated nerve	In vitro	AP	148	—	—	Tasaki and Sato (1951)
24.	Dog	4	Back	Magnetic coil	Muscle twitch	148	—	—	Bourland et al. (1990)
C. Theoretical model									
25.	SENN model	—	Myelinated nerve	Point elect.	AP	92–128	—	—	Reilly and Bauer (1987)
26.	SENN model	—	Myelinated nerve	Uniform field	AP	120	—	—	Reilly and Bauer (1987)

*Information not available.

—Information not applicable for table summary.

when the data of Parts B and C are included, the geometric mean is closer to 170 μs. The myelinated nerve SENN model indicates a range of τ_e depending on the distance between a point electrode and the nerve fiber. For uniform E-field excitation, the model indicates $\tau_e = 121$ μs; in comparison, a value of 140 μs was obtained under experimental conditions closely approximating uniform field stimulation, in which the internal F-field was induced by magnetic field exposure over a large area of the torso (Schaefer et al., 1991).

Experimental S-D time constants are not simply a property of the excitable tissue, but depend on the method of stimulation as well. For instance, time constants have been found to increase monotonically with electrode size in nerve and neuromuscular preparations (Pfeiffer, 1968; Davis, 1923), as well as in cardiac tissue (see Sec. 6.2). The area dependency is consistent with theoretical arguments, although the effect is expected to be much stronger in muscle than in nerve tissue (Jack et al., 1983).

For the data described in Table 7.1, absolute perception thresholds for short-duration stimuli and small electrodes converge to a constant charge value of 0.1 μC, and for long-duration stimuli to a constant amplitude value of about 0.25 mA. These values must be qualified by a variety of factors affecting sensitivity, such as body locus, electrode size, skin temperature, tactile masking, electrode contact, and stimulus waveform features. Additionally, individual differences in sensitivity will result in a distribution of threshold values.

Capacitive Discharges[†]

Exposure to transient electric shock is a common occurrence. We have all experienced shocks when we walked across a carpet on a dry day and then touched a grounded object. In such cases, our body acts as a capacitor that stores electric charges at levels of several thousand volts. Then, when we come sufficiently close to a grounded object, the stored charge is suddenly discharged at some discrete body location through a spark that may be felt, seen, and heard. The peak current of a carpet spark can be very large—typically more than an ampere—a level that could be lethal if sustained. Fortunately, the event is very brief, in the microsecond range. As a result, the shock is well below lethal intensity, but nonetheless can be annoying to many people.

There are also biomedical applications where capacitive discharges may be used beneficially. One potential research application takes advantage

---

[†]Portions of this section have been adapted from Larkin and Reilly (1984) and Reilly and Larkin (1984).

of the fact that a capacitive discharge has a sudden current onset, but a gradual decay (Accornero et al., 1977). This produces a stimulus that is effective primarily on its leading edge, while avoiding the possibility of excitation on the current break.

When the skin contacts an electrically charged object and so provides a ground path for a capacitive discharge, a variety of cutaneous sensations are possible. Near threshold, the sensations commonly resemble touch or mild pinprick. Above threshold, the pinprick may be followed by a burning sensation akin to the delayed pain related to C-fiber activity. Bishop (1943) reported other sense qualities as well, and suggested that a punctate "spark" from a capacitor could be used selectively on the skin to excite discrete neural channels. The capacitive discharge stimulus is monophasic (current flows in only one direction) and can be made brief in relation to the depolarization process in neural tissue. In these two respects, capacitive discharges resemble the short-duration, constant-current pulses often used in sensory and electrophysiological research.

In two other respects, capacitive discharge stimuli may differ from the current pulses commonly in use. First, the discharge waveform can have both a very high initial voltage and a very high current peak, even at threshold levels of stimulation. Furthermore, while constant-current stimulation requires an electrode in good contact with the skin, a high-voltage capacitive discharge can occur without contact, if the electrode is simply brought close enough to sustain an electric arc between it and the skin. These characteristics are also present in electrostatic "carpet sparks" of the sort familiar to all who live in dry, upholstered environments (Chakravarti & Pontrelli, 1976).

The capacitive discharge stimulus is typically associated with a high peak current density at a relatively high voltage. The electrical properties of the skin are much less significant with high-voltage capacitive discharge stimuli than with traditional constant-voltage or constant-current devices, because the skin impedance largely breaks down when presented with high-voltage and high-current densities of a capacitive discharge (see Sec. 2.5).

When a capacitance (C) discharges through a resistance (R), both current and voltage have simple exponential waveforms given by:

$$I(t) = I_0 e^{-t/\tau} \tag{7.2}$$

$$V(t) = V_0(1 - e^{-t/\tau}) \tag{7.3}$$

where V_0 is the initial voltage on the capacitor, $I_0 = V_0/R$, and τ is the discharge time constant given by the product RC. The amount of stored charge that may be released in the stimulus is simply

$$Q = CV_0 \tag{7.4}$$

While Eqs. (7.2) and (7.3) represent an ideal case, discharges to biological materials, including human skin, exhibit additional complexities such as nonlinear impedance (see Chapter 2). Despite these complexities, the capacitive discharge waveform is typically very close to Eqs. (7.2) and (7.3), so that an exponential approximation is sufficient. While the time course of the capacitive discharge current may be expected to follow the exponential form [Eq. (7.2)], discharges to intact skin are highly dependent on the initial voltage as well as the capacitance, as indicated by Figs. 2.28 and 2.34.

The author and colleagues have systematically explored the sensory and biophysical aspects of capacitive stimuli. As an example of this work, Fig. 7.4 illustrates mean perception thresholds for anodal capacitive discharges (Reilly and Larkin, 1987). Thresholds are shown for delivery of the stimulus by: (a) the finger tapping an energized electrode; (b) discharge to a large (1.27-cm-diameter) circular electrode held against the fingertip; (c) discharge to a small (0.11-cm-diameter) electrode held against the fingertip; and (d) discharge to a needle piercing the corneum of the forearm. The curve shapes and their relative displacements are based on intensive

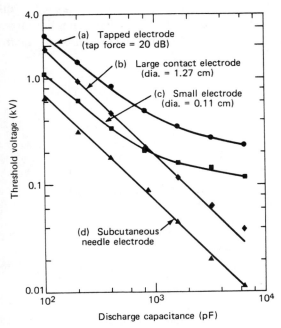

Figure 7.4   Threshold sensitivity contours for four methods of stimulation using capacitive discharges of positive polarity. (Measured data slightly adjusted to conform to large-sample results see text.) (From Reilly and Larkin, 1987.)

study of six expert subjects (Reilly and Larkin, 1985a), but their absolute positions have been slightly adjusted along the vertical axis to reflect measurements with a sample of 124 subjects (see Sec. 7.10). In Fig. 7.4, threshold charge at 100 pF is 0.25 μC for procedure (*a*); 0.19 μC for procedure (*b*); 0.11 μC for procedure (*c*); and 0.07 μC for procedure (*d*).

Two of the curves in Fig. 7.4 follow straight lines representing constant charge, that is, $CV =$ constant. The other two curves depart significantly from a constant-charge contour at the larger capacitances. Besides differences in curve shapes, there are significant vertical displacements among the curves. The contour shapes of Fig. 7.4 can be related to theoretical strength-duration (S-D) relationships if we account for stimulus discharge time constants (τ). Figure 7.5 presents capacitive stimulus thresholds (in normalized charge units) versus τ (adapted from Larkin and Reilly, 1984). The vertical axis has been normalized to the stimulus charge at the shortest value of τ corresponding to 100-pF discharges. The illustrated dependency is such that the threshold charge is reduced as τ is reduced, and converges to a minimum value for stimuli of very short duration. The S-D data of Fig. 7.5 are consistent with theoretical models describing electrical excitation of excitable neural tissue. The curve in Fig. 7.5 is a mathematical expression of an S-D curve for capacitive discharge stimuli given by Eq. (4.14). For undirectional (monophasic) current pulses that are brief relative to the neural-membrane response time, neuro-

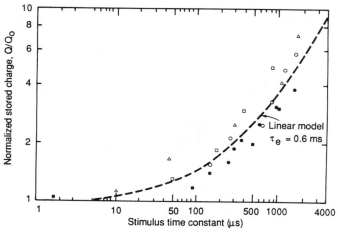

Figure 7.5   Strength-duration data—perception of cutaneous capacitive discharges for various procedures. (Thresholds normalized by minimum perception charge.) Adapted from Larkin and Reilly, 1984. in *Perception and Psychophysics*, vol. 36, pp. 68–78, reprinted by permission of Psychonomic Society, Inc.)

electric models show that excitation thresholds are governed by a minimum stimulus charge, and are not sensitive to the fine structure of the stimulus waveform. This result should not be generalized to brief oscillatory (biphasic) waveforms, where the waveshape can play a critical role.

It is possible to explain the contour shapes of Fig. 7.4 in the light of S-D relationships. For the range of parameters represented in the figure, stimulus time constants ranges from about $0.1\,\mu s$ to more than $1000\,\mu s$ (see Fig. 2.34). For procedures (b) and (d), discharge time constants were in all cases below $3\,\mu s$, a value much smaller than the time constants typical of excitable membranes. These brief time constants result from the relatively low impedance at the electrode/skin interface associated with the capacitive discharge. In the case of curve (b), the low impedance results from the large contact area of the electrode; in the case of curve (d) the electrode is small, but it bypasses the high-impedance corneal layer of skin. In contrast, discharge impedance associated with curves (a) and (c) becomes very large as the stimulus voltage is reduced. At small capacitance values, the time constants associated with all four curves in Fig. 7.4 are sufficiently short to be within the charge-dependent region of the S-D curve. In this region, the vertical displacements of the threshold curves can be explained by other factors discussed in subsequent sections of this chapter.

Capacitive discharge thresholds developed by Swiss and Austrian groups form the basis of acceptability criteria published by the IEC (1987) (see Chapter 11). The European measurements are shown in Fig. 7.6, as summarized by Biegelmeier (1986). Curves *a* and *c* apply to Swiss experiments in which subjects grasped with both hands a cylinder of 80-mm diameter and length 100 mm. Capacitive discharges were applied to the cylinder electrode. Curve *a* is for adults, and curve *c* is for children. Curves *b* and *d*, for adults and children, respectively, apply to a 5-mm-diameter rod electrode placed on the fingertip. For the Austrian data, curve *e* applies to an 80-mm cylinder grasped in the hands, and curve *f* applies to a 4-mm-diameter wire grasped in the hands. Referring to IEC criteria discussed in Chapter 11, perception thresholds are bounded by the Austrian data for adults, and the Swiss data for children.

There are notable similarities in the contact electrode thresholds of Fig. 7.5 (curves *b* and *c*), and European data if one allows for differences in plotting formats. In both sets of data, thresholds are defined by constant charge at small capacitances, and then approach constant voltage at large capacitances. Also, in both sets of data, thresholds for large and small electrodes cross over when plotted against capacitance: at small capacitances, thresholds are lower for the smaller electrodes; at larger capacitances, the reverse is true. The curve shapes, and their relative displacements, can be understood if we account for the effects of discharge time constant and electrode size (see Sec. 7.6).

7.4   Suprathreshold Responses

The determination of perception thresholds provides a valuable but incomplete account of human sensitivity. It is necessary to characterize

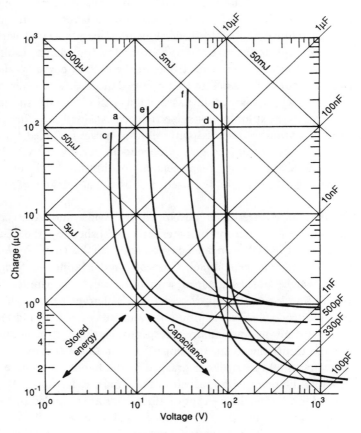

(a) 80 X 150 mm electrodes, adults
(b) 5 mm rod electrodes, adults
(c) As in (a), children age 4-14 yrs.
(d) As in (b), children age 4-14 yrs.
(e) 80 X 100 mm electrodes, adults
(f) 4 mm dia. wire electrodes

Figure 7.6   Perception thresholds for capacitor discharges. Curves *a–d* from Swiss measurement; curves *e* and *f* from Austrian measurements. (From Biegelmeier, 1986.)

sensation at suprathreshold levels in order to define levels at which people may be annoyed or disturbed by electric currents.

Magnitude Scaling

One method of characterizing subjective intensity is through magnitude-scaling experiments. In one procedure, a subject may be given a reference stimulus that is assigned some arbitrary numerical strength, and asked to rate other stimuli as a proportion or multiple of the reference. The absolute numbers themselves are meaningless, but their rate of growth with respect to the stimulus intensity reveals the subject's internal sensory interpretation.

In many cases, experimental data on subjective magnitude (M) versus stimulus intensity (I) is fitted to a simple power function of the form (Stevens, 1975)

$$M = \alpha I^{\beta} \tag{7.5}$$

where α and β are experimentally determined constants; α is a magnitude-scaling constant, and β indicates the rate of growth of sensory magnitude. Alternatively, the relationship is sometimes expressed in a form requiring three experimental parameters:

$$M = \alpha(I - I_0)^{\beta} \tag{7.6}$$

where I_0 is the threshold-level stimulus intensity. The advantages of the two forms have been debated over decades in a broad literature. The simpler form of Eq. (7.5) is preferred by many for electrical stimulation; it has been noted to give similar residual variance as compared with the more complex form of Eq. (7.6) (Reilly et al., 1982), and is argued to more closely convey the subjective impression of observers (Rollman, 1974). At any rate, the form of the two equations is similar for stimulus intensities sufficiently above the perception threshold. A variety of values for the exponent β have been reported for electrical stimulation (Tashiro and Higashyama, 1972); to some extent, these variations may be traced to features of the stimulus waveform, or to its locus or method of application.

Slopes of psychophysical power functions for electrical stimulation and two other modes of natural stimulation are illustrated in Fig. 7.7 (Reilly et al., 1982). The numerical values assigned to each curve represent the exponent β from Eq. (7.5). Values of β for capacitive discharge stimulation are significantly greater than that for heat or pressure, but not as great as with 60-Hz stimulation. Stimulation of the finger results in uniformly higher slopes (larger exponents) than for stimulation on the arm of leg. Additionally, it is noted that negative-polarity discharges produce lower exponents than positive-polarity discharges; on the average, the difference is 18%.

Categorical Scaling

Suprathreshold reactions may also be tested using *categorical* rating schemes, in which subjects choose from a list of affective (how unpleasant) or intensive (how strong) descriptors. While it might be expected that affective and intensive scales would be unfailingly linked, this is not always the case. Indeed, with the administration of narcotic analgesia, it is possible to reduce significantly the intensive aspects of painful electrical stimulation, without affecting its unpleasantness (Gracely et al., 1979).

Figure 7.8 provides an example of suprathreshold measurements including both subjective magnitude scaling and intensive descriptors (Reilly and Larkin, 1984). In these experiments, subjects were remarkably consistent in their use of the two scales, even though they were derived in separate procedures. As a result, it is possible to display both on a single graph. The suprathreshold data of Fig. 7.8 show that the growth of sensation magnitude is much greater than that of stimulus magnitude. When fitted by a power function, the data for perceived magnitude grows at about the 2.5 power of stimulus magnitude for stimulation of the finger, the 1.6 power for the arm, and the 1.4 power for the leg. This

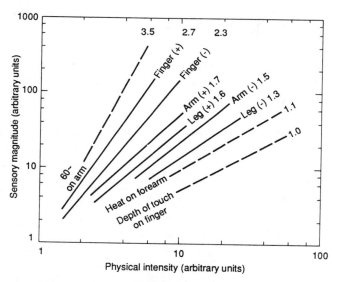

Figure 7.7   Psychophysical power function slopes for electrical, thermal, and mechanical stimulation. Solid lines represent spark discharges at negative (−) or positive (+) polarity. Dashed lines from Stevens et al. (1958, 1974). The relative positions of the functions are arbitrary on these coordinates. Slope values are given at upper right for each function. (From Reilly et al., 1982.)

range of exponents corresponds well with values reported in past studies using pulsed electrocutaneous stimulation (Sternbach and Tursky, 1965; Rollman 1974; Sachs et al., 1980).

The faster growth of sensation magnitude for the fingertip may be a consequence of its small volume relative to the arm or leg. Because of the volume constraint, current density becomes uniform along the finger beyond the stimulation point. This appears to result in a more spatially extensive sensory excitation: at suprathreshold levels, subjects report extended sensations along the finger.

A nonlinear trend is seen in Fig. 7.8, in which sensation magnitude tends to progress along a reduced slope at the larger stimulus levels. A similar description of a two-limbed power curve has been noted by others (Rosner & Goff, 1967). The reason for the deceleration may lie in the mechanism of neural response. Atkinson (1982) has proposed a general theory of interaction between excitatory and inhibitory neural processes, in which inhibition increases faster with stimulus level than does excitation. It is equally plausible, however, that the change in slope reflects a shift among populations of excitable neurons. The group of polymodal nociceptive fibers, for example, would respond only when their relatively

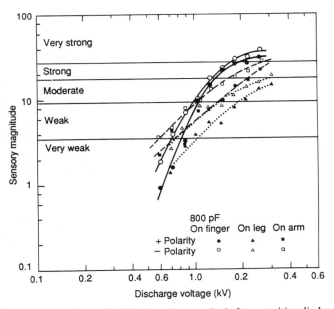

Figure 7.8 Growth of sensory magnitude for capacitive dischargers to three stimulation sites on the body. Vertical coordinate shows numerical magnitude judgments and ranges of adjectival rating categories. Composite data for eight subjects. (From Reilly and Larkin, 1984.)

high excitation thresholds are reached. Other groups of neurons may respond at various thresholds of the electrical stimulus. Because electrical stimulation has no specialized transduction mechanism, it is likely that the psychophysical function reflects a mixture of neural populations. The finding that both A and C fibers can participate in the process of pain transduction (Meyer and Campbell, 1981; Campbell et al., 1979) tends to support this hypothesis.

Stimulation levels corresponding to specific affective or intensive adjectives may be conveniently reported as multiples of perception thresholds. Suprathreshold multiples, if properly determined, are relatively consistent, despite wide variations in absolute thresholds due to factors such as electrode size, or intersubjective sensitivity variations. Table 7.2 lists affective response categories as multiples of the mean perception threshold as reported by investigators who used single monophasic pulse stimuli. In the data of Larkin and Reilly, for example, "tolerance thresholds" were determined by presenting the subjects with pairs of stimuli in an ascending sequence. The tolerance limit was reached when subjects indicated an unwillingness to accept a second stimulus, or to proceed to the next higher level. Tolerance limits determined in this manner are highly dependent on the context of the experimental procedure.

AC or Repetitive Stimuli

A brief electrical stimulus will elicit a single action potential (AP) on excited sensory neurons. The growth of sensory magnitude with increasing strength of a brief stimulus can be accounted for by the corresponding increase in recruitment of sensory neurons, including less-sensitive nociceptive fibers residing in deep subcutaneous strata. An additional factor, however, is involved in the growth of sensation for continued AC or repetitive pulse stimulation: sensory magnitude will also grow due to the higher CNS responses to repetitive APs (see Chapter 3). As the magnitude of an alternating current or pulse train is increased above the threshold level, there is a corresponding increase in the rate of AP production. As a result, continued AC or repetitively pulsed stimulation may be judged to be more intense than a single brief electrical stimulus of equal magnitude.

Table 7.3 lists suprathreshold multiples for AC or repetitive stimuli based on published experimental data. The dynamic range from perception to pain or tolerance for AC or repetitive pulsed stimulation is about one-half that for brief single-pulse stimuli (Table 7.2).

Figure 7.9 illustrates sensory power functions for stimulation by 0.1-ms pulses, delivered at a 60-Hz rate (from Rollman, 1974). The parameter n

Table 7.2. *Average suprathreshold multipliers for single monophasic pulses*

Body locus	N	Electrode	Stimulus[a]	Multiple above perception			Reference
				Unpleasant	Pain	Tolerance	
Finger	8	0.8-mm diam.	CD, 800 pF	2.3	3.5	7.1	Reilly and Larkin (1984)
Forearm	8	0.8-mm diam.	CD, 800 pF	3.5	5.5	11.0	Reilly and Larkin (1984)
Leg	8	0.8-mm diam.	CD, 800 pF	3.1	—	9.1	Reilly and Larkin (1984)
Finger	124	Tapped electrode	CD, 200 pF	2.6	—	—	Larkin and Reilly (1986)
Triceps	6	4-cm² carbon sponge	Pulse, 5–10 µs	—	4.3	8.9	Alon et al. (1983)
Triceps	6	4-cm² carbon sponge	Pulse, 20–1000 µs	—	7.4	10.5	Alon et al. (1983)

[a]CD, capacitive discharge stimulus.

Table 7.3. *Average suprathreshold multipliers for AC or repetitive pulse stimulation*

Body Locus	Subgroup	N	Electrode	Stimulus	Multiple above perception			Reference
					Unpleasant	Pain	Tolerance	
Fingertip		6	Tapped electrode	60-Hz AC	1.8	2.4	3.5	Reilly and Larkin (1987)
Forearm		40	1-cm diam.	Pulse train, 100 Hz	—	2.0	5.7	Rollman and Harris (1987)
Fingertip		2	*	100–10kHz AC	—	2.9	4.0	Hawkes and Warm (1960)
Fingertip	M	197	Tapped electrode	10-kHz AC	—	—	2.5[a]	Chatterjee et al. (1986)
Fingertip	F	170	Tapped electrode	10-kHz AC	—	—	2.1[a]	Chatterjee et al. (1986)
Forearm		12	Concentric ring	60-Hz AC	4.2	6.6	11.8	Higgins et al. (1971)

*Information sketchy or not available.
[a]Chatterjee threshold described as "pain," but functionally equivalent to "tolerance."
Note: N = number of subjects tested.

indicates the number of pulses in the train; β indicates the sensory magnitude power law, according to Eq. (7.5). It is seen that the power exponent grows with increasing n. More striking is the increase in sensory magnitude with increasing n for pulse trains at the same stimulus intensity. At 3.0-mA current level, for example, perceived sensory magnitude increases sixfold as the number of pulses increases from 1 to 30.

Dynamic Range of Electrical Stimulation

The data in Table 7.2 and 7.3 show that the dynamic range for electrocutaneous stimulation is very small. For stimulation by single pulses, the ratio of pain to perception threshold is only 3.5 or 7.4, depending on the locus of stimulation; for AC or repetitive stimulation the dynamic range is even less. We might contrast the measured dynamic range of electrical stimulation with that for hearing or pressure sensitivity, both of which have dynamic ranges of about 100,000 to 1. We can thus appreciate

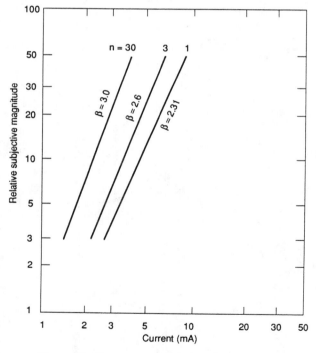

Figure 7.9   Sensory power functions for train of pulses (width = 0.1 ms), delivered at 60-Hz rate; n indicates number of pulses in train; β indicates power function exponent. (Adapted from Rollman, 1974, in *Conference on Cutaneous Communication Systems and Devices*, pp. 38–51, reprinted by permission of Psychonomic Society, Inc.)

the importance of perception threshold measurements in electrical sensitivity studies; if we can measure the perception threshold accurately, we know that a relatively small increase will result in a strong perceptual effect.

The ordering of tolerance values on the finger, arm, and leg has been found to be consistent among subjects who may differ widely in absolute threshold levels (Reilly et al., 1982). Overall, the fingertip is more sensitive than the leg by nearly a factor of 2, and the arm takes a midway position. This ordering contrasts with measurements taken at the perception threshold (see Sec. 7.7), in which electrical stimulation is more easily detected on the forearm than on the fingertip. One factor that may account for this difference is that current flow through the finger is volume-limited to a much greater degree than in the arm or leg. Consequently, at points distant from the introduction of current, current density may be higher in the finger than at equivalent distances on the arm or leg. Subjects experience suprathreshold stimulation as "radiating" through the finger and hand. This indicates that current density remains relatively high over a broad region, and that more than one population of afferent neurons may be involved. On the leg or arm, in contrast, the sensation appears to be more limited spatially.

There is a wide variation among individual tolerance limits. While part of this variation may reflect individual differences in sensitivity, it is clear that "tolerance" is not simply a sensory limitation. Subjects with high tolerance limits may have a detached attitude about the possibility of injury, while those with low limits may be more cautious or fearful. This variation does not seem tied to an individual's past experience or familiarity with electrical shocks. Indeed, the author has encountered subjects who had no prior experience, but were willing to accept levels substantially higher than those set by the highly experienced investigators. The acceptance of a large stimulus does not necessarily imply a lack of cutaneous sensitivity. Individuals who tend to have very nearly the same thresholds, and give numerical magnitude estimates in the same range, may have markedly different tolerance levels.

In view of the preceding comments, the tolerance limits in Tables 7.2 and 7.3 cannot be taken as absolute limits on an individual's capacity for painful stimulation. The concept of tolerance has empirical application only as a relative, context-dependent measure. Implied demand, or social expectations, influence a person's willingness to cooperate in a scientific endeavor. For some individuals, neither pain nor the potential for skin injury deters their voluntary exposure to electric shock, if they believe that there is some benefit to be gained.

7.5   Stimulus Waveform Factors[†]

We have seen in Sec. 7.3 that neural excitation thresholds are markedly sensitive to the duration of a monophasic stimulus current. Other features of the stimulus waveform, such as biphasic properties and pulse repetition patterns, can be equally important. A number of experimental studies have provided pertinent information on waveform effects. A general framework for evaluating such effects can be provided by appropriate neural excitation models, such as the SENN model discussed in Chapter 4.

Biphasic Stimuli

The current reversal of a biphasic pulse can suppress a developing AP that was elicited by the initial phase. To compensate for the reversal, a biphasic pulse must present a higher initial current. When thresholds are measured in terms of the initial current, the biphasic pulse may therefore have a higher threshold than a monophasic pulse. According to Fig. 4.15, the threshold elevation depends on the phase duration and the time delay for current reversal. Thresholds are elevated more when the pulse is short, and when the current reversal immediately follows the initial pulse. If the phase reversal is delayed by 100 μs or more, calculations show only small effects on the biphasic pulse threshold.

Experimental results with biphasic stimuli generally confirm the modeled threshold relationships of Fig. 4.15. One study tested supra-threshold sensory sensitivity of the hand to biphasic pulses having durations of 20 to 50 μs, and with delays from 10 to 50 μs (Bütikoffer and Lawrence, 1978). Although it is difficult to compare absolute thresholds with model results, we can compare the relative sensitivity for different waveforms. Table 7.4 lists the relative thresholds of biphasic pulse doublets having different phase durations (t_p) and interpulse delays (δ). The experimental and model results agree within a few percent. Other experimental data confirm that a biphasic stimulus has a reduced efficacy for neuromuscular stimulaton as compared with a single monophasic pulse of the same phase duration (Gorman and Mortimer, 1983; van den Honert' and Mortimer, 1979a).

The finding that thresholds are elevated for a single biphasic pulse should not be interpreted as implying that thresholds for all oscillatory stimuli are necessarily elevated above monophasic stimuli of the same phase duration. Indeed, if a biphasic stimulus is repeated as an oscillatory waveform, thresholds may decrease with successive oscillations to the

---

[†]Portions of this section have been adapted from Reilly (1989).

Table 7.4. *Relative tresholds of biphasic stimuli:*
comparison of model and experiment

Waveform parameters		Threshold ratio (Q_1/Q_2)	
$t_p(Q_1, Q_2)$ (μs)	$\delta(Q_1, Q_2)$ (μs)	Model	Experiment
20, 20	10, 50	1.20	1.19
50, 50	10, 50	1.10	1.12
50, 20	50, 50	1.17	1.23

Notes: t_p = phase duration; δ = interpulse delay.
Source: Experimental data from Bütikofer and
Lawrence (1978).

point that they eventually fall below the single pulse monophasic threshold
(see Sec. 4.6).

Repetitive Stimuli

We have seen in Chapter 4 that the threshold for exciting a single
AP by a sequence of pulses may be lower than the threshold for a single
stimulus pulse. Repetitive stimuli can also enhance sensory (and motor)
response if multiple APs are generated. In either case there is an integration
effect of the multiple pulses. In the first case, the integration takes place
at the membrane level. In the second case, response enhancement takes
place at the CNS level for neurosensory effects, and at the muscle level
for neuromuscular effects (see Chapters 3 and 8).

Multiple pulse threshold effects predicted by a neuroelectric model are
illustrated for two pulses in the lower section ($M < 1$) of Fig. 4.16. The
effects are most pronounced for short pulses and short interpulse delays.
The neuroelectric model also predicts that thresholds consistently fall
as the pulse number is increased, until a minimum plateau is reached.
Figure 4.21 suggests a temporal summation process that effectively sums
repetitive pulses, with an integration time that is roughly four times the
S-D time constant. At a delay of 500μs, the neuroelectric model shows no
measurable integration effect. Similar results with neuroelectric models
have been reported by others (Bütikofer and Lawrence, 1979).

Tasaki and Sato (1951) experimented with myelinated toad nerve to
examine the shift in threshold of a stimulus pulse caused by a prior conditioning pulse of either the same or the opposite polarity. The duration
of the two pulses was 10μs, and the interpulse delay varied from 0 to
200μs. Results from these experiments are shown in Table 7.5; tabulated
values express the threshold of a pulse doublet as a multiple of a single-

Table 7.5. *Threshold shifts for monophasic and biphasic pulse doublets* $(t_p = 10\,\mu s)$

Delay (µs)	Theory		Experiment ·	
	Mono-phasic	Biphasic	Mono-phasic	Biphasic
0	0.54	1.99	0.60	1.40
20	0.65	1.21	0.76	1.36
50	0.72	1.07	0.70	1.30
100	0.80	1.02	0.82	1.15
200	0.91	1.00	0.95	1.04

Source: Theoretical data derived from SENN model (see Chapter 4). Tabulated values are the threshold of a two-pulse doublet expressed as a ratio of the single-pulse threshold. Experimental data derived from Tasaki and Sato (1951).

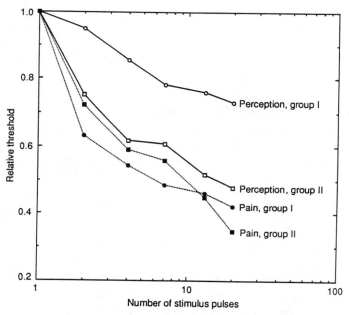

Figure 7.10   Relative thresholds versus number of stimulus pulses at 100-Hz rate; pulse width = 0.5 ms. Designations I and II refer to averages for two different groups of body loci. Data extracted from Gibson (1968).

pulse threshold. Also shown are theoretical predictions taken from
Fig. 4.16. In this instance, the experimental data and theoretical prediction agree quite well. Even for pulse delays as long as 5 ms, cutaneous
perception threshold reductions on the order of 7–8% have been reported
for continuous pulse trains with pulse durations from 100 to 400 μs
(Hahn, 1958).

The thresholds of perception as well as suprathreshold reactions both
decline with increasing number of pulses in a stimulus train. Figure 7.10
illustrates data of Gibson (1968) obtained with monophasic pulses
individually having durations of 0.5 ms, and delivered at a rate of 100 Hz.
The vertical axis indicates the threshold normalized by the value obtained
with $n = 1$. The plotted curves represent averages of thresholds obtained
for stimulation at two different groups of body loci: Group I applies to
the fingertip, lip, and ear; group II applies to the forearm and back. These
groupings were used because their thresholds appeared to have different
functional dependencies with pulse number. The coefficient of variation
of the data represented the figure was typically 0.1. Notermans (1966)
related the threshold of pain to the number of pulses, each having widths
of 5 ms, and delivered at a rate of 100 Hz. His results, when plotted on a
format similar to that in Fig. 7.10, nearly coincide with the pain data of
Gibson.

We have seen, in connection with Fig. 4.21, that the integration process
of a theoretical neuron model effectively sums pulsed stimuli over a
duration roughly four times the S-D time constant. The data of both
Gibson and Notermans demonstrate thresholds that continue to decline
with increasing n, out to the maximum duration of the pulse train (about
200 ms). This observation indicates a temporal integration time constant
of roughly 50 ms. It is difficult to justify a time constant that long solely
from the properties of a single neuron, whose S-D time constants are
typically a fraction of a millisecond (see Table 7.1). The precise mechanism
of temporal integration in electrical perception remains poorly
understood.

Besides lowering thresholds relative to a single pulse, multiple pulses
that are individually near or above threshold can generate a train of APs.
The consequences can be elevated sensory magnitudes for afferent
stimulation, or enhanced muscle contraction for efferent stimulation.
Sensory magnitude and unpleasantness in both A and C fibers are related
to several factors, including the AP frequency and burst duration
(Campbell et al., 1979; Gybels et al., 1979). Figure 7.11 illustrates the
effects of sensory enhancement due to multiple pulses that are individually
above threshold. The stimulus was designed to study 60-Hz electric field

induction effects, in which a train of capacitive discharges is produced on individual half-cycles of a 60-Hz waveform, i.e., 120-Hz repetition rate (Reilly and Larkin, 1987). Additional details concerning this figure are given in Sec. 9.4. The horizontal axis gives the number of stimuli, and the vertical axis gives the perceived magnitude (an arbitrary, but consistent, scale assigned by subjects). Also shown are regions of subjects' qualitative descriptors. In this experiment, sensory magnitude grows at about the 0.8 power of the number of stimuli. Notice that a stimulus that is perceptible, but qualitatively "neutral," can become highly unpleasant when presented as a repetitive train.

Another study (Rollman, 1974) found that sensory magnitude scaled approximately as the 0.5 power of the number of pulses up to $N = 30$ (pulse width $= 0.1$ ms, repetition interval $= 16$ ms; see Fig. 7.9). Experiments with biphasic pulse trains (Sachs et al., 1980) found that sensory magnitude scaled with N to the 0.8 power; these results were likely a combination

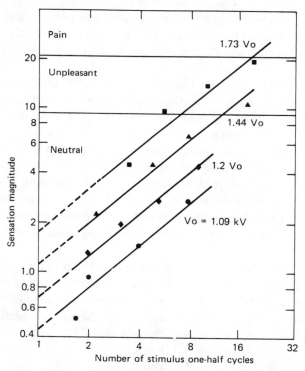

Figure 7.11   Effect of the number of stimulus cycles on sensation magnitude for 60-Hz AC electrical field-inducted stimuli, discharge capacitance $= 100$ pF. (From Reilly and Larkin, 1987.)

of membrane integration and multiple AP effects, because the experimental repetition interval (50 μs) was much too small to support on AP on each pulse.

Sinusoidal Stimuli

Our discussion thus far has primarily treated pulsatile stimuli. Acceptability standards for electric currents have traditionally been developed for sinusoidal stimuli. It is useful to relate thresholds for sinusoidal currents to pulsatile stimulus waveforms.

The sinusoidal stimulus is a special case biphasic waveform, which has been discussed in Chapter 4, and also in Sec. 7.4. We have seen the current reversal of a biphasic waveform can reverse the development of membrane excitation caused by the initial phase of the stimulus. As a result, a biphasic waveform may be associated with a higher excitation threshold than a monophasic current pulse of the same phase duration. The degree of desensitization caused by the biphasic current reversal is increased as the duration of the stimulus is reduced. This phenomenon partially accounts for the high-frequency upturn noted in strength-frequency (S-F) curves for neural excitation (Fig. 4.20). The S-F curve reaches a minimum plateau as the frequency is reduced. If the frequency is reduced further, the threshold begins to rise again as the time rate of change of the sinusoidal stimulus becomes small.

The overall shape of the frequency sensitivity curve has been developed in Chapter 4 using a neuroelectric model for myelinated nerve. A mathematical fit to the thresholds determined from the model expressed by Eqs. (4.35)–(4.37) is based on an analogy to the exponential S-D expression for monophasic stimulation. The expression includes terms to account for the threshold rise at high and low frequencies. By analogy with the S-D expression, a high-frequency parameter f_e specifies the frequency at which the threshold upturn occurs; a similar term f_0 applies to the low-frequency upturn. The low-frequency term is constrained to a maximum value, to account for the fact that stimulation is possible with direct current. A maximum value of $K_L \leqslant 4.6$ may be inferred from the ratio of the perception threshold for continuous DC to that for 60-Hz AC (Dalziel and Mansfield, 1950b).

If the sinusoidal waveform is presented as a continuous stimulus, we have seen that the degree of membrane depolarization may increase with each successive stimulation cycle because of the nonlinear action of the excitable membrane. With increasing stimulus duration, the threshold of an AC stimulus may eventually be reduced below that of a monophasic stimulus of the same duration.

Table 7.6. *Empirical strength/frequency curve parameters*

Species	Locus of stimuli	N	Electrode	Response	I_0 (mA)	f_e (Hz)	f_0 (Hz)	Reference
Human (M)	Hand	115	Hand-held wire	Perception	1.1	500	<60	Dalziel (1954)
Human	Arm	2	0.27 cm²; elect. paste (concentric)	Perception	0.11	1000	50	Anderson and Munson (1951)
Human	Thorax	8	1.0 cm²	Perception	0.2	100	<20	Geddes et al. (1969)
Human	Neck	8	25 cm²	Perception	0.5	150	<20	Geddes et al. (1969)
Human	Fingertip	2	*	Perception	0.4	500	<100	Hawkes and Warms (1960)
				Tolerance	1.0	400	<100	Hawkes and Warms (1960)
				Pain	1.8	460	<100	Hawkes and Warms (1960)
Human	Hand		Large-area grip	Tetanus		600	<100	Dalziel (1954)
Frog	Sciatic nerve	*	In vitro, elect. spac. = 2 mm	AP		1500	~30	Hill et al. (1937)
			elect. spac. = 25 mm	AP		533	~30	Hill et al. (1937)
Rat	Gast; tib. musc.	8	Subcutaneous	Muscle twitch		200	<100	LaCourse et al. (1985)
Toad	Mye. nerve	*	In vitro	AP		200	<100	Tasaki and Sato (1951)
Frog	Mye. nerve	*	In vitro	AP		500	28	Wyss (1963)

*Information incomplete or not available.

While we can suggest a functional form of an S-F curve from theoretical principles, the parameters of this functional relationship need to be determined experimentally. Table 7.6 summarizes experimental S-F data determined from neurosensory and neuromuscular responses to sinusoidal stimulation. The table lists experimental estimates of f_e and f_0 from Eqs. (4.36) and (4.37). The table also lists the minimum absolute threshold for electrocutaneous stimulation. For proper comparison of the absolute threshold, one should take into account the electrode size (see Sec. 7.6).

The values of f_e in Table 7.6 encompass a rather wide range, with a median (and also geometric mean) of 500 Hz. Part of the variation seen in f_e values might be due to the electrode size and its configuration. In the experiments of Hill et al. (1937), f_e changed from 1,500 to 533 Hz when a bipolar electrode spacing was changed from 2 to 25 mm. For the electrocutaneous sensory data reported in Table 7.6, there appears to be a tendency toward large f_e values with smaller electrodes, but there are insufficient data to establish a clear relationship. The direction of the effect of electrode size on f_e is consistent with the corresponding effect on the S-D time constant, τ_e, which increases monotonically with electrode size in neural stimulation (Pfeiffer, 1968; Davis, 1923), and also with cardiac stimulation (see Sec. 6.3).

The form of the S-F curve suggests that the perception threshold current ought to increase indefinitely as its frequency is increased. As a practical matter, however, if the electrical threshold is raised high enough, sensory thresholds will be dominated by thermal perception. Since heating effects are largely independent of frequency, perception thresholds cease to rise at that point. Experimental data show that, at a frequency of about 100 kHz, perception thresholds reach a maximum plateau due to thermal perception (Chatterjee et al., 1986; Daziel and Mansfield, 1950b). Data of Chatterjee and colleagues, shown in Fig. 7.12, indicate a threshold plateau due to thermal effects beyond 10^5 Hz.

Figure 7.13 illustrates S-F curves from several of the references listed in Table 7.6. Also drawn is a theoretical curve defined by Eqs. (4.35)–(4.37), using the empirical constants $f_e = 500$ Hz and $f_0 = 30$ Hz. In general, the analytic expression provides a good fit to the functional form of the S-F curves, provided that the constants f_e and f_0 are properly selected.

The analytic S-F expression provides a good fit to experimental data in most cases. However, that particular representation may not apply universally. In patients with diabetic neuropathy, frequency sensitivity reported by Katims et al. (1987) deviates significantly from the analytic form. In Katims' patients, thresholds were typically elevated above the values noted for normal individuals. It was especially notable that

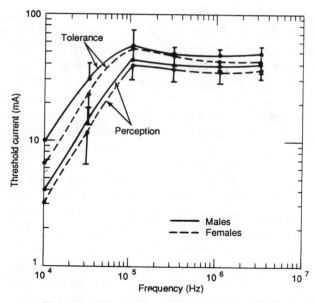

Figure 7.12   Perception and tolerance thresholds for AC stimulation of high frequency. Finger contact on 25-mm² plate. (Adapted from Chatterjee et al., 1986, © 1986 IEEE.)

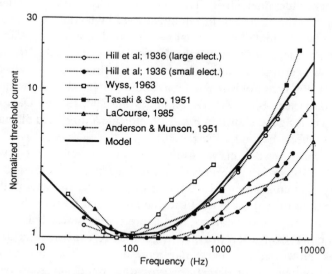

Figure 7.13   Strength-frequency curves for perception of sinusoidal currents applied cutaneously. Symbols represent experimental data from several investigators. Solid curve is analytic expression with $f_e = 500$ Hz and $f_0 = 30$ Hz.

perception thresholds on the hands and feet of the diabetic patients was significantly smaller at 2,000 Hz than at 250 Hz; the direction of this frequency sensitivity is opposite to that found in normal people. The electrophysiological basis for this finding is presently unknown.

Polarity Effects

Measurements taken under a wide variety of experimental conditions demonstrate lower thresholds for monophasic stimuli of negative polarity than with positive polarity. In the author's experiments, for example, perception thresholds for negative polarity stimuli averaged 23% lower than those for positive polarity (Reilly and Larkin, 1983). Similar polarity differences (about 30%) were found when the discharges were applied to corneum-piercing needles. It thus can be concluded that polarity-sensitive mechanisms do not depend on the properties of the corneum.

Similar polarity sensitivity has been found by other investigators. Gibson (1968), for example, reported cathodal perception thresholds to be about 50–75% of those for anodal pulses. Girvin and colleagues (1982) tested perception thresholds for monophasic pulses having durations ranging from 63 to 1,000 μs. At the shorter durations, the ratio of cathodal to anodal thresholds was 0.77; at the longer durations, the ratio was 0.63. Polarity sensitivity has also been noted for simulation of cardiac tissue (refer to Chapter 6), with polarity sensitivity similar to that found for sensory excitation.

We have seen in Chapter 4 that polarity sensitivity is predicted from theoretical considerations of neural excitation. With anodal stimulation, the distribution of current influx and efflux along the axon dictates that the nerve may be excited by current from an external electrode of either polarity, but that a cathodal electrode should be significantly more effective (Fig. 4.12). A model for excitation near the terminal structures of nerves also demonstrates that cathodal thresholds are lower than anodal thresholds, and that the magnitude of the polarity effect is similar to that noted in sensory experiments (Fig. 4.19).

Biphasic stimuli may also exhibit a sensitivity to the polarity of the initial phase, but the polarity effects are generally less than with single monophasic pulses. The neuroelectric model described in Chapter 4, for example, shows that with short-duration biphasic pulse doublets, the threshold of excitation is lower if the initial phase is cathodal rather than anodal. A similar effect was seen in the experimental biphasic sensory thresholds for short-duration biphasic pulses.

7.6   Electrodes and Current Density

In Chapter 4 we showed that the relevant force contributing to electrical excitation is the electric field within the biological medium. The field has the maximum effect when it is aligned along the long axis of an excitable cell (neuron, nerve ending, muscle fiber). The electric field, E, is simply related to the current density, J, by $E = J/\sigma$, where σ is the conductivity of the medium. It can be rather difficult to determine the spatial distribution of E or J in a typical experimental situation involving electrodes within or contacting biological materials. Frequently, one has only a gross estimate of the average current density at the surface of the stimulating electrode.

Cutaneous Electrodes

The relationship between electrode area and perception sensitivity was determined for capacitive discharges to contacts of different sizes (Reilly and Larkin, 1984; Reilly et al., 1983). To avoid artifacts of electrode placement, the precise point of contact was varied from one trial to the next, within a perimeter defined by the largest electrode. A low capacitance (100 pF) was used in these tests so that stimulus discharges would have time constants within the charge-dependent region of the strength-duration relationship, regardless of the electrode size. The electrode was either dry, or treated with electrode paste.

Results for untreated skin (Fig. 7.14) show that electrode size is a critical parameter only for diameters greater than about 1 mm. Below this point, sensitivity is nearly constant. We hypothesize that for dry skin, current is

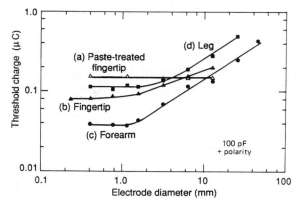

Figure 7.14   Effect of electrode contact size on sensitivity to discharge from 100-pF capacitor. Curve (*a*) applies to paste-treated skin; the others apply to dry skin. (From Reilly and Larkin, 1985a.)

conducted through discrete channels beneath a contact electrode, so that effective current density depends on the number and size of these channels and not simply on the electrode size. This hypothesis is consistent with the observation that current concentration is not homogeneous in the corneal layer of skin (Mueller et al., 1953). As discussed in Chapter 2, the number and size of these current channels are unknown, but tests with electroplating on the skin surface suggest a density of about one channel per square millimeter (Saunders, 1974). This estimate corresponds very well with the plateau in Fig. 7.14. For electrodes smaller than $1\,mm^2$, a single channel of excitation may be produced in dry skin. For larger electrodes, the discharge current may pass through the dry epidermis in more than one place. If so, current density would be constant for any electrode smaller than about $1\,mm^2$, and would decrease only when the electrode is made to cover at least two current channels.

If current density is a critical parameter for human sensitivity, then thresholds eventually must rise as electrode area is increased. This effect is expected whether the current is uniformly distributed or is concentrated in small current channels. But the rate at which thresholds increase may

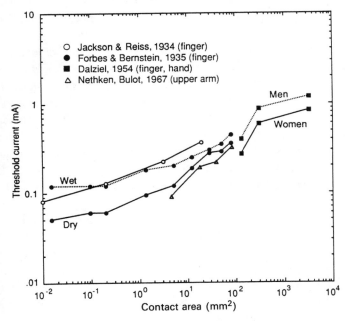

Figure 7.15   Sensory thresholds for 60-Hz stimulation versus area of contact. Data of Jackson and Reiss apply to pain threshold; other data apply to perception threshold.

be diagnostic of the spatial distribution of current. For the data shown in Fig. 7.14, thresholds above 1-mm diameter increase as the one-third power of area on the forearm and leg, and as the one-sixth power on the fingertip. Slopes falling between the one-sixth and one-third powers are evident in the data of previous investigators who used 60-Hz stimulation, as shown in Fig. 7.15 (Jackson and Riess, 1934; Forbes and Bernstein, 1935; Nethken and Bulot, 1967; Dalziel, 1954).

While cutaneous sensory thresholds rise as the electrode size is increased, the rate of rise is much less than would be expected if thresholds were simply inversely proportional to electrode area. The lack of area proportionally is most likely related to the nonuniform conduction of current beneath a cutaneous electrode. While electrode paste treatment or skin hydration would be expected to reduce this effect, the current distribution beneath even a treated electrode is not uniformly distributed (refer to Sec. 2.2). Furthermore, the presence of conduction "hot spots" in treated skin cannot be ruled out.

When the skin is treated with electrode paste, the area relationship is markedly altered. In Fig. 7.14, for example, thresholds on the paste-treated fingertip are elevated above the dry-skin thresholds by a factor of about 2 for small-diameter contacts. As the electrode size is increased, the thresholds of dry and paste-treated skin eventually cross each other. The data of Forbes and Bernstein shown in Fig. 7.15 also demonstrate the fact that paste treatment elevates thresholds by a factor of 2 above that for a small, dry electrode, and that the thresholds of dry and treated electrodes eventually converge as the electrode size is increased. The elevation of thresholds for hydrated skin can be explained by the fact that treatment would tend to result in a more even distribution of current beneath the contact electrode, thereby reducing (but not eliminating) current-density "hot-spots" and electrode-edge current enhancement. Furthermore, current is more readily conducted beyond the edges of the electrode if the skin is hydrated. The crossover of thresholds at larger contact areas is more difficult to explain.

The role of skin impedance breakdown in electrical sensation should not be discounted. It has been noted that, with rising applied voltage, the sensation of electric shock on the dry, untreated skin is accompanied by a dramatic reduction in impedance (Nute, 1985; Gibson, 1968; Mason and Mackay, 1976; Mueller et al., 1953). Such a decrease is likely to be caused by skin breakdown in discrete locations, leading to a sudden increase in current density in those locations.

Based on the preceding, we can postulate thresholds for brief monophasic stimuli delivered to the finger or hand as given in Table 7.7.

Table 7.7. *Calculated sensory thresholds for single monophasic stimuli to finger or hand (threshold in μC for short-duration currents)*

Electrode area (cm^2)	Finger, hand threshold (μC)				Forearm threshold (μC)			
	Percept.	Annoy.	Pain.	Tol.	Percept.	Annoy.	Pain.	Tol.
0.01	0.11	0.25	0.39	0.78	0.09	0.31	0.48	0.97
0.1	0.16	0.37	0.56	1.14	0.19	0.66	1.04	2.08
1.0	0.23	0.53	0.81	1.63	0.41	1.43	2.24	4.49
10.0	0.34	0.78	1.19	2.41	0.88	3.08	4.84	9.68
100.0	0.51	1.17	1.79	3.62	1.89	6.62	10.40	20.79

Notes: Median threshold for adults. Contact to finger or hand. Brief-duration (< 20-μs) currents. Area power law $= \frac{1}{6}$ for finger/hand; $\frac{1}{3}$ for arm. Suprathreshold multiples from Table 7.2.

The smallest listed "perception" threshold of 0.11 μC applies to a 0.01-cm^2 contact electrode; that value has been determined from the capacitive discharge measurements of Reilly and Larkin (1987), as indicated by curve (*c*) at 100 pF capacitance in Fig. 7.4. The corresponding value on the forearm has been taken from the body-location sensitivity data, as discussed in Sec. 7.7. Values at other contact areas have been calculated by assuming that thresholds rise as the one-sixth power on the fingertip or hand, and as the one-third power on the forearm, as discussed previously. The thresholds for unpleasantness, pain, and tolerance have been determined from Table 7.2. One postulate used to derive Table 7.7 is that the area power law determined for perception currents also applies to the suprathreshold categories of annoyance, pain, and tolerance. A second postulate is that the suprathreshold multiples of Table 7.2 are invariant with respect to electrode size. The perception threshold for the finger/hand in Table 7.7 compares favorably with the European data using capacitive discharges (Fig. 7.6). Note that in the Swiss large-electrode data, the contact area is probably on the order of 100 cm^2.

Current Density Considerations

In many instances, it is important to determine current density thresholds for peripheral nerve stimulation. For example, in magnetic resonance imaging (MRI) procedures, allowable magnetic field exposure guidelines have been related to the induced current density needed to stimulate peripheral nerves (refer to Chapter 9). Because of the afore-mentioned difficulties, estimates of threshold current density thresholds calculated from the average current density beneath a cutaneous electrode may be unrealistically low.

Current density thresholds have been calculated from a theoretical myelinated nerve model (Sec. 4.4). For fiber diameters of 5, 10, and 20 μm, the current density thresholds were calculated as 0.492, 0.246, and 0.123 mA/cm^2, respectively, for medium conductivity of 0.2 S/m (Table 4.2). In Chapter 9, these theoretical thresholds are compared with experimental values applying to stimulation by externally applied magnetic fields, or by in-vitro nerve stimulation by uniform electric fields. In these experiments, difficulties due to nonuniform electrode conduction are not present. If allowances are made for variations in stimulation parameters, the experimental values are found to be within the range predicted by the theoretical model. Accordingly, the value of about 0.1 mA/cm^2 appears, on the basis of available data, to represent a reasonable estimate of the current density threshold for stimulation of the most sensitive peripheral nerves.

7.7   Body Location Sensitivity

It is well known that different parts of the human body have different sensitivities to tactile and thermal stimulation. Figure 7.16, for example, displays the rank-orders of body-part sensitivity for three measures of tactile sensitivity: pressure, two-point discrimination, and point localization (Weinstein, 1968). The rank-ordering has been tabulated for a mixed population of male and female subjects. Females were found to have consistently lower pressure thresholds on all body parts than males, as illustrated in Fig. 7.17; the vertical scale expresses the logarithm of the sensitivity relative to 0.1 mg. The gender difference is appreciable, reaching 10:1, for example, on the back, forearm, and leg. No such difference appears in Weinstein's data for spatial discrimination. We shall return to a discussion of male/female differences in sensitivity in Sec. 7.10, where we show that, on the average, females are more sensitive to electrical stimulation as well, and we advance a hypothesis for these differences. Relative pressure sensitivity in absolute units varies in Fig. 7.17 by nearly 10:1 from the most sensitive regions (the face), to the least sensitive (the foot).

Electrical sensitivity also varies considerably across different body parts, as demonstrated in the experiments conducted by the author and colleagues (Reilly et al., 1982; Reilly and Larkin, 1984). In these experiments, sensitivity thresholds were measured for five individuals at each of seven body loci: forehead, cheek (both right and left), tip of third finger, thenar eminence (portion of the palm at the base of the thumb), underside of forearm, back of the midcalf, and a point 3 cm above the ankle bone. The stimuli were discharges from a 200-pF capacitor. To minimize the influence of concomitant tactile stimulation, a 1.6-mm-diameter contact electrode was used, flush-mounted in a large plastic holder. At each body locus, the electrode contacted a slightly different spot in order to avoid an average measurement that might have been unduly influenced by spots of unusually high or low sensitivity.

Figure 7.18 illustrates the results, ordered in decreasing sensitivity from left to right. Results have been normalized in relation to the fingertip threshold for each subject. The fingertip was chosen for this purpose because it showed smaller variability than the other points, both among and within subjects. Electrical sensitivity across body loci is seen to vary by about 4:1, with points on the face being the most sensitive, the lower leg the least sensitive, and the finger about in midrange.

Five of the seven locations tested in this experiment were included in Weinstein's study of cutaneous tactile sensitivity; a sixth, the thenar eminence, can be compared with Weinstein's measurement on the palm. By comparing Figs. 7.17 and 7.18, it can be seen that the rank-ordering of pressure and electrical sensitivity are similar. On the basis of these

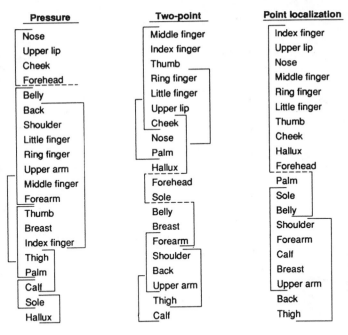

Pressure

- Nose
- Upper lip
- Cheek
- Forehead
- Belly
- Back
- Shoulder
- Little finger
- Ring finger
- Upper arm
- Middle finger
- Forearm
- Thumb
- Breast
- Index finger
- Thigh
- Palm
- Calf
- Sole
- Hallux

Two-point

- Middle finger
- Index finger
- Thumb
- Ring finger
- Little finger
- Upper lip
- Cheek
- Nose
- Palm
- Hallux
- Forehead
- Sole
- Belly
- Breast
- Forearm
- Shoulder
- Back
- Upper arm
- Thigh
- Calf

Point localization

- Index finger
- Upper lip
- Nose
- Middle finger
- Ring finger
- Little finger
- Thumb
- Cheek
- Hallux
- Forehead
- Palm
- Sole
- Belly
- Shoulder
- Forearm
- Calf
- Breast
- Upper arm
- Back
- Thigh

] Items grouped within a bracket not significantly different.

- - - - All upper items significantly different from all lower items.

Figure 7.16   Rank-order of tactile sensitivity across various body parts, for three measures of sensitivity. (From Weinstein, in D. R. Kenshalo, ed., *The Skin Senses*, 1968. Courtesy of Charles C Thomas, Publisher, Springfield, Ill.)

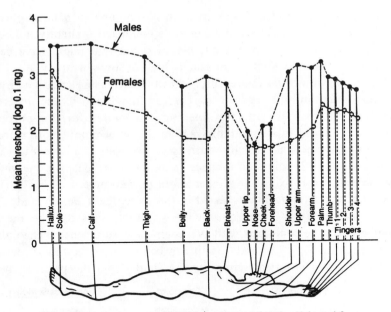

Figure 7.17   Pressure sensitivity for males and females (Adapted from Weinstein, in D. R. Kenshalo, ed., *The Skin Senses*, 1968. Courtesy of Charles C Thomas, Publisher, Springfield Ill.)

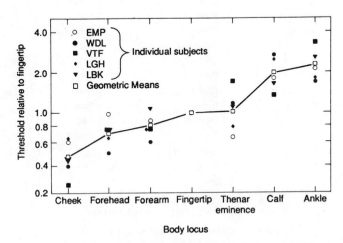

Figure 7.18   Threshold data for seven body loci, normalized to the threshold for the fingertip, flush-tip contact electrode, 200-pF capacitance. (From Reilly and Larkin, 1984.)

results, it is reasonable to expect that electrical sensitivity at other points of stimulation will correlate highly with tactile pressure sensitivity.

Although the ankle is about one-half as sensitive to electrical stimulation as the fingertip, this difference should not be construed to mean that the lower extremities are in all conditions protected from accidental exposure that cannot be felt on the fingertip. There may be situations in which it is natural to use some force in touching an electrical source with the fingertip, but not with the ankle. In such cases, the force of the finger tap can mask an electrical sensation (see Sec. 7.9).

As a general observation, we observe that the further the site of stimulation is from the brain, the higher the threshold. Whether this observation implies some connection with the synchronous timing of AP volleys, or differences in receptor type or density at various body locations, can only be guessed. Lacking sufficient histological and experimental data, any explanation for electrical sensitivity differences at different body locations will remain speculative.

7.8   Skin Temperature[†]

That human sensitivity to electrical stimulation depends on skin temperature is not surprising in view of the fact that the responsiveness of the underlying neural fibers depends on their temperature. But there is a major difficulty in extrapolating from the temperature-dependent behavior of a single neuron to the temperature-dependent behavior of human skin sensitivity: The skin, together with its blood supply, is a very efficient thermoregulator which protects underlying tissues from the stress of changing temperatures at the body surface. It is important, therefore to measure human sensitivity under controlled conditions of ambient temperature. Larkin and Reilly (1986) performed experiments for this purpose, using an air-immersion technique that mimics exposure to environmental heat and cold. The experiments tested both perception thresholds, and suprathreshold responses for capacitive discharges on the fingertip and on the forearm. An electrode-tapping procedure was used for the fingertip, and a contact electrode was used for the forearm; the stimulated arm was placed inside an insulated, temperature-controlled chamber.

Perception threshold measurements, illustrated in Fig. 7.19, demonstrate that the effects of skin temperature are similar on the fingertip and forearm. Warming the skin affects sensitivity very little in either case, but there is

[†]Portions at this section have been adapted from Larkin and Reilly (1986) and from Reilly and Larkin (1985a).

a drastic effect with cooling. The somewhat higher rate of climb for the fingertip at low temperatures is consistent with its relatively lower priority in thermoregulation compared with the forearm. Figure 7.19 also shows normalized punctuate-pressure thresholds adapted from a study of temperature effects on pressure thresholds (Stevens et al., 1977; Stevens, 1980). The temperature dependence of pressure and electrical thresholds is similar: Pressure sensitivity degrades if the skin is cooled, but shows very little dependence on warming, up to 43°C.

In a suprathreshold procedure, subjects adjusted a stimulus on the cooled or heated right hand to match the perceived intensity of a capacitive discharge stimulus delivered to the left hand that was maintained at room temperature. The stimuli presented to the left hand ranged from just perceptible levels to the threshold of pain. Results of the sensation-matching procedure are given in Fig. 7.20. The horizontal axis plots the voltage presented in each sample to the left fingertip. The vertical axis shows the stimulus charge needed on thermally treated skin to achieve an equivalent level of sensation. The most striking aspect of these data is that the matching charge falls along straight lines parallel to the diagonal;

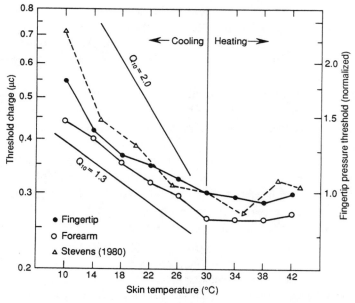

Figure 7.19   Effect of skin temperature on cutaneous thresholds. Solid lines with circles show detectability of electrical pulse on fingertip and forearm. Broken line shows punctate-pressure sensitivity on fingertip [from Stevens (1980), Fig. 1]. Reference lines of constant slope are shown for $Q_{10} = 1.3$ and 2.0. (From Larkin and Reilly, 1986.)

thus, the change in sensitivity is uniform over the range of stimuli tested, and can be represented by a multiplicative constant for each temperature. The multipliers for the matching experiment are listed on Fig. 7.20. They indicate that, for example, charge must be raised by about 60% to reach the same sensation magnitude on the skin at 10°C, compared with skin at normal temperature (30°C). These multipliers fall very close to similar numbers derived from the threshold data in Fig. 7.19. Thus, skin temperature affects perception sensitivity and suprathreshold sensitivity nearly identically.

The fact that heating or cooling affects electrical sensitivity differently is not surprising in view of the body's more efficient thermoregulation against heat than against cold. The tolerance of pain to skin temperature occurs around 45°C, at which point tissue damage begins to occur (Chapter 10), and nerve conduction is irreversibly blocked (Klump and Zimmerman, 1980). On the other hand, the body can withstand cooling well below normal temperatures. Below 30°C, the skin loses heat faster than it can be restored metabolically (Tregear, 1966). It can therefore be expected that neural responsiveness would be affected by cooling of the skin much more so than with heating.

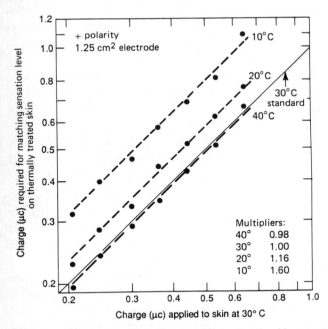

Figure 7.20   Effect of heating and cooling the skin on suprathreshold sensation levels. Measurements for capacitive discharges at 200 pF to the fingertip. (From Larkin and Reilly, 1986.)

The temperature dependence of neural activity is frequently reported in terms of a Q_{10} coefficient, which is the ratio of a given index of activity at one temperature to that when the temperature is shifted by 10°C. For a variety of indices of neural responsiveness, Q_{10} coefficients for isolated neural preparations typically range from about 1.3 to 3.0 (Larkin and Reilly, 1986). The data of Fig. 7.19 indicate a Q_{10} value of 1.3 for cooling on the arm. On the fingertip, a Q_{10} value around 1.3 applies to temperatures down to about 14°C, below which a value around 2.0 is indicated.

7.9   Tactile Masking

The mechanical stimulation of touch can have a masking effect on electrical sensation. Most of us learn, for example, that a carpet spark is less bothersome if we grasp the grounded object firmly, rather than tentatively. The attenuating effects of contact force are well known among electricians and others who are called upon to touch "hot wires." The explanation of this phenomenon is not fully understood, but is likely to include the fact that electrical and mechanical stimulation excites the same neural pathways in the skin. At low voltages, people commonly report that capacitive discharges feel very much like tactile pressure, mild pricks with a needle, or pinpoint touches. Furthermore, we have seen that tactile and electrical perception share several features in common, including similar rank-ordering with respect to body-part sensitivity, temperature sensitivity, and male/female differential sensitivity.

The author and colleagues studied the relationship between tactile force and electrical threshold by measuring perception thresholds when subjects tapped an energized electrode with calibrated force (Reilly and Larkin, 1983). Subjects were trained to tap with a force within 2 dB of a target position on an accelerometer scale. The scale ranged from 0 dB (the lightest force for repeated tapping) to 50 dB. Repeated tapping at a force of 60 dB was painful. Steps of 10 dB on this scale represent approximately equal intervals of subjective sensation, as well as of physical intensity. Thus, 50 dB is a "strong" tap, 40 dB is "firm," 30 dB is "moderate," 20 dB is "light," and 10 dB is "very light." Tap force was calibrated in a mechanical arrangement, in which a force of 20 dB was registered by dropping a steel ball with 11 μJ of kinetic energy (Reilly et al., 1982).

Figure 7.21 illustrates results from an experiment in which capacitance was held fixed while tap force was changed randomly from one trial to another. The middle set of curves represents results for four subjects using a tip electrode (1-mm-diameter tip, elevated 0.5 mm above an insulating holder) at 600 pF. The threshold voltage at the highest tap force, 50 dB, is approximately double the voltage at the minimum tap force. A second

set of experiments was performed with capacitor discharges at 200 and
3,200 pF. Results for one individual are plotted in the top and bottom of
Fig. 7.21. The data represent trials using a variety of electrodes of different
shapes and sizes. Two general conclusions were made from these
experiments. First, the electrode size and shape has almost no effect in
the tapping procedure. Second, the masking effect of tap force is greater
for electrical stimuli with shorter durations. The masking of electrical
thresholds by touch can be large and should not be discounted in any
assessment of human sensitivity. Accordingly, tap force should be
monitored or controlled in any test involving active touching.

Rollman (1974) tested the masking effects of a conditioning pressure
stimulus on electrical thresholds and the reverse (electrical masking of
pressure stimulus). The degree of masking depended on the time sequence
of two stimuli. For coincident pressure and electrical stimuli matched to
the same subjective intensity (determined by cross-modal matching to an
audible tone), Rollman reported a 1- to 3-dB threshold shift in the electrical

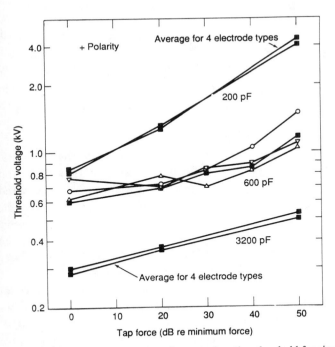

Figure 7.21   Effect of tap force on detection threshold for single-spark
discharges into the fingertip. Each symbol represents an individual subject.
Except where noted, a 1-mm raised-tip electrode was used. (From Reilly
and Larkin, 1985a.)

stimulus if the mechanical stimulus is the conditioning event; for the reverse situation (detection of pressure, masking by electrical), the pressure threshold was shifted by 12 dB.

A concurrent pressure stimulus has been observed to raise thresholds of painful 60-Hz electrical stimulation (Higgins et al., 1971). This observation was offered as evidence supporting Melzack and Wall's (1965) so-called "gate theory" in which stimulation of myelinated fibers is said to inhibit pain perception from nearby unmyelinated nociceptors.

7.10   Individual Differences in Electrical Sensitivity[†]

It is not uncommon for one person's detection threshold to be as much as twice of another person, though the measurement procedure is in all respects identical. These individual differences persist from day to day, and from week to week: they are not temporary artifacts of the measuring procedure. In spite of these differences, electrical sensitivity data exhibit a high degree of parametric invariance, in the sense that curves of sensitivity (for example, psychometric functions or equal-sensation contours) commonly have the same shape, regardless of who is picked as the test subject. Curves for different test subjects can frequently be related to one another by a simple multiplicative transformation. Consequently, basic studies of sensitivity need not require a large number of participants. However, none can serve as a basis for estimating the distribution of sensitivity variation in the general population. For that purpose, a large sample of test participants is needed.

Potential Sources of Variability

Variations in sensory measurements fall into three general categories: trial-to-trial variations, long-term sensitivity drifts, and subject-to-subject differences. Trial-to-trial variations in measured thresholds are observed when an individual is tested repeatedly in a single session. These variations cannot be eliminated in practice. For capacitive discharge stimuli, these variations depend on the method of applying the stimulus (Larkin and Reilly, 1984). For stimuli delivered by actively tapping an energized electrode with the fingertip, perception thresholds exhibit a coefficient of variation (ratio of standard deviation to mean) of about 0.12; for capacitive discharges delivered to an electrode held in contact with the skin, the observed coefficient of variation is about 0.05. In the contact-electrode case, the coefficient of variation is comparable to that observed in other studies using constant-current pulses (Rollman,

[†]Portions of this section have been adapted from Larkin and Reilly (1984).

1974). In the tapped electrode case the electrical sensation can be masked by tactile stimulation, raising the threshold of detectability and increasing its variance. Gradual threshold drifts reflect a second source of variability: Even highly practiced subjects do not give precisely the same results from day to day. We examined individual sensitivity drifts in a longitudinal study in which four subjects were tested once weekly over a period of 8 weeks (Reilly et al., 1984). Treating the session means for each subject as a statistical variable, the coefficient of variation across sessions averaged 0.11 for the four subjects. These long-term drifts cannot be separated from intersubjective comparisons without enormous experimental effort. Such an effort would not be repaid by any substantial increase in the precision with which two individuals can be compared, because the drifts appear to account for only a small fraction of the total intersubjective variance. Individual differences in sensitivity are a much greater source of variability in sensory measurements. These differences persist over time, and greatly outweigh the fluctuations observed from trial to trial or from one week to the next.

It is known that elevated sensory thresholds exist in individuals with neuropathological conditions (Friedli and Meyer, 1984; Katims et al., 1986; Notermans, 1967; Pruna et al., 1989), as well as psychopathological conditions (Hare, 1968). Neuromotor excitation threshold shifts are also known to exist in individuals with neuropathy (Parry, 1971; Harris, 1971). Among individuals who appear to be normal, wide variations in electrical sensory thresholds are also found. The basis for these variations is not well understood.

Distribution of Sensitivity

The author and colleagues studied electrical sensitivity of 124 subjects (Larkin et al., 1986). Rather than make an extensive study of people chosen completely at random—a practical impossibility in any case—the study focused on specific subgroups whose sensitivity might be expected to be either high or low. Subgroups consisted of college students (25 women and 24 men), 25 female office workers and 50 men engaged in skilled trades (electricians, carpenters, plumbers, sheet-metal workers). The sampling strategy allowed direct comparisons between men and women, and between occupational groups. Among the groups, there was a wide distribution of ages and body sizes, so that a statistical estimate could be made of their relationship to electrical sensitivity. A variety of physical attributes were quantified, namely: age, height, weight, temperature of fingertip and forearm, circumference of fingertip and forearm, index of habitual physical activity, and index of prior shock experience.

Table 7.8. *Geometric averages and ratios of thresholds from the large-sample study*

Group	N	Perception thresholds		Annoyance levels		Ratios		
		(A) Arm-contact electrode, 200 pF	(B) Fingertip tapping, 200 pF	(C) Fingertip tapping, 200 pF	(D) Fingertip tapping, 6400 pF	B/A	C/B	C/D
Female students	25	0.083	0.252	0.576	1.91	3.0	2.3	9.7
Male students	24	0.107	0.277	0.664	2.14	2.6	2.4	9.9
Female office workers	25	0.074	0.266	0.648	1.95	3.6	2.4	10.7
Male maintenance workers	50	0.113	0.314	0.924	2.94	2.8	2.9	10.0
Total sample	124	0.097	0.284	0.734	2.33	2.9	2.6	10.1

Notes: Data given in μC of charge for positive-polarity stimulation; values of capacitance are listed at heads of columns.

Statistical distributions of sensitivity were determined separately for the subgroups. Geometric means are summarized in Table 7.8. One indication is that there are sensitivity differences related to sex and occupation: males appear to have higher thresholds than females, and maintenance men have higher thresholds than college men. As shown subsequently, the apparent group differences are thought to be artifacts of a body-size dependency.

Table 7.8 also provides the geometric means of ratios among individuals' sensitivity values. The column labeled B/A gives the ratios of perception for finger versus arm stimulation. This ratio not only reflects the greater sensitivity of the arm relative to the finger (see Sec. 7.7), it also arises because tactile masking and electrode size effects were included in the method of stimulation used at these two body loci. The effect of tactile masking is to elevate thresholds for fingertip stimulation relative to a contact electrode condition (see Sec. 7.9). The combined effects of these factors result in an average fingertip-to-arm threshold ratio (B/A) of 2.9. The column labeled C/B indicates the multiple above the perception threshold needed to produce the rating "definitely annoying." These "perception-to-annoyance" ratios agree well with results from other experiments designed specifically to study suprathreshold reactions (Sec. 7.4).

The column labeled C/D indicates the relative sensitivity of 200- versus 6,400-pF discharges from a tapped electrode: On the average, 3.2 times as much charge is required at 6,400 pF to produce the same annoyance rating obtained at 200 pF. This capacitance effect stems from differences in discharge time constants at the different capacitance values (see Sec. 7.3).

Threshold distributions are well represented by the log-normal distribution, both for subgroups and for pooled data. Figure 7.22 gives an example of sensitivity distributions for 200-pF discharges, separated into male and female subgroups. Voltage levels are indicated on the horizontal axis. The vertical axis plots the relative number of individuals reporting perception [curves (a) and (b)] or annoyance [curves (c)]. The horizontal axis is logarithmic and the vertical axis is subdivided according to the standard Gaussian probability distribution. The data are nearly linear on these plots, indicating the log-normal distribution. The straight lines through the data are minimum chi-square fits to the log-normal model. This depiction omits some outliers, and therefore slightly underrepresents the total variability in the sample. The voltage ratio spanning the middle 90% of the fingertip perception thresholds (that is, from the 5th to the 95th percentile) is approximately 2:1. The ratio is closer to 3:1 in the case of forearm thresholds and the annoyance ratings.

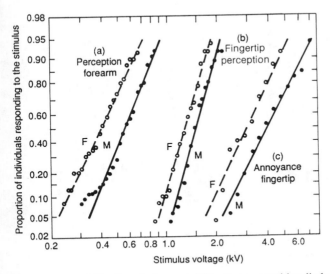

Figure 7.22 Cumulative probability that a capacitive discharge (+ polarity, 200 pF) exceeds perception or annoyance thresholds; samples of 74 men (M) and 50 women (F). Discharges on the fingertip via a tapped electrode; discharges on the arm via a 0.5-cm-diameter contact electrode. (From Larkin et al., 1986.)

Table 7.9. *Statistical distributions of sensory thresholds*

Body locus	Subgroup	N	Stimulus	Measurement	Normalized threshold[a]		Reference
					10%	90%	
Forearm	M + F	124	200-pF cap. discharge	Perception	0.54	1.67	Larkin and Reilly (1984)
Fingertip	M + F	124	200-pF cap. discharge	Perception	0.71	1.43	Larkin and Reilly (1984)
Fingertip	M + F	124	200-pF cap. discharge	Annoyance	0.70	1.81	Larkin and Reilly (1984)
Forearm	M or F	74, 50	200-pF cap. discharge	Perception	0.61	1.52	Larkin and Reilly (1984)
Fingertip	M or F	74, 50	200-pF cap. discharge	Perception	0.78	1.32	Larkin and Reilly (1984)
Fingertip	M or F	74, 50	200-pF cap. discharge	Annoyance	0.74	1.55	Larkin and Reilly (1984)
Forearm	Weight-adjusted	124	200-pF cap. discharge	Perception	0.73	1.36	Larkin and Reilly (1984)
Fingertip	Weight-adjusted	124	200-pF cap. discharge	Perception	0.79	1.26	Larkin and Reilly (1984)
Forearm	M + F	40	1-ms pulses @ 10 pps	Perception	0.34	1.69	Rollman and Harris (1987)
Forearm	M + F	40	1-ms pulses @ 10 pps	Pain	0.43	1.57	Rollman and Harris (1987)
Hand (grip)	M	115	Continuous DC	Perception	0.63	1.50	Dalziel and Mansfield (1950b)
Hand (grip)	M	115	60-Hz AC	Perception	0.73	1.27	Dalziel and Mansfield (1950b)

[a]Thresholds normalized by median value.

Table 7.10. *Normalized perception threshold distribution for log-normal model of adult population*

Percentile rank (%)	M + F		Weight-adjusted	
	Forearm	Finger	Forearm	Finger
99	3.11	1.91	1.78	1.51
95	2.24	1.58	1.50	1.33
90	1.85	1.41	1.36	1.25
75	1.40	1.21	1.18	1.12
50	1.00	1.00	1.00	1.00
25	0.72	0.83	0.85	0.89
10	0.54	0.71	0.73	0.80
5	0.45	0.63	0.67	0.75
1	0.32	0.52	0.56	0.66
0.5	0.29	0.49	0.53	0.63

Note: Data express multiples of median threshold.

The statistical distribution will depend, to some extent, on the subgroup that is represented. As explained subsequently, electrical thresholds are correlated with measures of body size. Consequently, subgroups that are correlated with body size will tend to have predictable differences in thresholds and, consequently, will exhibit a threshold distribution that is less dispersed than a more heterogeneous group. This point is illustrated in Table 7.9 by the data of Larkin and Reilly. For comparison, statistical parameters from other experimenters are also shown. In the first subgroup of the table, adult males and females are combined in presumed equal numbers. The second subgroup applies to either males or females. Although these subgroups are associated with significantly different thresholds, the distributions are virtually indistinguishable when their thresholds are normalized with respect to their respective medians (see Fig. 7.22). The third subgroup, assumed to be of homogeneous body weight, has weight-adjusted thresholds in accordance with a regression formula presented subsequently in this section.

Much of the experimental data in this chapter has been given in terms of median or mean values. Frequently, for analysis of safety or electrical acceptability, we would like to know the distribution of thresholds across a population of exposed individuals. For example, one may wish to select a criterion of exposure that protects some percentile rank of sensitive individuals. Table 7.10 indicates thresholds for a representative log-normal model, based on the author's measurements, as indicated by the 10 percentile values in Table 7.9. The data listed under (M + F) apply to a group of adults, with males and females in equal proportion. The "weight-adjusted" data apply to calculated thresholds, using body-weight regression formula. The data in this table have been normalized by the median values of the respective subgroups.

Correlates of Sensitivity

The cumulative distributions plotted in Fig. 7.22 show an unmistakable difference between the two sexes. For a homogeneous group of students, for example, the forearm thresholds for males is 1.3 times that for females (column A of Table 7.8). The direction of this difference is not surprising, in view of the research literature on sex differences in cutaneous sensitivity. Women reportedly are more sensitive than men to pressure (see Sec. 7.8), to vibratory stimulation (Verillo, 1979), and to pain (Notermans and Tophoff, 1975). Women are also reported to be more sensitive to a variety of electrical stimuli, including pulse trains (Rollman and Harris, 1987), 60-Hz currents (Dalziel, 1972), and sinusoidal currents of frequency from 10^4 to 10^7 Hz (Chatterjee et al., 1986). The reported ratio of the thresholds for males versus females is typically consistent with the author's findings (about 1.3:1).

To the author's knowledge, no physiological hypothesis has ever been advanced to account for these differences, though there have been suggestions that cultural and social factors may account for them—especially in the case of pain tolerance. Our data do not support a cultural explanation, since detection thresholds were determined in a manner that effectively eliminates all sources of bias in response tendencies, and the ratio between annoyance level and threshold level is remarkably constant across the two sexes (see the column labeled C/B in Table 7.8).

The data of Table 7.8 also suggest that higher electrical thresholds might be associated with occupational groups involved in strenuous labor. There is a common belief that those who lead more strenuous lives acquire a more stoic or tolerant attitude toward physical discomfort and pain. The physical demands of daily life may contribute to this effect if skin sensitivity is altered by injury to the epidermis, or by the development of hard callused layers. Whether pain tolerance is culturally acquired, as many have suggested (Zborowski, 1952; Sternbach and Tursky, 1965; Craig and Weiss, 1971; Wolff and Langley, 1975; Clark and Clark, 1980), or reflects an underlying physiological adaptation, is not at issue in the present study, but we might have expected college students and office workers to differ greatly from electricians, carpenters, and sheet-metal workers, who are more frequently exposed to strenuous physical labor and harsh weather. If the threshold and annoyance levels of the maintenance workers are compared with the other groups (Table 7.8), there is a difference in the expected direction. Observe, however, that the difference is nearly as large for forearm sensitivity, where calluses do not develop, as it is for the fingertip. This observation suggests that skin thickness alone does not account for the result.

Much the same conclusion applies in the case of physical exercise and the subjects' reported levels of previous exposure to electric shock. Neither of these factors is well correlated with sensitivity; they do not explain why maintenance workers have higher thresholds and annoyance levels than the other subgroups.

It might seem that a factor related to sex and occupation could partially account for individual sensitivity differences. But multiple regression analysis shows that the apparent sex and occupation differences appear to be artifacts of an underlying body-size dependency: Woman in our sample group tended to be smaller than the men, and maintenance workers were larger than college men (Larkin and Reilly, 1986). When these size differences were taken into account, no statistically significant differences in sensitivity remained. A regression equation based on body weight was determined:

$$V = ae^{bW} \qquad (7.7)$$

where V is the predicted median perception or annoyance voltage, W is the subject's weight in kilograms, and a and b are empirical parameters. Best-fit values for a and b that apply to discharges to the fingertip from a 200-pF capacitor were, for perception and annoyance, respectively, $a = 971$, 1904; and $b = 0.0053$, 0.0093. Statistical variations about the median indicated by Eq. (7.7) follow the log-normal distribution, with a 90th percentile value at a factor of approximately 1.5 times the median and a 10th percentile at approximately 0.7 times the median.

Equation (7.7) provided a better overall fit (greatest r^2) to the data than other two-parameter functions. However, from theoretical considerations (Schmidt-Nielsen, 1984), one might expect a power function of body weight. In this case we can represent the data by

$$V = cW^d \qquad (7.8)$$

where c and d are empirical parameters. The least-squares fit of Eq. (7.7) was marginally better than that of Eq. (7.8). The estimated parameter values that apply to discharges to the fingertip from a tapped 200-pF capacitor were for perception and annoyance, respectively: $c = 204$, 121; and $d = 0.386$, 0.678. With 200-pF discharges to a 0.5-cm-diameter electrode held against the arm, perception parameter values were $c = 13.8$ and $d = 0.702$.

Table 7.11. *Calculated thresholds[a] for various body-weight classes, discharge to the fingertip from a 200-pF capacitor, tapped electrode delivery*

Weight		Perception		Annoyance		Annoyance multiple	
(kg)	(Ib)	(7.7)	(7.8)	(7.7)	(7.8)	(7.7)	(7.8)
23	50	0.22	0.18	0.47	0.35	2.15	1.87
45	100	0.25	0.24	0.58	0.55	2.45	2.29
68	150	0.28	0.28	0.72	0.73	2.57	2.58
91	200	0.31	0.31	0.88	0.88	2.81	2.81
113	250	0.35	0.34	1.09	1.03	3.11	3.00

[a]Thresholds are expressed in μC of charge. Corresponding voltages are calculated according to Eq. (7.7) or (7.8) as indicated in column head.
Source: From Reilly and Larkin (1987).

Table 7.11 illustrates weight-dependent fingertip thresholds for discharges of 200 pF, based on Eqs. (7.7) and (7.8). The two formulations are in close agreement over the weight range of the individuals who were tested (41 to 128 kg). Note that the dynamic range of sensitivity, represented by the perception-to-annoyance multiple, increases from low to high body weights.

Downward extrapolation of Eqs. (7.7) and (7.8) yields rather different estimates of sensitivity at the lower body weights. A representative weight for children (23 kg) is included in Table 7.11. In this case, the power function gives a value 16% lower for perception, and 26% lower for annoyance, compared with the exponential function. Without further testing, it is not possible to make a secure choice between these alternatives, nor to be confident that extrapolation to children is valid.

The regression formulas presented above used body weight as a matter of convenience. Other relationships with body size (e.g., forearm diameter) yield equally valid correlates of sensitivity, since various measures of body size are strongly correlated with one another. Based on these findings, it would be interesting to examine whether male/female differences in tactile pressure sensitivity might also reflect an underlying body-size effect. The fact that pressure sensitivity of women's breasts is more closely related to the inverse of breast size, rather than to the overall size of the women (Weinstein, 1963), suggests that the size of the stimulated body part is relevant.

An intriguing question remains: why is electrical sensitivity related to body size? Conceivably, the relationship might be explainable if cutaneous receptor density were inversely related to the size of the individual.

However, until further histological and experimental evidence becomes available, the etiology of the body-size role in electrical sensitivity can only be guessed.

7.11   Startle Reactions[†]

Muscle movement that is initiated through reflex activity involves both sensory and motor responses (see Sec. 3.6). Reflex activity operates outside conscious control to regulate many vital body functions. When this activity is in response to a surprise sensory insult, the reaction is frequently called a "startle reaction." A startle reaction to electrical stimulation could potentially pose a hazard to an individual who was engaged in certain inherently dangerous activities, such as climbing a ladder, handling hot liquids, or handling power cutting tools.

It is very difficult to design a meaningful experiment to define startle reactions. One difficulty arises from the fact that it is necessary to introduce the element of surprise in the context of informed consent on the part of experimental subjects. Expectations in an experimental setting are likely to introduce a bias in the measured reaction threshold or the extent of the reaction.

To the author's knowledge, the only published data on startle reactions to electrical stimulation were obtained from experiments carried out at the Underwriters Laboratories in connection with the development of ANSI standards on leakage current (Smoot and Stevenson, 1968a 1968b; Stevenson, 1969). These studies were designed to help define levels of currents that could pose a potential hazard due to a startle reaction. Men and women volunteers were asked to transfer rice from one container to another using a small metal cup held in the fingers. Random levels of currents were applied at unexpected times to various parts of the hand, wrist, or lower arm, including the small cup itself. The other electrode was provided by the other arm being immersed in salt water. The subjects' reactions were given numerical ratings according to the distance that the hand moved during a startle reaction, the speed of movement, and the amount of rice spilled. A panel consisting of approximately 240 people viewed the volunteers' reactions via videotape and made judgements as to what reaction might pose a hazard in the home, such as due to spillage of hot grease or falling from a ladder. Most panel members were engineers and scientists concerned with electrical safety, although 11 homemakers were also represented on the panel. A borderline hazardous reaction was

[†]This section has been adapted from Reilly (1978b).

typically judged as one corresponding to between 3 and 12 in. of motion at a moderately fast rate, with some rice spillage (usually less than 10%).

Initial tests established that women were somewhat more sensitive than men. As a result, succeeding tests were carried out with women in order to establish a conservative estimate of hazardous thresholds. Based on 20 women tested, the average reaction judged as borderline hazardous was elicited at a conducted current[†] of approximately 2.2 mA when applied to the lower arm and 3.2 mA when applied to the cup held between the fingers. These average reaction currents may be compared with the median perception levels for women of 0.73 mA for gripped conductors, 0.59 mA for pinched contacts (Thompson, 1933), and 0.24 mA for lightly touched contacts (Dalziel, 1954). Thus, it appears that average reaction currents are at least a factor of 3 above average perception currents.

Based on the UL reaction tests, a limit of 0.5 mA was adopted as the ANSI standard for leakage current from portable appliances (see Chapter 11). We would like to determine the probability that the specified leakage might cause a reaction fitting into the "borderline hazardous" category defined by the UL observers. Even lacking the statistical distribution of reaction thresholds, it is still possible to make some estimates by assuming that the distribution of reaction thresholds has the same coefficient of variation (standard deviation to mean ratio) as do perception thresholds. The coefficient of variation of the 60-Hz perception threshold for touched or gripped contacts is about 0.37 (Dalziel, 1954). Applying that value to a normal distribution for reaction currents along with the mean reaction values mentioned above, it is estimated that an unexpected current of 0.5 mA applied to a woman's lower arm will have a 2% probability of causing a reaction fitting into the UL "borderline hazardous" category. If the contact is pinched between the thumb and fingers, the probability is 1%.

Acknowledgment

I wish to acknowledge the contributions of Willard D. Larkin, who is with the Department of Psychology at the University of Maryland. Dr. Larkin has participated for several years as co-investigator with me in a research project concerning human reactions to transient electric currents. Results and conclusions from that project have been used extensively in this chapter. Dr. Larkin is co-author of many of the papers referred to in this chapter.

[†]Because of body impedance, conducted current was on the average 0.72 times the available short-circuit current.

8

Skeletal Muscle Response to Electrical Stimulation

JAMES D. SWEENEY

8.1 Introduction

In the eighteenth and nineteenth centuries, many seminal investigations into the basic properties of electricity were electrophysiological in nature and focused on neuromuscular phenomena. In Galvani's commentary (1791) a frog nerve-muscle preparation was used to propose a theory of "animal electricity" in which muscle fibers were thought to store charge. Volta (1800) later demonstrated that the bimetallic electrodes used by Galvani were a source of electrical stimulation current, rather than a pathway for discharge of animal electricity. In modern society electrical stimulation of skeletal muscle may occur unintentionally as one possible effect of an accidental electric shock, or intentionally in medical devices for artificially induced muscle exercise or control.

This chapter presents the fundamental principles and effects of electrical stimulation of skeletal muscle. We first introduce the principles of skeletal muscle electrical stimulation in general, and then focus on electrically induced skeletal muscle responses that may occur during intentional stimulation or electric shock. To achieve an understanding of the skeletal muscle responses that may be elicited by any form of electrical stimulation, it is necessary to consider the basic structure and function of muscle and the nerves that innervate muscle.

8.2 Neuromuscular Structure and Function

Skeletal Muscle Innervation

The myelinated nerve fibers that innervate skeletal muscle fibers derive from α-motor neurons (see Fig. 3.1), whose cell bodies lie within the ventral horn of the spinal cord. These efferent axons (approximately 9 to 20 μm in diameter) leave the spinal cord via spinal-nerve ventral roots and project out to muscles through peripheral nerve trunks. Skeletal muscles may contain thousands of individual muscle fibers. However, a one-to-one relation does not generally exist between α-motor neurons and muscle fibers. Rather, as an α-motor neuron fiber nears and enters a skeletal muscle, it branches into a number of collateral axons each of which innervates a single skeletal muscle fiber. An α-motor neuron and the skeletal muscle fibers it innervates are known as a *motor unit*. These skeletal muscle fibers, which upon contracting contribute to total muscle force, are known as *extrafusal* muscle fibers. *Intrafusal* muscle fibers produce only a minute fraction of total muscle force output. They are innervated by efferent myelinated γ-motor neuron fibers (averaging 5 μm in diameter) and afferent sensory myelinated Ia and II fibers (averaging about 17 and 8 μm in diameter, respectively). The sensory receptors within skeletal muscle, the *muscle spindles*, incorporate a small number of intrafusal fibers. These receptors sense muscle length and the rate of change of muscle length, and are responsible for the well-known *stretch reflex* of skeletal muscle—that is, when a skeletal muscle is stretched, excitation of muscle spindles elicits reflex activation of the muscle. The second type of skeletal muscle receptors, the *Golgi tendon organs*, lie in series with muscle fibers as they enter tendons. These receptors are innervated by afferent myelinated Ib nerve fibers (averaging 16 μm in diameter) and report muscle tension information to the central nervous system.

Peripheral Nerve Trunk Structure

Peripheral nerve trunks in general carry motor, sensory, and sympathetic nervous system information between the central nervous system and the periphery (or vice versa). Axons are not simply "packed" into nerve trunks. Peripheral nerve trunks usually contain several fascicles (i.e., bundles, also called "funiculi"), each of which contains many axons as well as endoneurial connective tissue (see Fig. 8.1). A thin perineurial connective tissue sheath encloses each fascicle. The surface of the nerve trunk is covered by epineurial connective tissue. The fascicular structure of a nerve trunk is not constant throughout its course. On the contrary, regrouping and redistribution of fascicles (and the axons they contain) occurs throughout the nerve trunk. For example, in the far periphery some fascicles tend to exit from main nerve trunks and provide innervation to one or two skeletal muscles. They therefore must at this point contain all of the appropriate muscular efferent and afferent axons. As peripheral nerve trunks near the spinal roots, however, axons that innervate particular muscles must regroup into a multisegmental arrangement of fascicles. Sunderland (1978) provides an excellent review of this and related material. An implication of this aspect of trunk structure is that more central electrical stimulation of nerve trunks tends to elicit more diffuse muscular responses. Electrical stimulation of nerve trunks or branches in the far periphery can give quite specific responses.

Muscle Action Potentials and Excitation-Contraction Coupling

Action potentials of α-motor neurons are generated by the cell body given sufficient dendritic input. Each action potential then propagates along a myelinated nerve fiber until it reaches a synapse at the neuromuscular junction (or motor end plate) (see Fig. 3.1). At the neuromuscular junction, neurotransmitter (acetylcholine or ACH) is released

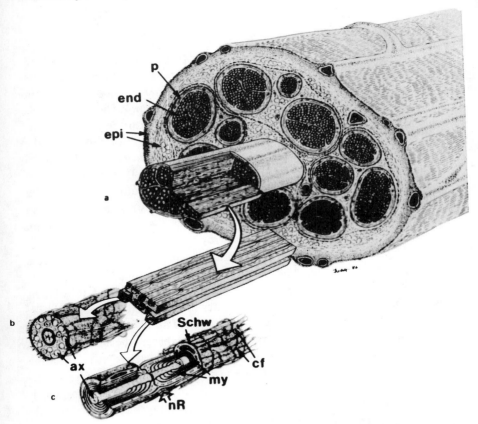

Figure 8.1 Microanatomy of a peripheral nerve trunk and its components. (a) Fascicles surrounded by a multilaminated perineurium (p) are embedded in a loose connective tissue, the epineurium (epi). The outer layers of the epineurium are condensed into a sheath. (b) and (c) illustrate the appearance of unmyelinated axons and myelinated nerve fibers, respectively. Schw, Schwann cell; my, myelin sheath; ax, axon; nR, node of Ranvier (components not to scale). (From Lundborg, 1988, p. 33.)

into the synaptic cleft in response to the presynaptic action potential. The postsynaptic response on the muscle fiber membrane (sarcolemma) is, in general, a muscle action potential (MAP), which propagates away bidirectionally given that the motor end plate tends to be located toward the middle of the fiber. Muscle action potentials penetrate each fiber through the transverse-tubule (T-tubule) system. Propagation of an MAP into the T-tubule network of a fiber elicits release of calcium ions by the sarcoplasmic reticulum system into the proximity of the myofilaments. These calcium ions, upon binding with troponin C, initiate muscle fiber contraction (see below). This process, whereby muscle action potentials lead to contraction, is known as *excitation-contraction coupling.*

Skeletal Muscle Force Production

The skeletal muscle filaments (i.e., myofilaments) of Fig. 3.23 consist of relatively thick myosin filaments and thinner actin filaments (see Fig. 5.5). Each myosin filament contains many individual myosin molecules, each of which possess a "head" and a "tail." The myosin tails are wrapped together to form a myosin filament, with the myosin heads protruding outward. (Each head and the "arm" of myosin attaching the head to the myosin filament is known as a *cross-bridge.*) Each myosin head possesses strong ATPase activity. That is, ATP is readily bound and cleaved (into ADP and phosphate). An actin filament consists of a double-helix F-actin molecule as well as tropomyosin strands, which in resting muscle act to cover "active" sites on the actin, and troponin complexes, which are attached to the tropomyosin molecules.

During muscle contraction, the excitation-contraction coupling process described above results in the release of calcium ions near the myofilaments. Calcium binding to troponin is thought to elicit movement of tropomyosin so that actin "active" sites are uncovered. Myosin heads, upon binding ATP, produce ADP and phosphate (which remain bound to the head) and achieve a conformational state wherein the heads protrude toward surrounding actin filaments. When actin "active" sites become exposed, myosin heads attach to these sites. This binding is thought to elicit a new conformational change in the myosin cross-bridge, which "pulls" the actin relative to the myosin—providing the force of muscle contraction. The myosin head then releases from the actin, frees the bound ADP and phosphate, and can bind a new molecule of ATP. At this point the cycle can repeat if sufficient free calcium and ATP are present at the myofilaments. A calcium pump in the sarcoplasmic reticulum acts continuously to decrease the calcium concentration near the myofibrils. (Thus, in response to a single action potential, a brief surge in myofibrillar calcium concentration, and therefore a brief contractile response, occurs.) To prevent rapid depletion of ATP concentrations within a contracting muscle, ADP molecules must be rephosphorylated. Some energy for this process is immediately available from phosphocreatine. However, anaerobic and aerobic metabolic processes must provide the energy for contractions that last longer than a few seconds. Glycolysis of stored carbohydrates (such as glycogen) is a relatively fast, but metabolically inefficient, anaerobic process by which new ATP can be formed. Oxidative (aerobic) metabolic processing of carbohydrates, fats, or proteins can provide slower, lower, but steadier levels of energy for ATP production.

The contractile output of a single motor unit receiving a single action potential input is a muscle twitch (actually, such a twitch represents a summation of responses from the individual muscle fibers within the motor unit). As we have seen in Chapter 3, two fundamental physiological means exist by which the nervous system can modulate force production within a skeletal muscle: temporal summation (also known as rate modulation), and recruitment. In temporal summation (Fig. 3.24) the rate of action potential input to the fiber is controlled so that twitch summation results in a desired force level. Note that this summation process is in general nonlinear and can be quite time-dependent. Recruitment in this context refers to the activation of greater or lesser numbers of motor units within the skeletal muscle (i.e., α-motor neuron fibers).

Differentiation of Skeletal Muscle Fiber Types

The muscle fibers of motor units that innervate a given skeletal muscle can be categorized with respect to physiological, ultrastructural, and metabolic characteristics. Table 8.1 presents a highly simplified, but useful, differentiation among fiber types based on three common classification schemes (see Burke, 1981, for a detailed review). It should be noted that, in general, muscle fiber and motor unit properties lie within an overlapping continuum—and that classification schemes serve mainly as aids in relating structure-function properties of the muscle. The classification method of Brooke and Kaiser (1970) places muscle fibers into three categories, I, IIA, and IIB (based on histochemical measurements of ATPase reactivities). Peter et al. (1972) differentiated fiber types based on a combination of physiological and metabolic properties (SO, slow oxidative; FOG, fast oxidative-glycolytic; FG, fast glycolytic). The method of Burke et al. (1973) differentiates motor units on the basis of their twitch speed and fatigue resistance (S, slow; FR, fast-resistant; FF, fast-fatigable). An "intermediate" type of fast muscle fiber also exists in some muscles (type IIAB or FI, fast intermediate resistance to fatigue).

Table 8.1. *Correlation of skeletal muscle major fiber type physiological, ultrastructural, and metabolic properties*

Fiber type	IIB, FG	IIAB	IIA, FOG	I, SO
Corresponding motor unit type	FF	FI	FR	S
Myofibrillar ATPase, pH 9.4	High	High	High	Low
Glycogen	High	High	High	Low
Phosphorylase	High		High	Low
Neutral fat	Low		Medium	High
Capillary supply	Sparse		Rich	Rich
Aerobic metabolism capacity	Low	Medium	Medium-High	High
Anaerobic metabolism capacity	High	High	High	Low

Source: Adapted from Burke, 1981, p. 349.

In general, we may think of slow-twitch skeletal muscle fibers as being optimized for maintenance of relatively low force levels, at relatively low speeds, but for prolonged periods of time. Fast-twitch fibers are optimized for quick production of relatively high force levels for briefer time intervals. We see from Table 8.1 that the physiological, ultrastructural, and metabolic properties of the major skeletal muscle fiber types reflect a trade-off between the ability to produce force quickly and powerfully, or slowly and steadily. Type I fibers generate force relatively slowly (myofibrillar ATPase activity at high pH is relatively low), yet can maintain force well via a relatively high aerobic metabolic capacity (high neutral fat content, rich capillary supply, etc.). In the other extreme case, type IIB fibers can generate force quickly (myofibrillar ATPase activity is relatively high) via the use of anaerobic metabolic mechanisms (high glycogen storage, high phosphorylase). However, muscle force fatigue occurs relatively rapidly in these fibers. Type IIAB and IIA fibers represent varying degrees of compromise between the design requirements of force generation, speed/power, and fatigue resistance.

Recruitment and Firing Patterns

In normal physiological use, motor unit recruitment and firing patterns also appear to be optimized for performing desired motor tasks (see Burke, 1981, 1986). In 1965 Henneman and colleagues advanced a theory known as the "size principle" of motor neuron recruitment (Henneman et al., 1965a, 1965b), in which it was proposed that small α-motor neurons innervating slow motor units are recruited at lower "functional thresholds" than large α-motor neurons, which innervate fast motor units. The term "functional threshold" is analogous to "excitability level" in this context and essentially reflects the level of synaptic input necessary to excite the motor neuron (Burke, 1981).

The logical implication of this theory is that, in "mixed" skeletal muscles containing a distribution of muscle fiber types, low-force tasks requiring fatigue resistance (such as force generation to maintain posture) are met by recruiting slower motor units. Only when greater and/or faster force levels are desired are the more readily fatigued fast motor units brought to bear. While this theory has, to a great extent, stood the test of time and further experimentation (see Henneman and Mendell, 1981; Burke, 1981, 1986), it is clear that some exceptions do occur in normal physiological use and that motor neuron size alone does not completely determine functional excitability. For example, in rapid, powerful "ballistic" movements, relatively synchronous activation of both slow and fast motor units, and even preferential activation of fast units, can occur.

In general, it has been found that α-motor neurons innervating slow-twitch motor units tend to have average firing rates that are lower than those that innervate fast-twitch units (see Burke, 1968, 1981). This seems sensible in that twitches generated by slow motor units start to fuse at approximately 5–10 Hz and reach a tetanic state by 25–30 Hz, while fast motor units may require 80–100 Hz to reach total fusion (see Burke, 1981; Wuerker et al., 1965). In typical physiological use, however, motor units rarely exhibit sustained firing frequencies that even approach these tetanic fusion limits. For example, in the human extensor digitorum communis muscle, Monster and Chan (1977) observed that virtually all motor units began firing at about 8 Hz regardless of functional threshold (i.e., motor unit type) and increased their firing rates to a maximum of 16–24 Hz at maximal voluntary force levels.

Burke (1986) has suggested that recruitment patterns that fit the size principle probably meet the normal functional demands placed on muscles well while maintaining an acceptable metabolic cost. In instances where recruitment deviates from the "orderly" pattern dictated by the size principle, metabolic cost is sacrificed in order to achieve less frequently needed fast and/or powerful motor demands.

Fatigue in Normal Physiological Use

Muscle fatigue is often defined as an inability to generate the required or expected force (Edwards, 1981). Bigland-Ritchie et al. (1986) have pointed out that neuromuscular fatigue might more generally be defined as "any reduction in the maximum force generating *capacity*, regardless of what type of work is being done." It should also be emphasized that fatigue should be regarded as a normal protective measure for survival, rather than as a deficit or abnormal phenomenon. If fatigue mechanisms did not exist, it would be possible to exercise muscles to the point where ATP levels were completely exhausted and muscle rigor (i.e., irreversible binding of cross-bridges to actin) occurred.

We have seen that voluntary skeletal muscle contractions depend on a sequence of events involving the motor neuron (dendritic input, cell body excitability, action potential propagation, motor end-plate presynaptic transmission), excitation-contraction coupling (motor end-plate post-synaptic transmission, muscle action potential propagation along the sarcolemma and into the T-tubules, calcium release from the sarcoplasmic reticulum), and the actual contractile machinery (ATP-dependent cross-bridge cycling, ATP rephosphorylation through metabolic processes). Fatigue may be induced due to impairment of any of these processes (Bigland-Ritchie et al., 1986). It is generally accepted that both short-lasting and long-lasting components exist in physiological fatigue. Quickly reversible effects include sarcoplasmic electrolyte shifts and pH changes that may impair MAP propagation through the T-tubule system, sarcoplasmic reticulum calcium ion release, and/or force development at the myofilaments. Force development is obviously also dependent on metabolic processes and products. At the motor unit level a clear correlation between aerobic (oxidative) metabolic capacity and resistance to fatigue exists. As mentioned above, the fatigue-resistant fiber types S (SO) and FR (FOG) can be associated with the high aerobic capacity I and IIA categories; while easily fatigued type FF (FG) units are associated with type IIB low aerobic capacity (high anaerobic capacity). Longer-lasting fatigue may be due to depletion of metabolic energy stores (such as glycogen) or excitation-contraction coupling impairment [i.e., "low-frequency fatigue" (Edwards et al., 1977)].

In normal physiological use, motor unit firing patterns appear to be adaptable to match fatigue levels. Bigland-Ritchie et al. (1983) have reported that, in a series of brief maximal voluntary contractions of the human pollicis muscle, the average firing rate of more than 200 motor units (from five subjects) dropped from 29.8 Hz ($\pm$ 6.4) to 18.8 Hz ($\pm$ 4.6) between 30 and 60 s after contraction onset, and to 14.3 Hz ($\pm$ 4.4) after 60–90 s. In the same study, twitch contraction times before and immediately after a 60 s maximal voluntary contraction were not significantly different. However, twitch relaxation times were prolonged by approximately 50%. This has the effect of lowering the firing frequency needed to achieve fused tetanic contractions. Thus, firing frequency decreases during fatigue may in part occur because higher rates are unnecessary to achieve fused force output.

Dynamic Nature of Skeletal Muscle Fiber Types

A great deal of evidence exists that the skeletal muscle fiber "type" of an individual motor unit is not fixed, but is instead a dynamic quantity that depends on the neural input to and motor usage of the unit (see reviews of Pette and Vrobova, 1985; Burke, 1981). Prolonged endurance training, or constant electric stimulation at low frequencies (10 Hz), can result in a transformation of fiber type populations within mixed skeletal muscles. This transformation is in the direction of faster to slower twitch, and increased aerobic metabolism capacity. Increases in fatty acid metabolic capacity, fiber cross-sectional areas, whole-muscle fatigue resistance, and capillary density can be elicited within weeks to months after training onset. At moderate levels of exercise demand, oxidative capacity increases are most dramatic as the result of apparent conversion of type FF fibers to type FR. With constant electrical stimulation, actual conversion of type II (fast) myosin to type I (slow) eventually can be brought about in addition to the metabolic and ultrastructural changes already noted. Conversely, extended disuse of skeletal muscle tends to result in fiber atrophy, decreases in tetanic force output per fiber, and aerobic capacity loss in fatigue-resistant fibers.

The properties of muscle therefore appear to be optimized for normal use, as determined by overall activity level, and to some extent activity and force production patterns. Fast-twitch fibers, which normally see relatively low levels of phasic activity, appear capable of adapting to new high, tonic levels of activity. Conversely, muscle fiber inactivity appears to result in an opposite shift to metabolic profiles more appropriate for infrequent, phasic levels of activation.

8.3   Fundamental Principles of Skeletal Muscle Electrical Stimulation

Strength-Duration Relations for Neuromuscular Excitation

The most basic requirement for electrical stimulation of a skeletal muscle fiber is that an action potential(s) must be elicited on the α-motor neuron innervating the fiber, or within the membrane of the muscle fiber itself. As has been discussed in Sec. 3.5, the electric excitability of myelinated nerve is generally much greater than that of skeletal muscle (see Parry, 1971; Walthard and Tchicaloff, 1971; Mortimer, 1981). Figure 8.2 depicts strength-duration (S-D) curves (see Sec. 4.2) for transcutaneous stimulation of human skeletal muscles via regulated-voltage waveforms. We see that, in this example, direct electrical stimulation of denervated muscle requires substantially higher rheobase (i.e., the stimulus strength needed for large pulse widths) and chronaxy (i.e., the stimulus pulse width

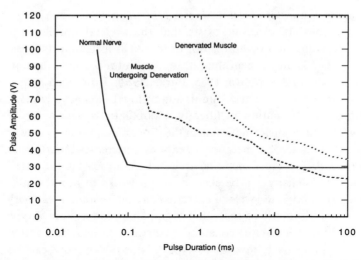

Figure 8.2   Strength-duration curves for transcutaneous stimulation of human Peroneus Longus skeletal muscles via regulated voltage waveforms. The left-hand curve is typical of normally innervaed muscle. The middle curve was obtained following a left lateral popliteal nerve lesion when the Peroneus Longus was undergoing denervation atrophy. The right-hand curve, measured six weeks after the middle record, is typical of denervated muscle. (Adapted from Sunderland, 1978, p. 259.)

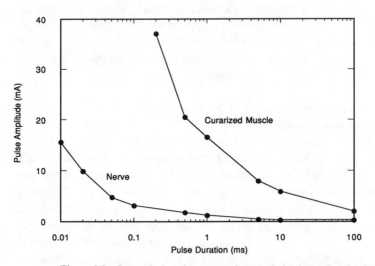

Figure 8.3   Strength-duration curves for cat skeletal muscle stimulation with an intramuscular electrode. In these experiments, evoked muscle response was held constant at a small fraction of total possible muscle force. The left-hand curve represents data for direct nerve stimulation. The right-hand curve represents data for direct muscle fiber stimulation (following administration of curare). (Adapted from Mortimer, 1981, p. 170.)

at twice rheobase) values than for normal stimulation, where α-motor neuron axons can be activated.

Strength-duration curves for cat skeletal muscle stimulation with an implanted intramuscular electrode (see Sec. 8.4) and regulated-current monophasic stimuli are shown in Fig. 8.3. In these experiments on cat skeletal muscle (Mortimer, 1981), evoked isometric muscle response was held constant at a small fraction of total possible force. Once again, we see that the curve for direct muscle stimulation (in this case, following administration of the motor end-plate blocking-agent curare) shifts up and to the right in comparison to the curve for neural stimulation. Crago et al. (1974) conducted similar experiments on intramuscular stimulation of curarized and noncurarized cat skeletal muscle. They found that, over a range of stimulus pulse durations and amplitudes, direct (curarized) muscle stimulation accounted for only 3–7% of isometric muscle force evoked by neural (uncurarized) stimulation with identical parameters.

An S-D curve time constant τ_e can be determined for both neural and direct muscle stimulation by performing a mathematical fit of experimental data to Lapicque's form of the strength-duration curve as expressed by Eq. (4.17). Such data indicate S-D time constants for denervated skeletal muscle of 4.3 to 57.6 ms, and for intact muscle of 20 to 700 μs (Oester and Licht, 1971). In that just-suprathreshold intact muscle stimulation is accomplished indirectly through excitation of α-motor neuron axons, we would expect that the S-D time constant values for intact muscle and large myelinated nerve stimulation would be similar. Indeed, for mammalian peripheral myelinated nerve fibers, experimentally obtained values from 29 μs to over 800 μs can be determined (Ranck, 1975; Li and Bak, 1976).

We have seen in Chapter 4 that the S-D time constant is a property not just of the excitable membrane, but also of the method of excitation. In general, τ_e becomes greater as the stimulus current crossing the cell membrane becomes more gradually distributed along the longitudinal dimension. The absolute minimum values of the S-D time constants for neural and direct muscle monophasic rectangular stimulation are the actual *membrane* time constants (τ_m) for α-motor neuron nodes of Ranvier, and skeletal muscle sarcolemma.[†] Based on data obtained from rabbit peripheral myelinated axons by Chiu and colleagues (Chiu et al., 1979, Chiu and Ritchie, 1981), the author has estimated that node of Ranvier

---

[†]This situation occurs, for example, when a regulated current stimulus is applied directly across a small patch of excitable membrane. The S-D time constant τ_e then becomes equal to the membrane time constant τ_m [i.e., Eq. (4.4) simplifies to Eq. (4.5)].

τ_m values would lie in the range of 10 to 20 μs. Albuquerque and Thesleff (1968) found that, for innervated rat skeletal muscles, sarcolemmal τ_m was approximately 1.7 ms. While interspecies differences are to be expected, these values indeed fall just below the lowest experimentally obtained measurements of S-D time constants mentioned above.

Regulated-Current versus Regulated-Voltage Stimulation

Many typical stimulation waveforms are rectangular in shape. A regulated-current stimulus (of magnitude I) with pulse width t will deliver a fixed total charge Q per stimulus, such that $Q = It$. Regardless of electrode impedance and potential shifts (either during a stimulus, or between stimuli), because the current of the stimulus is regulated, a reproducible applied electric field can be created within the tissue(s) to be stimulated (see Sec. 2.1). For this reason, many recently designed implanted neuromuscular stimulation systems utilize regulated-current stimulus output stages. Transcutaneous neuromuscular stimulation systems, i.e., systems that use surface electrodes, often produce regulated-voltage waveforms. Despite the drawback that stimulation results may be less reproducible, in this case regulating the voltage of a stimulus is safer—a regulated-current stimulus passed through a dislodged or broken surface electrode could produce skin irritation or even burns due to the resultant high current density.

Biphasic Stimulation

Balanced-charge biphasic (BCB) regulated-current stimuli are used with many implantable electrode neuromuscular stimulation systems (see Sec. 8.4 below). With such waveforms a cathodal rectangular pulse is followed by an anodal phase of equal charge (either rectangular in shape, or with the form of a capacitive discharge)(see Robblee and Rose, 1990). While the "primary" cathodal phase serves to stimulate excitable tissue(s), the "secondary" anodal phase performs charge balancing mainly for electrochemical reasons. Without some form of electrochemical reversal, deleterious cathodic reactions such as alkaline pH swings, hydrogen gas evolution, and oxidizing agent(s) formation can occur at the electrode-tissue interface [see Sec. 8.4 and Robblee and Rose, (1990)]. However, while the intent of BCB stimulation is to minimize adverse electrochemical reactions, we have seen in Sec. 4.6 that the anodic current reversal of a biphasic stimulus can act to abolish an action potential developing in response to the cathodic phase. This physiological effect has been studied in vitro on isolated Xenopus Laevis myelinated nerve fibers, and in vivo using a cat nerve-muscle preparation, by van den Honert and Mortimer

(1979a). It was found that a delay between the primary and secondary phases of approximately 100 μs effectively prevented this effect from occurring. Modeling of myelinated nerve BCB electrical stimulation also predicts this result (see Fig. 4.16). Gorman and Mortimer (1983) found that the effect could be used potentially to advantage for separating the stimulation thresholds of differing diameter α-motor neuron axons.

Fatigue and Conduction Failure in Response to Electrical Stimulation

Trains of electrical pulses at rates of 10 to 20 Hz and at magnitudes that activate α-motor neuron axons can elicit a long-lasting depression in skeletal muscle force output in addition to metabolic deficit fatigue. This "low-frequency fatigue" phenomenon (Edwards et al., 1977), which also occurs in normal physiological use (see Sec. 8.2), is thought to occur (at least in part) due to a deficit in excitation-contraction coupling. That is, in response to each muscle action potential, less calcium in released by the sarcoplasmic reticulum. With increased stimulation frequencies of 20 to 100 Hz, "high-frequency fatigue" may be elicited. Motor neuron propagation failure at axon branch points, neurotransmitter depletion at the motor end-plate, and MAP propagation failure all may contribute to this effect. Bigland-Ritchie et al. (1986) feel that occurrence of high-frequency fatigue is avoided in normal physiological use because motor neuron firing rates remain below the frequencies needed for propagation failure.

Bowman and McNeal (1986) have studied in vivo the response characteristics of single α-motor neuron axons of cat sciatic nerves when very-high-frequency trains (100–10,000 Hz) of balanced-charge biphasic stimuli were applied. At frequencies of 100 to 1000 Hz, axonal firing rates were generally equal to or were subharmonics of the stimulation frequency. Over a 3-min stimulation period, an axon often typically initially fired at the same rate as the stimulus train for a very brief time, shifted to firing with every other stimulus, gradually shifted to firing with every third stimulus, and sometimes then shifted to every fourth. Increased stimulus amplitudes tended to delay the downward shift in firing frequency with time. Axonal firing frequencies in this range tend to rapidly deplete supplies of ACH at the neuromuscular junction. This type of "junction fatigue," first studied by Wedensky (1884) and Waller in 1885 (see Bowman and McNeal, 1986), rapidly results in contractile force loss (see McNeal et al., 1973; Solomonow et al., 1983).

At stimulation frequencies of 2000 to 10,000 Hz, Bowman and McNeal found that axons would briefly respond with firing rates of several hundred hertz, but stopped firing completely within seconds after stimulation initiation. This type of "electrical conduction block" could be maintained for periods up to 20 min, with recovery following stimulation termination occurring within 1 s. The mechanism involved in this type of very-high-frequency conduction block, which has also been studied by a number of other investigators using rectangular stimuli or AC sinusoidal trains (e.g., Cattell and Gerard, 1935; Katz, 1939; Tasaki and Sato, 1951; Tanner, 1962), is not completely known. Polarization buildup at nodes of Ranvier due to differing conductivities for inward and outward currents (see Sec. 4.6) may act to depolarize excitable membranes to the point where strong sodium channel inactivation results.

Sinusoidal Stimulation Strength-Frequency Effects

Nonlinear membrane polarization buildup is also thought to account for the upward deflection in AC sinusoidal stimulation strength-frequency curves above a minimum level that typically occurs at about 150 Hz for myelinated axons (see Fig. 4.20, right). At stimulation frequencies below about 100 to 150 Hz, rectangular stimulation thresholds for α-motor neurons are independent of stimulation frequency, and each stimulus tends to produce one action potential. AC sinusoidal trains, however, exhibit an upturn in their strength-frequency behavior (see Fig. 4.20, left). That is, higher-amplitude sinusoidal trains are needed to excite the nerve. The well-known "accommodation" of excitable membrane thresholds (see Sec. 4.6) in response to slowly increasing stimuli is thought to be responsible for this effect. For example, Sato and Ushiyama (1950) demonstrated in an in-vitro toad motor nerve preparation that accurate mathematical prediction of the form of strength-frequency curves below 40 Hz could be performed through consideration of membrane accommodation properties [see Tasaki (1982) for an excellent review of this and related topics].

As sinusoidal frequencies approach DC, it becomes quite difficult to stimulate peripheral nerves. Stimulation may be elicited at the onset of direct current as with any rectangular pulse. However, following the onset of direct current, additional stimulation of α-motor neurons or skeletal muscle is not easily achieved except at very high stimulus amplitudes. Extremely high-amplitude stimuli, at very high frequencies (2000 Hz) or at DC, have been found to produce "tetanoid" responses directly in isolated (curarized) frog skeletal muscles (Koniarek, 1989). After initiation of this type of response, a portion of each muscle fiber appears to reach a constant depolarization level.

8.4   Functional Neuromuscular Stimulation Systems

Electrically induced stimulation of skeletal muscle for artificially controlled exercise and restoration of motor function (lost due to injury or disease) are the objectives of functional neuromuscular stimulation (FNS) systems. A number of excellent reviews of this field, and the general area of functional electrical stimulation (FES), already exist (e.g., Vodovnik et al., 1981; Mortimer, 1981; Peckham, 1987; Agnew and McCreery, 1990). In this section we will briefly present the current state of FNS applications, methods, models, and electropathology, and will introduce basic principles that can provide greater understanding of accidental stimulation in electric accidents.

Objectives and Applications of Functional Neuromuscular Stimulation Systems

It has been estimated that there are over 200,000 spinal cord-injured citizens in the united states. Spinal cord injuries occur mostly in younger people, with the most common age at injury being 19 years and the mean age at injury being 29 years (Young et al., 1982). A conservative estimate of the annual cost of care for spinal cord-injured persons is $1.5 billion. In the paraplegic person (spinal cord damage in the thoracic or lumber region), the freedom to walk, run, or simply stand from a chair to reach a shelf is restricted or lost. Quadriplegic individuals (spinal cord damage in the cervical region) will have lost, in addition, some or all of the ability to manipulate and control their environment. Functional neuro-muscular stimulation of α-motor neurons that remain intact below the level of spinal cord damage provides a means by which some control of lost motor function can be reasserted. FNS can also be used for motor restoration in other disabled populations where α-motor neurons remain intact, but central control has been lost or impaired (e.g., stroke, cerebral palsy). At this time a number of research centers around the world are developing systems designed to provide upper-extremity control (e.g., Peckham, 1983; Hoshimiya et al., 1989) and lower-extremity control (e.g., Waters et al., 1985; Holle et al., 1984; Kralj et al., 1986; Marsolais and Kobetic, 1987; Graupe, 1989; Jaeger et al., 1989).

Electrical Excitation of Skeletal Muscle by Functional Neuromuscular Stimulation

Excitation of a given motor nerve fiber can be achieved, in general, if an electric field is introduced (by electrodes) that involves a focalized change in the extracellular voltage gradient. Equations such as (4.31) imply that a second-difference function involving extracellular potentials at nodes of Ranvier [see Sec. 4.4 and Rattay (1986)] characterizes the "driving force" of electric field distributions. In Eq. (4.31) the term $(V_{e,n-1} - 2V_{e,n} + V_{e,n+1})$ represents a second-difference sampling of the longitudinal extracellular potential distribution along the fiber. Focal changes in the extracellular electric field can therefore cause depolarization of the axonal membrane at nodes of Ranvier and generation of action potentials that will propagate toward a muscle.

In spinal cord injured people, most motor neurons in the spinal cord will remain intact if they are below the level of the injury. Electrical stimulation of their axons can therefore be used to cause muscle contraction. The restoration of functional movement in general requires the fine gradation of force so that adequate control can be achieved, and damage to the body or to objects in the environment can be avoided. Just as skeletal-muscle force output is controlled in normal physiological use by recruitment of α-motor neurons and by rate modulation of each motor neuron's firing, electrical stimulation can be used to control the same variables artificially. FNS system designers often intuitively strive to replace the body's normal control strategies; however, major limitations still exist in the present ability to achieve such control.

Electrodes for Functional Neuromuscular Stimulation

An electric field can be generated within the body using a variety of electrode types. Stimulation using electrodes located on the skin surface is the most straightforward possible approach. It was determined as early as 1833 by Duchenne de Boulogne that skeletal muscles could be stimulated by passing current percutaneously via cloth-covered electrodes (see Licht, 1971; Walthard and Tchicaloff, 1971). Duchenne found that contractions were most easily obtained by stimulating at a discrete location on the skin overlying each muscle. He called such areas "points of election" and theorized that the muscle beneath each point was hyperexcitable. Remark and von Ziemssen subsequently showed that such locations, now known as "motor points," actually overlie sites where nerve trunks penetrate and branch out into the deep face of muscles. Coers (1955) refined this essentially correct viewpoint. He found that the greatest concentration of motor end-plates in a superficial muscle is located in the superficial part of the muscle beneath the skin surface motor point. "Motor lines" reflect linear projections onto the skin surface of underlying superficially located motor nerve trunks. Walthard and Tchicaloff (1971) provide extensive maps and tabulations of approximate motor point and motor line locations.

The stimulus parameters for surface-electrode motor point stimulation will obviously vary from one muscle to another, and will depend on muscle/nerve anatomy, whether voltage or current is regulated, the waveform utilized, as well as electrode(s) type, size, placement, and configuration. Vodovnik et al. (1967) found that for regulated-voltage rectangular stimuli, while threshold parameters for stimulation of upper extremity muscles were quite variable, typical threshold values (for 200-μs pulses at 50 Hz) were on the order of 20 V. Voltages in excess of 100 V may sometimes be necessary to achieve high activation levels (Vodovnik et al., 1981). Figure 8.4 presents typical S-D curves for surface-electrode regulated-current stimulation of four upper-extremity muscles. These curves were obtained by placing an electrode over the motor point of each muscle and measuring, for each pulse duration, the just-threshold pulse amplitude. Crago et al. (1974) have reported similar results (e.g., for stimulation of the flexor digitorum sublimis muscle in a typical subject: 9.9 mA threshold, for 0.5-ms-duration stimuli at 50 Hz). Bajzek and Jaeger (1987) found that, for triceps surae stimulation, similar or somewhat higher charge parameters were necessary (e.g., on the order of 50 to 100 μs pulse duration thresholds, for 60 to 80 mA stimuli at 25 Hz).

An increase in surface-electrode stimulus intensity above the threshold level will, of course, tend to result in activation of a larger volume of the underlying muscle and therefore increased force output until maximal muscle activation is achieved. Equation (4.31) implies that, for a given electric field distribution beneath a surface electrode, a "shell" will exist

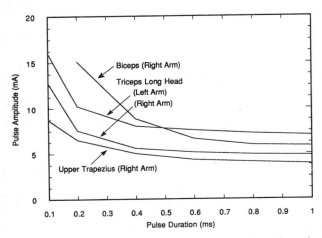

Figure 8.4   Strength-duration curves for surface-electrode threshold stimulation of four upper-extremity muscles. (Frequency = 50 Hz; C-3 level spinal cord injured patient "H.K.") (Adapted from Vodovnik et al., 1967.)

under the electrode that separates (for each size nerve fiber) regions where excitation will occur or will not occur. Figure 8.5 presents a family of S-D curves for varying levels of torque about the elbow given surface-electrode stimulation of the biceps muscle. Each curve essentially represents a constant degree of biceps muscle activation (i.e., a constant level of neural excitation penetration into the muscle body). Prediction of the relationship between surface-electrode stimulation parameters and underlying muscle activation level is nontrivial. Rattay (1988) has modeled the excitation of nerve fibers under disk-shaped surface electrodes using the "activating function" approach. He found, as have others, that the tendency for high current densities to arise at the edge of a surface electrode will ultimately limit the depth to which excitation can be elicited (i.e., at sufficiently high edge current densities, burns can occur).

A fundamental difficulty with surface motor point stimulation is that, while electrodes are applied to the *skin*, the motor point is inherently dependent on anatomical properties of the underlying *muscle*. Thus, as shortening or lengthening of a muscle occurs during a motor task, and as limb positions or joint angles change, the location of a motor point will shift with respect to fixed locations on the skin surface. Figure 8.6 illustrates this effect for surface electrode stimulation of the medial head of the triceps. In these experiments Crochetiere et al. (1967) estimated that the 2.5-cm shift in the motor point skin surface location (over an elbow

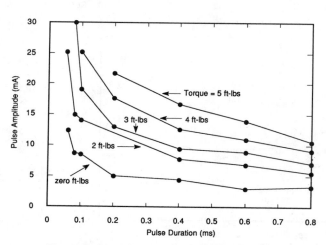

Figure 8.5   Strength-duration curves for surface-electrode tetanic stimulation of the biceps muscle at five levels of isometric torque about the elbow. Curve for zero ft-lbs reflects the just-threshold response. (Frequency = 50 Hz; elbow angle = 75°; subject "P.C.") (Adapted from Vodovnik et al., 1967.)

angle range of 15–90°) corresponded well with a theoretical expected excursion of 2.5 cm in the underlying muscle.

In a recent study of joint angle and electrode position effects on surface stimulation characteristics, McNeal and Baker (1988) found that, in activation of the hamstring muscles, a large inter-subject variability existed in predicting surface regions of greatest excitability. Changes in regions of excitability also generally occurred as the knee was flexed. Excitability regions for surface stimulation of the quadriceps muscles showed less dependence on joint angle. Surface electrode stimulation also tends to be relatively non-selective because the electric field created is necessarily widespread. For example, stimulating the innervation of deep muscles without also stimulating overlying superficial muscles is impossible. Surface electrode systems are in general cosmetically undesirable for chronic use, may cause skin irritation, and are difficult for a handicapped individual to maintain. Despite these drawbacks, however, the non-invasiveness of surface stimulation has great appeal, and a number of surface electrode based FNS systems have been reported. Patients using surface electrode systems for lower-extremity FNS have been able to stand, to walk at speeds approximately one-quarter of normal, and to climb stairs (Kralj et al., 1983; Petrofsky and Phillips, 1983). In stimulation systems intended purely for artificially generated muscle exercise (as opposed to function restoration), surface stimulation is clearly preferred.

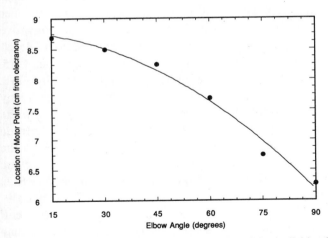

Figure 8.6 Location of the motor point of the medial head of the triceps muscle versus elbow angle for surface stimulation using a 1-in.-diameter electrode (pulse amplitude = 15 mA; pulse duration = 0.2 ms; frequency = 50 Hz; subject "W.C.") (Adapted from Crochetiere et al., 1967.)

Intramuscular electrodes (Caldwell and Reswick, 1975; Cargo et al., 1980; Handa et al., 1989), which can be placed percutaneously or fully implanted into skeletal muscles, and implanted epimysial electrodes, which are placed on skeletal muscle surfaces (Grandjean and Mortimer, 1986), enable selective activation of individual muscles (or even muscle regions). FNS research systems that use intramuscular or epimysial electrodes are presently limited by the need to implant, maintain, and use a large number of electrodes if many muscles are to be controlled (e.g., see Marsolais and Kobetic, 1987).

So-called "neural" electrode systems place electrodes directly into or adjacent to peripheral nerve trunks (Naples et al., 1990) and offer the possibility of controlling multiple muscles with a single implant. As presented in Sec. 8.2, in the distal segments of peripheral nerve trunks, individual fascicles (bundles) tend to contain the innervation for one or two skeletal muscles (Sunderland, 1978). Intraneural stimulation, which utilizes implantation of coiled-wire electrodes into peripheral nerve trunks, has been tested in animals (Bowman and Erickson, 1985). Epineural stimulation (i.e., surgical implantation of electrodes onto the epineurium of a nerve trunk) has been used by one research group (Holle et al., 1984). Both of these electrode systems are potentially suitable for selective activation of individual nerve trunk fascicles and, because they are not introduced into or onto muscles, may be more reliable and in some instances easier to implant than intramuscular or epimysial electrodes. Nerve cuff electrodes, which involve placement of electrodes (within insulating tubes or sheaths) near but not within the peripheral nerve, appear to offer similar advantages as intra- or epineural electrodes but without invasive penetration of the nerve trunk. A peroneal nerve cuff electrode, developed by Medtronic, Inc., and McNeal and colleagues (Waters et al., 1975, 1985; McNeal et al., 1977), has been used for correction of footdrop in a large patient population, with some implants functioning well after many years. McNeal and Bowman (1985) tested in animals multiple-electrode cuffs intended to provide greater selectivity of muscle activation. Electrode position and good contact (i.e., close apposition of the electrodes to the nerve trunk) were found to be important factors in successfully achieving selectivity of stimulation with this design. Agnew and co-workers (e.g., Agnew et al., 1989) have developed a helical cuff electrode intended to reduce possible mechanical damage to nerve trunks. A spiral nerve cuff (Naples et al., 1988) suitable for fascicle selective stimulation (Sweeney et al., 1990) is under development that may enable direct apposition of electrodes to a nerve trunk surface without eliciting the mechanical damage normally caused by tightly fitting cuff systems (Naples et al., 1990).

Recruitment and Rate Modulation in
Functional Neuromuscular Stimulation

With regulated-current BCB rectangular stimuli, recruitment levels are in general controlled by altering the charge within the primary stimulus (i.e., by changing the amplitude or pulse width). Recruitment selectivity in FNS involves two separate aspects—spatial selectivity and nerve-fiber diameter selectivity. In the above discussion we have seen that implanted electrodes offer greater selectivity in activating different muscles than does surface stimulation. This enhanced "spatial selectivity" occurs because the electric field introduced can be focused closer to the α-motor neuron fibers of interest.

An aspect of FNS not yet discussed, fiber diameter selectivity, relates to the tendency to stimulate sub-populations of nerve fibers (at the same spatial location) based on their size. Consider again the form of the spatially extended nonlinear nodal (SENN) myelinated nerve model equation (4.31). In Eqs. (4.27)–(4.29) and (4.32a), the terms C_m, G_a, and $I_{i,n}$ are all proportional to fiber diameter. These terms occur in Eq. (4.31) only as ratios. As pointed out by McNeal (1976), the only effect of changing fiber diameter is therefore in the calculation of the extracellular potential terms. Because internodal length is proportional to fiber diameter, the extracellular potentials at the nodes will change with fiber diameter. In one idealized situation, where an extracellular potential distribution rises linearly along a fiber, peaks over a node of Ranvier, and then declines linearly, Eq. (4.31) predicts that fiber excitation threshold will be exactly inversely proportional to fiber diameter. This theoretical dependence of excitability given a linear field along a nerve matches the experimental results of Tasaki (1953) for studies on single myelinated fibers from frog sciatic nerve. In that the focalized electric fields generated by actual stimulation systems rarely approximate such an ideal distribution, more complex dependencies of stimulus threshold on fiber diameter are generally found. McNeal's original model predicted that, for regulated-current stimuli of 100-ms duration from a monopolar point cathode located 1 mm above a node of Ranvier, large-fiber-diameter thresholds were inversely proportional to the square root of fiber diameter. In the small-diameter range, threshold approached an inverse-square relationship with fiber diameter.

Recall from Sec. 8.2 that, in most normal physiological use, the "size principle" predicts an orderly recruitment of motor units from slow (small-diameter axons from small motor neurons) to fast (large-diameter axons from large motor neurons). In most electrical stimulation of myelinated fibers, however, there will be a tendency to recruit large axons at small stimulus magnitudes, and then smaller axons with increased stimulus levels. Such "reversed recruitment" of myelinated nerve fibers with extracellular electrical stimulation is a well-known phenomenon (e.g., Blair and Erlanger, 1933; Petrofsky, 1978; Fang and Mortimer, 1987). A major problem in FNS for motor control restoration is that, even if sufficient spatial selectivity is achieved (primarily through electrode placement), reversed recruitment of motor units will inappropriately utilize fast, more readily fatigued muscle fibers for low-force tasks. Slower, fatigue-resistant muscle fibers will be recruited only at higher stimulus levels. This also results in an undesirable, steep relation between force output and stimulus magnitude. Exacerbating this problem is the fact that with the muscle disuse that tends to follow spinal cord injuries, many fatigue-resistant muscle fibers tend to shift their metabolism toward less oxidative, and more anaerobic, more readily fatigued mechanisms. To circumvent these problems in clinical use of FNS systems, electrical stimulation "exercise" procedures are sometimes employed to alter the metabolic and physiologic properties of type IIB (FF) fibers and type IIAB (FI) toward the less readily fatigued type IIA (FR) and type I (S) profiles.

Exceptions to the reversed recruitment principle of electrical stimulation probably occur at very small electrode–nerve separation distances. Consider the idealized situation where an electric field generated by a point source very close to a node of Ranvier is so highly focused that, at adjacent nodes of Ranvier, *regardless of fiber diameter* (and therefore internodal spacing), no appreciable extracellular potentials occur. In this case the extracellular potential second-difference function of Eq. (4.31) simply reduces to the term $V_{e,n}$ for all size fibers. We would therefore expect that all size fibers would be recruited at about the same current threshold. Modeling studies of Veltink et al. (1988, 1989a), Altman and Plonsey (1989), and Sweeney et al. (1989) predict that as electrode–nerve separations decrease below the order of internodal lengths, strict reversed recruitment order disappears. This effect also has been studied experimentally by Veltink et al. (1989b).

Rate modulation in FNS differs from that in normal physiological use in one important respect. FNS systems will, by their nature, activate all motor units synchronously in phase with each stimulus at a predetermined frequency. In normal physiological use, the nervous system asynchronously activates different motor units. The result is that, at moderate average firing rates, smoothly controlled force production is possible. In synchronized FNS activation of motor units, however, smoothly fused contractions are possible only at higher average firing rates. As we have seen in Sec. 8.2, however, higher firing rates more rapidly fatigue the muscle fibers.

A variety of control strategies have been implemented, tested, and/or proposed for FNS systems (Chizeck et al., 1988). Recruitment modulation (with a fixed stimulation frequency) is generally used to achieve low to moderate force output control. Rate modulation can then be used to produce high force levels after all motor units have been recruited. Simultaneous modulation of both recruitment levels and rate modulation is much more complex, but offers the possibility of finer, more rapid control of force and of changes in force.

Electropathology of Functional Neuromuscular Stimulation Systems

The mechanisms by which electrical stimulation might damage tissues fall into two general categories: those deriving from electrochemical processes that enable charge transfer at electrode–tissue interfaces, and those that occur directly or indirectly as a result of current flow within the tissues [for a review, see McCreery and Agnew (1990)]. Of course, any implanted neuromuscular stimulation system must meet general biocompatability requirements (beyond those related to electrical stimulation) as well (Yuen et al., 1990).

Recall that current flows in metals by electron movement, while current in the body consists of the flow of charged ions. This flow of charge can be induced by an electrode by capacitive or conductive mechanisms [see Robblee and Rose (1990) for a review of stimulating electrode electrochemistry]. In general, capacitive mechanisms (which involve charging and discharging of the electrode "double layer") are not sufficient to deliver the current densities needed for nerve or muscle stimulation. Conductive mechanisms (which are also known as *Faradaic* and involve electron transfer across the electrode–tissue interface via electrochemical redox reactions) are therefore almost always necessary. "Reversible" conductive reactions involve chemical species that, ideally, are bound at the electrode surface and can be quantitatively reversed by alternately passing equal and opposite amounts of charge. Examples of such reactions are oxide layer formation/depletion and hydrogen atom plating. Balanced-charge biphasic (BCB) regulated-current stimuli (see Sec. 8.3) can therefore use reversible conductive reactions for electrical stimulation without introducing new electrochemical species into the tissue. "Irreversible" conductive reactions necessitate formation of new species, and in general may elicit variable amounts of tissue damage depending on the body's ability to buffer or suppress deleterious effects. Examples of such processes are electrode corrosion, the oxidation of chloride ion, and the electrolysis of water. Such reactions can bring about the formation of biologically toxic products and may generate large swings of pH near the electrode–tissue interface.

While electrical stimulation of axons within their normal physiological range does not in general elicit any damage, Agnew et al. (1989) have reported that prolonged BCB electrical stimulation (in which only reversible conductive reactions were presumably utilized) of cat peripheral nerve could elicit damage in a fraction of large myelinated nerve fibers. This result depended on the stimulus parameters (current, pulse width, and frequency) and the total duration of stimulation. For the peroneal nerve trunk, supramaximal stimulation of $A\alpha$ nerves for 8 h or more at 50 Hz consistently produced this *early axonal degeneration* (EAD) in some fibers. Depending on the stimulation parameters, damage could sometimes be found with frequencies as low as 20 Hz. In a recent series of experiments, Agnew and colleagues (see McCreery and Agnew, 1990) found that local anesthetic applied to peroneal nerves near the stimulation site prevented EAD. This complete abolition of damage by local anesthesia (which will prevent electrical excitation of the nerve fibers) strongly suggests that hyperactivity of axons may be at least partially responsible for EAD generation.

8.5   Skeletal Muscle Stimulation in Electrical Accidents[†]

The neuromuscular effects of accidental electric shocks have been a subject of study since the first development of commercial electric current sources. In particular, two possible neuromuscular responses to accidental electrical stimulation have received a great deal of attention because they can potentially contribute to injurious or fatal outcomes: the generation of involuntary movements by skeletal muscle activation, and arrest of respiration. In this section we will review existing experimental data on skeletal muscle stimulation in electrical accidents. Section 8.6 analyzes contractile movements of the hand contacting a relatively low-voltage electrical source, and the tetanizing condition of the so-called "let-go" threshold. In Sec. 8.7 we will review the possible effects of electric shock on respiration.

History and Overview

Virtually since the introduction of commercial electricity, it has been recognized that alternating currents introduced into or onto the

[†]Much of the material in this section is extended from an excellent review of 60-Hz contact current safety levels by Banks and Vinh (1984).

surface of the body could produce tingling and pain (see Chapter 7), sensations of tissue heating, stimulation of muscle contractions, respiratory difficulty or arrest, and even cardiac fibrillation (see Chapter 6). At current levels where voluntary skeletal muscle control is lost, and involuntary movements occur, "freezing" to a live conductor held in the hand is possible. As C. F. Dalziel, who performed many landmark investigations in this area, noted in 1956, electric stimulation of the forearm and hand flexors can generate very powerful grasping contractions that are not easily overcome. Dalziel reported that "...in 1942, a shipyard workman in the San Francisco Bay area attempted to obtain a better view of an in-plant Christmas Eve party by climbing around an overhead crane. He slipped and, to keep from falling, grasped the 440-volt crane trolly wires. He immediately froze to the trolley wires, and his body swung back and forth until someone opened the switch. He then fell to the ground, fractured his skull and died. A similar accident occurred in 1947 at a brass works in the Los Angeles area. A laborer lost his footing when climbing down from the roof of the factory. As he fell he grasped the 220-volt trolley wires of an overhead crane, and froze to the trolley wires. His body dangled from the wires until one employee opened the switch while another employee caught him as he fell. He received severe burns but recovered" (Dalziel and Massoglia, 1956).

Early investigators of this effect defined the current level where an individual could, through remaining voluntary control, just barely free himself from a hand-held current source as the *let-go value*. Experimental study of this value and its dependence on a number of variables [subject sex and age, contact(s) size and location, skin conductivity conditions, stimulus waveform and parameters, etc.] has dominated investigations on accidental neuromuscular stimulation effects.

Let-Go Thresholds for Hand Contact with Electrical Stimuli

In the 1920s and the 1930s a number of studies on let-go thresholds (or similar tests) for 50- or 60-Hz stimulation were carried out (Grayson, 1931; Thompson, 1933; Dalziel, 1938; Gilbert, 1939; Whitaker, 1939). A summary of the results of these studies is presented in Table 8.2 (after Banks and Vinh, 1984). Despite the widely varying methods and end points used, the average values for humans fell within a fairly reasonable range of about 8 to 15 mA rms.

From 1941 to 1956 Dalziel and his colleagues reported in a series of papers on human let-go currents and voltages. Despite some experimental and statistical methods that would not be acceptable by today's standards (Banks and Vinh, 1984), these reports still form the foundation

Table 8.2. *Summary of data from early let-go current investigations*

Report	Subjects	Current path	Active electrode	Endpoint	Let-go current (mA rms)			Frequency (Hz)
					Minimum	Average	Maximum	
Gilbert (1939)	25 Both sexes	Hand–hand	20-mm brass rod	Release grip	NR	15	NR	50
Whitaker (1939)	13	Hand–hand	Pliers	Release grip	6.0	7.8	10	NR
Thompson (1933)	42 men	Hand–hand	Rotatable metal handle	Rotate handle	NR	8.35	20.0	60
	27 women				NR	5.15	8.8	60
Grayson (1931)	42 men	NR	NR	Not "serious discomfort"	NR	8	NR	NR
Dalziel (1938)[a]	56 men	Hand–hand	Flat metal disk	Tolerate sensation	NR	13.9	22	60
	18 men	Hand–foot	Flat metal disk	Tolerate sensation	NR	13.5	NR	60
Dalziel (1938)[b]	42 men	Hand–hand	No. 10 AWG copper wire	Release grip	NR	12.6	18	60

NR = not reported.
[a] First series.
[b] Second series.
Source: Adapted from Banks and Vinh, © 1984, IEEE.

of our knowledge about the let-go phenomenon and have had great impact in setting safety standards for line-frequency contact currents (see Chapter 11). The adult experimental subjects in these studies were predominantly healthy, young, male volunteers aged 21 to 25 years (with an upper age tested of 46 years). A smaller number of female subjects in their "late teens to the early 20's" were also tested (Dalziel and Massoglia, 1956). Dalziel described his female subjects as probably representing a fairly sedentary population. A small number of children were also tested. In all experiments "...an individual's let-go current [was] defined as the maximum current he [could] tolerate when holding a copper conductor by using muscles directly affected by that current" (Dalziel et al., 1943). A No. 6 or No. 7 AWG polished copper wire was grasped as the active electrode in most studies. Dalziel and colleagues felt that conductor size and shape did not have a large effect on threshold measurements. The hands of the subjects were kept moist with a "weak salt" solution, and any of three indifferent electrode sites/types were used (usually a brass plate under the opposite hand, but sometimes a plate under one or both feet, or a conducting band around the upper arm), apparently without strongly affecting the results. After allowing each subject to become accustomed to the sensations that would be encountered, the current was raised slowly to a certain value, at which point the subject was commanded to drop the electrode. If he or she succeeded, the current values were increased by steps until the subject was no longer able to let go of the wire; if the subject failed, the test was repeated at a lower value. It is important to note that, at the let-go threshold current value, "...failure to interrupt promptly the current is accompanied by a rapid decrease in muscular strength caused by pain and the fatigue associated with the severer involuntary muscular contractions" (Dalziel and Massoglia, 1956). The end point of experiments was checked by several trials thought to be insufficient in number to fatigue the individual but enough to determine the let-go value accurately (Dalziel et al., 1941). This process was undoubtedly quite unpleasant (see Fig. 8.7). Dalziel and colleagues recognized that psychological state and motivation levels certainly affected results. "Psychological factors, especially fear and competitive spirit, were the most important causes for the variations" (Dalziel et al., 1943). Figure 8.8 and Table 8.3 summarize results obtained for the 60-Hz let-go currents of 134 men and 28 women. The 99.5 percentile rank levels (9 mA rms for men, 6 mA rms for women) were considered to represent a reasonable criterion for safety (Dalziel and Massoglia, 1956).

Great interest existed at this time in also establishing 60-Hz safety levels for small children. In general, Dalziel and colleagues found that obtaining

Figure 8.7   An experimental subject undergoes let-go current threshold testing. (From Dalziel, © 1972, IEEE.)

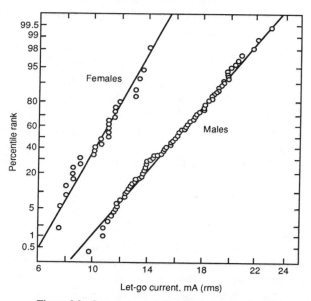

Figure 8.8   Let-go current distribution curve for adults given hand contact with a 60-Hz Ac source. Average value for 134 males = 15.9 mA rms; extrapolated 0.5 percentile = 9 mA rms. Average of 28 females = 10.5 mA rms; extrapolated 0.5 percentile = 6 mA rms. (Adapted from Dalziel, © 1972 IEEE.)

Table 8.3. *Summary of Dalziel's 60-Hz let-go current threshold data*

Subjects	0.5 Percentile rank	Minimum	Average	Maximum
134 men	9.0	9.7	15.9	22.0
28 women	6.0	7.4	10.5	14.0

Note: Values of let-go current in mA rms.
Source: Adapted from Banks and Vinh, © 1984, IEEE.

Table 8.4. *Dalziel's 60-Hz let-go threshold data for three boys*

Subject	Age	Let-go current (mA rms)
1	5 years	7
2	9 years, 3 months	7.6
3	10 years, 11 months	9
Mean		7.9

Source: Adapted from Banks and Vinh, © 1984, IEEE.

let-go currents for young children was extremely difficult. In addition to the reluctance of parents to permit experimentation on their children, Dalziel's limited experiences with children indicated that children were likely to "...just cry at the higher values" (Dalziel, 1972). Despite these problems, results were reported for three small boys (Table 8.4), and Dalziel recommended a "reasonably safe" current level for children as 50% of the 99.5 percentile value for normal adult males, namely, 4.5 mA rms.

The frequency dependence of tetanizing sinusoidal current thresholds was reported in 1943 (Dalziel et al.) for groups of adult male subjects ranging in number from 26 to 30 at frequencies from 5 to 10,000 Hz, as well as at DC. These results were compared with the 60-Hz study on 134 men. Figure 8.9 depicts the percentile curves for frequencies ranging from 5 to 10,000 Hz, as well as at DC.[†] Tests on steady or gradually increasing DC currents "...produced sensations of internal heating rather than

---

[†]A somewhat dubious statistical method of "correcting" these curves, and DC data, for variations in the 60-Hz let-go threshold mean value was used. For each group of subjects tested at a given frequency, a 60-Hz let-go threshold mean was also found. A "corrected mean" for each frequency was calculated as (the actual mean of the sample group at the given frequency) × (the 60-Hz mean for the group of 134 men)/(the 60-Hz mean for the current sample group). These corrected means were also used in calculation of percentile ranks at each frequency. The correction factors ranged from 1.00 (at 60 Hz) to 1.09 (at 5 Hz).

muscular contractions. Sudden changes in current magnitudes produced muscular contractions, and interruption of the current produced a very severe shock.... The maximum [level of discomfort] a subject could take and release was termed the release current. It represents the limit of voluntary endurance rather than the let-go limit." The DC release current mean of 28 men was 73.7 mA. After "correction" for variation in the 60-Hz let-go threshold for this group, the 99.5 percentile rank "reasonably safe" DC current value for men was 62 mA. Because at 60 Hz the mean value for let-go threshold values of women was about 66% of that for men (Table 8.3), it was felt that the 99.5 percentile male values at 5 to 10,000 Hz, and at DC, could be scaled by 0.66 to obtain estimates of reasonably safe values for women.

8.6   Analysis of the Let-Go Phenomenon

Analysis of factors that may be involved in the let-go phenomenon first requires consideration of forearm/hand functional anatomy (especially with regard to power grip generation), and the current paths and

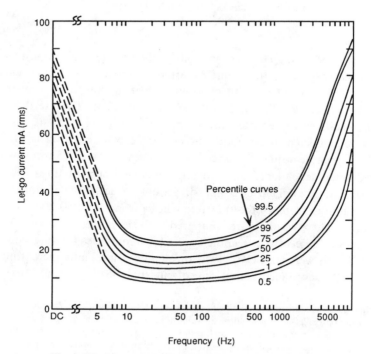

Figure 8.9   The effect of stimulus frequency on the let-go current threshold for adult males. Values on the left-hand axis are for DC. (Adapted from Dalziel, © 1972, IEEE, using some additional data from Dalziel et al., 1943.)

impedances of hand-contact accidents. The fundamental principles of reuromuscular electrical stimulation presented previously in this chapter can then be utilized.

Functional Anatomy of the Hand and Wrist

A brief introduction to the anatomy and kinesiology of the hand and wrist is essential to an understanding of how powerful gripping actions occur. More detailed descriptions can be found in *Gray's Anatomy* (Gray, 1989), Kaplan's excellent functional anatomy text (1984), and Rasch (1989). The forearm (antebrachial) muscles are divided functionally into flexors and extensors (anterior and posterior), although often complex interactions are possible. Foremost among the flexors of this group that contribute to powerful gripping are the flexor digitorum profundus muscle (the medial part of which is innervated by the ulnar nerve, and the lateral by the anterior interosseous branch of the median nerve) and the flexor digitorum superficialis muscle (innervated by the median nerve) (see Figs. 8.10, 8.11, and 8.12). In a transverse section through the middle of the forearm (Fig. 8.12), the relatively large fraction of cross-sectional area occupied by these two muscles (52% of the area taken up by muscle, nerve, and blood vessels) attests to their potential power. Maximal flexion of the hand thus overcomes maximal extension. The intrinsic muscles of the hand are categorized into three groups—those of the thumb, those of the little finger, and those in the middle part of the hand. In gripping an object, the four small lumbrical muscles (innervated by the median and ulnar nerves) (Figs. 8.11 and 8.13) in particular provide flexion of the proximal segment of each finger (and under some conditions, extension of the distal finger segments) (see Gray, 1989). We can therefore hypothesize that, until the large flexor muscles of the forearm are activated, electrically activated gripping actions using only the intrinsic muscles of the hand can probably be overcome by use of the forearm extensors. The let-go threshold should occur at about the point where strong involuntary contraction of the forearm flexors has been initiated.

Body Impedances and Current Paths in Hand-Contact Accidents

We have seen in Chapter 2 that the impedance properties of dry skin are highly nonlinear and time-varying. It is likely that the major source of skin nonlinearity is the corneum. At the level of electric line voltages, a skin impedance breakdown occurs fairly rapidly. As depicted in Fig. 2.12, for example, 50-Hz AC voltage stimulation results in an initial rapid drop in impedance within a fraction of a minute, followed by a

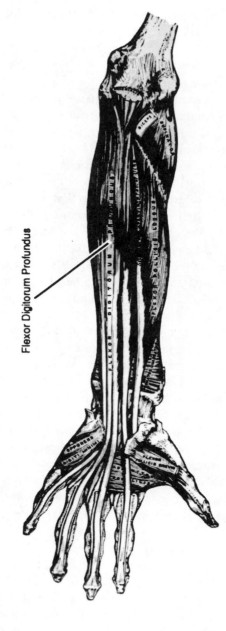

Flexor Digitorum Profundus

Figure 8.10　Anterior aspect of the left forearm. Deep muscles. (Adapted from Gray, 1985, p. 533.)

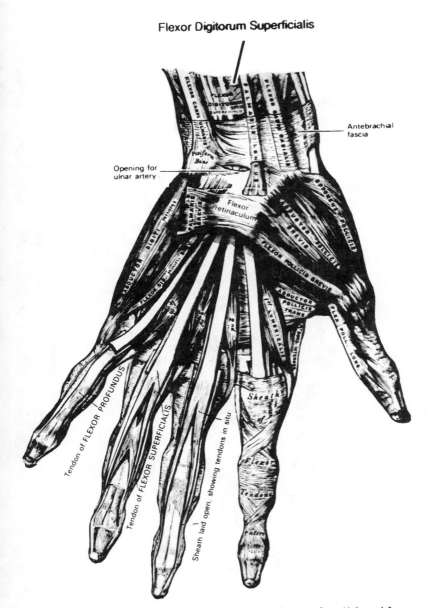

Figure 8.11 The muscles of the left hand. Palmar surface. (Adapted from Gray, 1985, p. 553.)

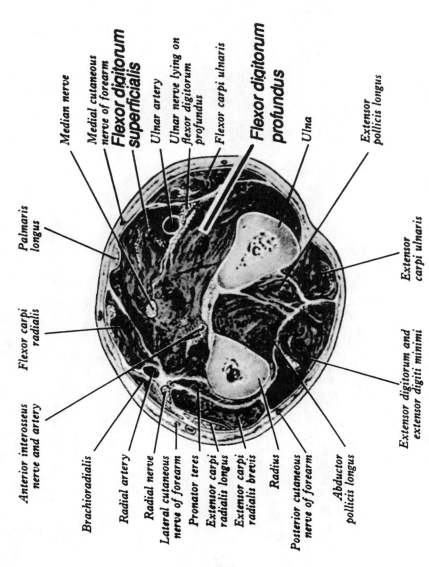

Median nerve

Medial cutaneous
nerve of forearm

**Flexor digitorum
superficialis**

Ulnar artery

Ulnar nerve lying on
flexor digitorum
profundus

Flexor carpi ulnaris

**Flexor digitorum
profundus**

Ulna

Extensor
pollicis longus

Palmaris
longus

Flexor carpi
radialis

Anterior interosseus
nerve and artery

Brachioradialis

Radial artery

Radial nerve

Lateral cutaneous
nerve of forearm

Pronator teres

Extensor carpi
radialis longus

Extensor carpi
radialis brevis

Radius

Posterior cutaneous
nerve of forearm

Abductor
pollicis longus

Extensor digitorum and
extensor digiti minimi

Extensor
carpi ulnaris

Figure 8.12 Transverse section through the middle of the left forearm. (Adapted from Gray, 1989, p. 620.)

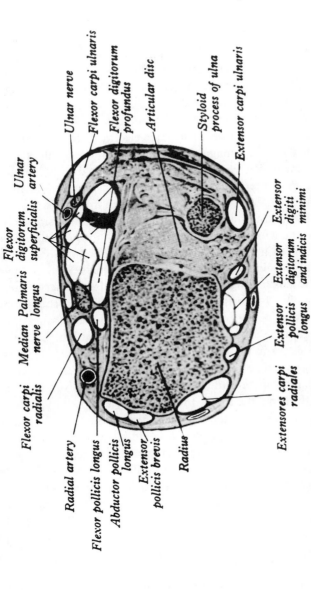

Figure 8.13   Transverse section adjacent to the wrist, passing through the distal end of the right radius and the styloid process of the right ulna, made with the hand in full supination; distal (inferior) aspect. (Adapted from Gray, 1989, p. 625.)

slower drop in the first minute, and then more gradual decreases in the next 7 to 8 min until a plateau value is achieved [see Sec. 2.2 and Freiberger (1934)]. This phenomenon may partially explain Dalziel's observation that, near the let-go threshold, failure to interrupt current flow promptly rapidly results in an inability to do so.

While the total internal body impedance, pathways, and current densities that occur in sinusoidal-voltage electrical stimulation are highly dependent on the voltage magnitudes and frequencies as well as electrode(s) size and placement, some useful generalities can be drawn. Hand-to-hand and hand-to-foot internal impedances for line voltages and frequencies are in the range of 500 to 750 Ω. As can be seen from Figs. 2.20 and 2.21, about 50% of the total internal body impedance in this situation can be attributed to the wrists and/or ankles. Cross sections through these areas (e.g., Fig. 8.13) reveal that relatively poorly conducting bone, tendon, fat, and ligament dominate. Current densities in these regions should be highest in the existing higher-conductivity soft tissues (nerve trunks) and blood vessels. We would therefore expect that, in hand-contact electrical accidents, current density will be very focused directly beneath the electrode in the muscle and nerve of the palm, and within the median and ulnar nerve trunks in the wrist.

To a first approximation, current densities beneath a cylindrical object held in the hand can be estimated by considering the surface area of the electrode (although if the rod is not completely enclosed by tissue, current densities will be highest at the "edges" of the conductor). The No. 6 AWG wires used by Dalziel and colleagues would have an approximate surface area of 11.1 cm^2 when held by an average adult male, assuming a typical hand breadth of 8.6 cm (McConville et al., 1980) and No. 6 AWG diameter of 0.41 cm. Given a total current I (in mA), a current density of $0.09I$ (mA/cm^2) would therefore occur close to the electrode surface. For adult females, average hand breadth is approximately 7.8 cm (Young et al., 1983), so current density would be $0.10I$ (mA/cm^2). Typical hand width of a 5-year-old boy is about 5.7 cm (Snyder et al., 1975), so current density would be $0.14I$ (mA/cm^2). An estimate of current densities within the median and ulnar nerve trunks near the wrist can be obtained from Fig. 8.13. Tissues other than bone, cartilage, fat, and tendon in Fig. 8.13 comprise less than 5% of the wrist cross-sectional area at this level. From anthropometric estimates of wrist circumferences [17 cm for adult males (McConville et al., 1980); 15.7 cm for adult females (Young et al., 1983); about 11 cm for small children], we can roughly calculate wrist cross-sectional areas making the simple assumption that all wrists have the elliptical "shape" of Fig. 8.13. Performing these calculations yields estimates of 21.3, 18.2, and 8.9 cm^2 for adult males and females, and for small children, respectively. For total current I, current densities of $0.94I$, $1.10I$, and $2.25I$ (mA/cm^2) result for men, women, and children making the simple approximations that most current will flow in the nerve trunks and blood vessels at this level, that these tissues have roughly the same conductivity, and that such tissues comprise approximately 5% of the total cross-sectional wrist area regardless of sex or age. This is in contrast to densities for the more proximal forearm. From Fig. 8.12 we see that skeletal muscle dominates the cross section at the mid-forearm, and highly conductive tissues (muscle, nerve, and blood vessels) comprise about 72% of the cross-sectional area. This relatively high percentage decreases rapidly as the wrist is approached [see cross sections in Morton (1944)]. Using this information from Fig. 8.12, and anthropometric data on typical mid-forearm circumferences [23.4 cm for adult males (McConville et al., 1980); 21.2 cm for adult females (Young et al., 1983); 16 cm for 5-year-old children (Snyder et al., 1975)], we can again estimate current densities within the relatively conductive tissues at this level. Assuming that the elliptical shape of Fig. 8.12 is typical regardless of sex or age, current densities of $0.032I$, $0.039I$, and $0.069I$ (mA/cm^2) result for men, women, and children, respectively.

Excitation Threshold Estimates

Recall from Secs. 4.7 and 8.4 that, to a first approximation, changes in the extracellular electric field (i.e., the change in voltage gradients, or second difference of voltage values) along an excitable fiber provide the driving force for eliciting transmembrane potential shifts. Given current density estimates, J, we can calculate voltage gradient values, $\Delta\Phi$, using the relation $\Delta\Phi = J\rho$, where ρ is an estimate of bulk tissue resistivity.

Making use of the current density estimates derived in the previous section, we can therefore predict analogous voltage gradients if we assume a certain tissue resistivity value. While the conductivities of both skeletal muscle and peripheral nerve trunks are anisotropic, with both a spatial and temporal dependence (see Chapter 2), we will use 200 Ω cm as an approximate value for the lumped resistivity for muscle, nerve, and blood vessels with longitudinally oriented current at power-line frequencies [see Table 2.1 and Sances et al. (1983)]. Voltage gradients immediately adjacent to a No. 6 AWG wire (passing a current I in mA) held in the hand of adult males and females, or small children (5-year-old boys), become $18I$, $20I$, and $28I$ (mV/cm), respectively. In the nerve trunks of the wrist we would expect voltage gradients of $188I$, $220I$, and $45I$ (mV/cm), and at the mid-forearm 6.4I, 7.8I, and 13.8I (mV/cm) (for men, women, and small children, respectively).

As we have seen in Chapter 4, not only must a longitudinally oriented voltage gradient exist along myelinated nerve fibers for electric stimulation to occur, but a *change* in voltage gradient is necessary. The terms $(V_{e,n-1} - 2V_{e,n} + V_{e,n+1})$ in Eq. (4.31) imply that a second-difference sampling of nodal extracellular voltages provides the driving force for excitation. That is, a change (difference) in two first-difference terms ($\{V_{e,n-1} - V_{e,n}\}$ and $\{V_{e,n+1} - V_{e,n}\}$) is needed. Each first-difference term is essentially a linear sampling of the extracellular voltage gradient between two nodes of Ranvier. If the electric field along a nerve fiber is truly uniform, then in theory excitation cannot occur. Realistically, in this situation spatial orientation changes (e.g., bends) in nerve fibers tend to introduce apparent second-difference effects. Likewise, terminations of motor myelinated nerves at end plates, or sensory axons at receptors, allow excitation within a uniform field (see Sec. 4.5).

The peripheral myelinated nerve fibers of α-motor neurons range in diameter from approximately 9 to 20 μm (see Sec. 8.2). If we assume a factor of 100 difference between fiber diameter and internodal length (Table 4.1), then large (20-μm) fibers will have internodal lengths equal to about 2 mm, while small (10-μm) fibers will have internodal lengths of about 1 mm. As we have seen in Sec. 8.4, large fibers tend to be the most easily stimulated by fairly diffuse extracellular electric field distributions, leading to the phenomenon of "reversed recruitment." With either the myelinated nerve model described in Chapter 4, or compartmental cable models of mammalian myelinated nerve (Sweeney et al., 1987), the author has found that second differences of approximately 50 mV are needed to bring lengths of nerve fibers to threshold with long-pulse width (rheobase) stimuli. A 20-μm-diameter nerve fiber lying within a uniform voltage gradient of 25 mV/mm would experience a second difference of 50 mV if the nerve turned abruptly by 90° with respect to the field (similarly, 50 mV/mm will excite a 10-μm-diameter fiber). In Sec. 4.5 it is reported that in a terminated nerve model (e.g., roughly representative of axons ending at a motor end plate) a voltage gradient of 6.2 mV/mm suffices to excite a 20-μm-diameter myelinated nerve (12.3 mV/mm for a 10-μm fiber). These lower values presumably result due to the abrupt impedance transition at the cable end. All of the above voltage gradients apply for square-wave stimuli having durations several times greater than the chronaxie. For 60-Hz sinusoidal stimuli, the peak threshold current amplitudes are about 17% greater than those of square-wave stimuli (Fig. 4.15).

Table 8.5. *Analysis of Hand-Contact skeletal muscle stimulation*

	Current density (mA/cm²)	Voltage gradient (mV/cm)	20-μm Fiber threshold (mA rms)	10-μm Fiber threshold (mA rms)	60-Hz let-go current (mA rms)[a]
Directly beneath No. 6 AWG electrode in lumbrical muscles of the hand					
Adult males	0.09I	18I	2.9	5.7	—
Adult females	0.10I	20I	2.6	5.1	—
Small children	0.14I	28I	1.9	3.6	—
At median and ulnar nerve trunks within the wrist					
Adult males	0.94I	188I	1.1	2.2	—
Adult females	0.10I	220I	0.94	1.9	—
Small children	2.25I	450I	0.46	0.92	—
At the large flexor muscles of the mid-forearm					
Adult males	0.032I	6.4I	8.1	15.9	15.9
Adult females	0.039I	7.8I	6.7	13.1	10.5
Small children	0.069I	13.8I	3.8	7.4	7

[a]Estimates of Dalziel for adult males and females (Dalziel and Massoglia, 1956); single study on a 5-year-old boy as reported by Dalziel (1943).
Note: I is an arbitrary hand-contact current in milliamperes.

Directly under a hand-held electrode, or within the mid-forearm, we will therefore hypothesize that a voltage gradient of approximately 7.3 mV/mm (117% of 6.2 mV/mm) or 5.2 mV rms/mm is necessary to achieve threshold stimulation of the largest (20-μm-diameter) myelinated axons terminating at motor end plates of the lumbrical muscles in the hand or the large flexor muscles of the forearm (10.2 mV rms/mm for 10-μm fibers). For initial excitation of the median and ulnar nerve trunks within the wrist we will hypothesize that a rheobase voltage gradient of 29.2 mV/mm (117% of 25 mV/mm) or 20.6 mV rms/mm is needed, assuming that a "bending" of the nerve trunks with respect to the electric field occurs as the nerve trunks spread into branches, and the current distributes within the hand (41.2 mV rms/mm for 10-μm fibers).

Table 8.5 lists the current density and voltage gradient estimates derived above for an arbitrary current I (in mA), and the resultant estimates of absolute current levels needed to achieve threshold excitation in 20-μm- and 10-μm-diameter nerve fibers. Dalziel and colleague's mean values for the 60-Hz let-go threshold currents of adult males and females, and even Dalziel's single value for the let-go threshold of a 5-year-old boy, lie within the bounds of our estimates for threshold stimulation of. large and small α-motor neuron fibers innervating the forearm flexors (the flexor digitorum profundus and flexor digitorum superficialis muscles) (compare the last three columns of the table). This is in keeping with our hypothesis that the let-go threshold should reflect the point at which forearm flexor activation reaches a level where it cannot be voluntarily overcome.

The results in Table 8.5 predict that, at current values on the order of 1 mA (1.1 mA rms for males, 0.94 mA rms for females), typical adult males and females should experience initial, relatively small contractions of intrinsic muscles in the hand (such as the lumbricals) due to excitation of 20-μm-diameter fibers in the nerve trunks of the wrist. These estimates can be compared with Dalziel's studies on perception current thresholds for hand-contact electrical stimulation. [Recall from Chapter 7, and Sec. 8.2, that large-diameter $A\alpha$ sensory nerve fibers will also be present in the nerve trunks of the wrist. For example, muscle spindle primary endings are innervated by type IA fibers, which average about 17 μm in diameter and can be 20 μm (see Fig. 3.12).] Our predictions correspond quite well with the actual 60-Hz perception threshold mean values (1.1 mA rms for adult males, and 0.7 mA rms for females) (Dalziel, 1972). Somewhat surprisingly, current densities directly beneath a No. 6 AWG electrode should result in initial muscle activation only at somewhat higher thresholds (e.g., 2.9 mA for adult males). These values probably reflect overestimations of the actual values because, in reality, the edges and proximal surface of the electrode will carry the majority of the current.

Due primarily to our anthropometric calculations (above), our estimates predict that adult females should have let-go thresholds about 82% of those of adult males. Recall that Dalziel and colleagues estimated that female thresholds would be about 66% of those of males. Our overestimate may be due in part to the fact that females tend to have a higher percentage of body fat than males (e.g., see Malina, 1975); a factor that we have not taken into account in estimating cross-sectional areas of high conductivity in the forearm. Our estimate for the let-go threshold range in a 5-year-old child (3.8 to 7.4 mA rms) is reasonably close to the value of 7 mA rms for a 5-year-old boy as reported by Dalziel (1943). This current range estimate is 47% of our predicted adult male range, a factor in good agreement with the 50% factor originally proposed by Dalziel (1943). More accurate modeling of let-go thresholds for children could incorporate known relations between anthropometric measures and the child's sex and age. It is also known that, during development, the relative composition of bone, fat, and muscle in the arm tends to change (see Johnston and Malina, 1966). As Banks and Vinh (1984) have emphasized, a general safety standard for children would have to take into account that children range in age and size from infants to adults.

Thus, while Dalziel and Massoglia (1956) found that their attempts to correlate let-go currents with physical measurements of the forearm, wrist, strength of grip, etc., were inconclusive, our relatively simple estimates of let-go thresholds based on information from functional anatomy, anthropometry, bioelectricity, and electrical stimulation theory appear to indicate that such correlations should be possible. Measurements of let-go thresholds, or at least single-sinusoidal-pulse electrical stimulation thresholds, in individuals where radiographs or magnetic resonance imaging (MRI) scans of the forearm are also available, should enable more significant investigation of such correlations.

Strength-Frequency Considerations

The above analysis ignores the effects of sinusoidal frequency on stimulus thresholds (Sec. 8.3), and will be most accurate at the minimum plateau in the strength-frequency curve (Fig. 4.20). In general, the frequency dependence of let-go sinusoidal current thresholds reported by Dalziel et al. (1943) for groups of adult males matches well the form of the frequency dependence of general myelinated nerve excitation as studied experimentally (see Sec. 8.3) and as modeled using SENN approaches (Fig. 4.15). In theory it would be possible to extend the above analysis to model the frequency dependence of let-go stimulation by introducing frequency-dependent estimates of threshold voltage gradients, and average tissue conductivities.

Hart (1985) has pointed out that, in common hand-contact accidents, the electrical source is voltage (not current) regulated. Tests were performed on five adult subjects (four males and one female) utilizing the "involuntary squeeze test." With this method, a reference level of electric shock somewhat below the let-go threshold is determined, where the hand first starts to squeeze a spring-loaded probe. Hart found that, with the hand-held electrode placed against dry skin, the plateau of the strength-frequency curve for sinusoidal stimulation shifted toward higher frequencies (about 500 Hz to 3 kHz). Hart proposed that this effect occurred due to filtering of the electrical input by the skin capacitance. Therefore, at least when the skin is very dry and before any "breakdown" of the skin capacitance occurs, minimum let-go thresholds for sinusoidal regulated-voltage stimulation may occur at significantly higher frequencies than the 50- to 150-Hz level observed with regulated-current stimuli.

8.7   Effects of Electric Shock on Respiration

It has been long recognized that accidental electric shocks could inhibit or prevent the ability to breathe. In 1913, Jex-Blake (as cited by

Lee and Zoledziowski, 1964) listed three processes that could lead to death from electric shock: ventricular fibrillation, prolonged respiratory arrest that outlasts the shock itself, and asphyxia resulting from tetanic contraction of the respiratory muscles during the shock. Normal low-intensity breathing is performed almost entirely by movement of the diaphragm for inspiration, and by passive elastic recoil for expiration. The diaphragm is composed of skeletal muscle and dense collagenous connective tissue. It is innervated by the phrenic nerve, which derives from C3-C5 cervical plexes.[†] Inspiration can also be achieved by contraction of skeletal muscles that raise the rib cage (sternocleidomastoids, anterior serrati, scaleni, external intercostals). The abdominal recti and internal intercostal skeletal muscles may be used during forced expiration to depress the chest cage. The neural "respiratory centers" of the brainstem provide the control and drive for respiration.

In several of his review papers on the dangers of electrical shock, Dalziel reported anecdotally that in his group's experiments on the let-go phenomenon, they had observed that 60-Hz electrical stimulation with hand-held electrodes could inhibit or stop respiration. Dalziel and Lee (1968) reported that "...the muscular reactions caused by commercial frequency alternating currents in the upper ranges of let-go currents, typically 18 to 22 mA or more, flowing across the chest stopped breathing during the period the current flowed, and in several instances caused temporary paralysis of the middle finger. However, normal respiration resumed upon interruption of the current, and no adverse after effects were produced as a result of not breathing for short periods." This possible danger of accidental electric shock, as well as that of cardiac fibrillation (Chapter 6), are often cited as justifying the importance of let-go thresholds and the establishment of safety standards at currents well below the let-go range. An individual who is "frozen" to a conductor by his inability to release it cannot easily save himself. "Prolonged exposure to currents only slightly in excess of a person's let-go limit may produce exhaustion, asphyxia, collapse and unconsciousness followed by death" (Dalziel, 1959).

In the case of small children, as we have seen in Secs. 8.5 and 8.6, let-go thresholds can be on the order of 7–9 mA or less (see Table 8.4). In 1940 it was reported that a 4-year-old boy had died after contacting

---

[†] Electric stimulation of the phrenic nerves with implanted electrodes can be used to artificially ventilate patients who, due to high-level spinal cord damage, cannot breath unassisted (e.g., Glenn and Phelps, 1985). Geddes and colleagues have studied the possibility of electrically eliciting respiration via transthoracic stimulation with surface electrodes located in the axillary region (e.g., Geddes et al., 1990). It is thought that this technique also primarily stimulates the phrenic nerves.

an electric fence wire (*IAEI News Bulletin*, 1940). It was assumed that his hand "froze" to the conductor, although Dalziel recognized that small children exposed to painful electrical stimulation might sometimes not even attempt to extricate themselves from an electric source (Dalziel, 1972). Tests performed on the electric fence power supply indicated that a current of 9 mA was delivered when a load of 100 Ω was placed across the terminal (8.4 mA for 1000 Ω; 8 mA for 2000 Ω; 6.8 mA for 5000 Ω).

In 1961, Lee reported a clinical study of 104 electrical accidents. In 30 cases, individuals were "frozen" to the conductor. "The longer the victim was held on to the circuit the greater appeared to be his chances of developing heart and chest symptoms suggestive of impending asphyxia, and of losing consciousness. Artificial respiration was administered in two cases, one of whom was 'held on' and was being asphyxiated." Greenberg (1940) found that, in forelimb to hindlimb electrical stimulation of dogs, currents of up to 50 mA caused arrest of respiration during the period of current flow; however, respiration restarted as soon as the current was removed. Lee and Zoledziowski (1964) studied the effects of electric shock on the respiration of rabbits. With currents up to about 200 mA, respiratory arrest appeared to be due to muscular contraction in the chest. Only with exceptionally high limb-to-limb current levels, or with current paths that include the head, does it appear possible to electrically bring about an arrest of respiration that persists after the current is switched off (see review of Lee, 1964). It is presumed that this effect is caused by electrically induced damage to the brainstem respiratory centers.

9

Stimulation via Electric and Magnetic Fields

9.1 Introduction

Electromagnetic (E-M) fields are ubiquitous in modern society. Radio frequency fields, for example, are produced by a variety of communications devices. Extremely low-frequency fields are produced by power transmission and distribution lines, as well as by home wiring and appliances. Occasionally, these exposures can result in measurable short-term reactions. The electric fields from high-voltage transmission lines, for example, can provide the opportunity for electric shock under some circumstances. Recent developments in the design of magnetic resonance imagers are pressing the limits where perceptible effects might be caused by their time-varying gradient fields. Numerous research efforts have focused on biological effects associated with chronic exposure to low-level E-M fields. While chronic effects studies are necessary to assess the acceptability of long-term exposure to low-level fields, they are beyond the intended scope of this book.

This chapter concentrates on short-term human reactions related to fields whose wavelengths are long in comparison with the dimensions of the human body. The phenomena discussed here are related to charges and currents that are induced in the body by electric and magnetic fields. By concentrating on long-wavelength phenomena, it is possible to treat electric and magnetic field effects separately. This approach differs from the study of high-frequency radiation fields, where the electric and magnetic components are intimately connected. In the long-wavelength regime, the term "electromagnetic field" is often replaced by "electric and magnetic fields."

9.2 Electric Field Induction Principles

A steady electric field produces surface charges on conducting objects, including the human body. In a simplified view, positive charges are accumulated on the side of a conducting body nearest to the negative source of a steady field, and negative charges on the side nearest the positive source. If the field is alternating, the positive and negative charges alternate in position, resulting in an alternating current within the biological medium. The physical effects of charge distribution and current induction can result in short-term perceptible effects.

It is important to distinguish between *direct* and *indirect* E-field induction effects. The former refers to charges and currents induced directly in the body. The latter refers to current or charge transfer between a person and a conducting object within the field. Before proceeding further, consider physical induction mechanisms.

Induction Mechanisms

Consider a conducting object situated above a ground plane in a vertically oriented electric field. If the object is electrically connected to ground, it will acquire a net charge, which, for simple geometric shapes (flat plate, horizontal cylinder, sphere), can be expressed by (Deno, 1975a; 1975b)

$$Q = V_s C_0 \tag{9.1}$$

where C_0 is the capacitance between the object and ground, and V_s is the space potential that would be present at the centroid location of the object if it were not present, i.e., the undisturbed space potential. In a low-frequency alternating field where the electrical wavelength is much greater than the dimensions of the object, one can ignore propagation effects and resort to the so-called quasi-static solution (Kaune, 1981). In that case, the induced alternating current is given by the time derivative dQ/dt, which, for a sinusoidal field, is

$$I_s = j2\pi f V_s C_0 \tag{9.2}$$

where f is the frequency of the field, and j is the complex phaser operator (indicating 90° phase shift with respect to the E-field). In a uniform field, an equivalent expression for Eq. (9.2) is obtained by noting that $V_s = Eh$:

$$I_s = j2\pi f EhC_0 \tag{9.3}$$

where E is the electric field strength, and h is the height above the ground plane at the electrical centroid of the object. The terms in Eq. (9.3) can be conveniently regrouped as

$$I_s = (2\pi f \varepsilon_0 E)\left(\frac{hC_0}{\varepsilon_0}\right) \tag{9.4}$$

where ε_0 is the permittivity of free space. The phasor operator j has been dropped, since we will be dealing with magnitudes. The second term in

Eq. (9.4) expresses the area of a parallel-plate capacitor with small separation and having capacitance C_0. This leads to the *equivalent-area* concept for describing induced current (Deno, 1975b):

$$I_s = 2\pi f \varepsilon_0 E A_e \tag{9.5}$$

For irregularly shaped objects, A_e is understood to be the area of a flat plate that would result in the equivalent induced current. Methods for evaluating Eq. (9.5) have been determined for a variety of objects and shapes (Deno, 1975a, 1975b; Deno and Zafanella, 1982; Reilly 1979a, 1979b, 1982; Reilly and Cwicklewski, 1981).

Another parameter of interest is the *open-circuit voltage* V_0. It is the voltage that would be measured between the object and the ground plane in the absence of current conduction to ground. The open-circuit voltage can be determined from the short-circuit current by

$$V_0 = I_s Z_0 \tag{9.6}$$

where Z_0 is the impedance measured between the induction object and ground. For an object that is perfectly insulated from ground, Z_0 is determined by its coupling capacitance to ground. In this case, V_0 achieves its maximum possible value, given by

$$V_0(\text{max}) = \frac{I_s}{2\pi f C_0} \tag{9.7}$$

where V_0 and I_s apply to magnitudes, without regard to phase [the phasor operator j has been dropped from Eq. (9.7)].

Table 9.1 provides examples of the capacitance, normalized induced current (I_s/E), and normalized open-circuit voltage (V_0/E) for several objects in a 60-Hz AC field of strength E. References cited in the table discuss measurements involving these objects in electric fields.

The value of I_s given by Eqs. (9.2)–(9.5) expresses the so-called *short-circuit current*, i.e., the current that would flow in a grounded conductor that contacts an otherwise insulated object. When a grounded person provides the ground path, the current that flows at the points of contact may result in a perceptible electric shock. An insulated person touching a grounded connection would also result in current flow, but here the induction object is the person.

We can gain some insight into the induction on a person by using the theoretical expression for short-circuit current in an upright cylinder (Reilly, 1978a):

$$I_s = \frac{2\pi^2 \varepsilon_0 f h E}{\ln\{(h/r)[(4d+h)/(4d+3h)]^{1/2}\}} \tag{9.8}$$

where h is the length of the cylinder, r is its radius, d is the distance from

Table 9.1. *Example induction parameters of various objects in 60-Hz electric fields (object on insulated surface)*

Object	I_s/E (A V^{-1} m)	V_0/E (m)	C_0 (pF)
Person (1.8 m tall)	1.8×10^{-8}	0.48	100–150
Auto (midsize)	9.1×10^{-8}	0.21	200
Auto (large)	1.2×10^{-7}	0.16	2000
School bus	3.7×10^{-7}	0.26	3700
Commercial bus	4.9×10^{-7}	0.45	2900
Fence wire (ht = 1 m, length = 100 m)	2.7×10^{-7}	1.0	720

Source: Data from Bracken (1976), Reilly (1982), Reilly (1979a).

the bottom of the cylinder to the ground plane, f is the frequency of the field, and E is the strength of a vertically oriented electric field. [Units in Eq. (9.8) are amperes, Hertz, meters, and volts/meter]. The author has verified Eq. (9.8) with measurements on cylinders placed on a ground plane within a 60-Hz electric field.

If we model a person as an upright cylinder with a height-to-radius ratio of 12, and standing on a ground plane ($d = 0$), then Eq. (9.8) predicts a value of short-circuit current given by:

$$I_s = 9.0 \times 10^{-11} h^2 f E \tag{9.9}$$

When evaluated at 60 Hz, Eq. (9.9) is identical to the empirical formula determined by Bracken (1976) for people within transmission-line electric fields.

Whereas Eq. (9.9) expresses the short-circuit current at the point of ground contact, an investigator may wish to know the current density in a person who is immersed in an alternating electric field. Experimentors, using conducting models of the human body, have mapped the distribution of internally generated currents for both grounded and ungrounded humans standing within vertically oriented 60-Hz electric fields (Deno, 1977; Kaune and Forsythe, 1985). Calculations over a wider frequency range and for more complex field environments relevant to display terminals have also been carried out (Takemoto-Hambleton et al., 1988). Consider, for example, a person standing within a vertically oriented AC electric field, and grounded through the feet. In this case, the amount of current passing through any given horizontal cross section of the body increases as the cross section is moved from the head toward the feet, with the greatest current passing through the feet as illustrated in Fig. 9.1

(adapted from Deno, 1977). The horizontal axis expresses the fraction of maximum current, given by Eq. (9.9), passing through a cross section at the indicated height. The outermost curve applies to a person grounded through the feet; the other curve applies to an ungrounded person standing 13 cm above a ground plane. The maximum current in the grounded person flows through the feet. For an ungrounded person, maximum induced current flows through the midsection of the body.

The current density in a homogeneous model is the induced current divided by the cross-sectional area. One can infer from Eq. (9.9) that the current density induced in the upright body would not depend on a person's height, as long as the ratio h/r is held constant. This inference can be drawn by noting that the induced current and cross-sectional area of the model cylinder both grow in proportion to the square of height.

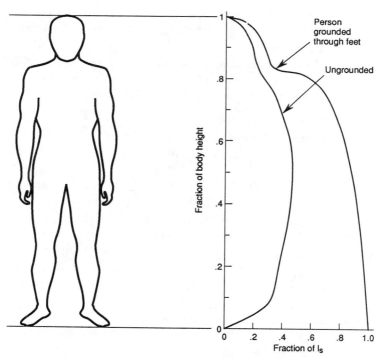

Figure 9.1   Distribution of current flowing in a person standing in a vertically oriented AC field. Horizontal axis describes fraction of maximum short-circuit current passing through cross section at indicated height. (Adapted from Deno, 1977.)

Body Surface Effects

Environmental electric field strengths are typically reported as *undisturbed* field levels, i.e., the intensity that would be measured in the absence of a biological subject. The presence of a conducting object will distort the undisturbed field, such that the field incident on the surface of the object will be significantly modified. The effect is illustrated in Fig. 9.2, which shows the distortion of a vertically oriented uniform field in the presence of a person. The orientation of the electric field is indicated by the direction of the *flux lines*, and its intensity by their density (not drawn to scale in the figure). The flux lines are normal to the surface of the body. The lines drawn within the body roughly represent the internally induced current.

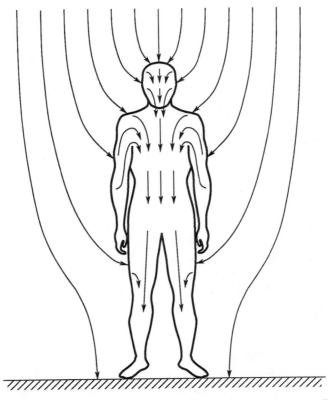

Figure 9.2   Perturbation of environmental electric field by a person. External electric field indicated by "flux lines" outside body. Arrows inside body roughly indicate direction and intensity of induced internal electric field and body current. Direction of arrows reverses every one-half cycle. Field lines are illustrative, and not drawn to scale.

Table 9.2. *Electric field enhancement and induced current for several species*

Species	Field enhancement		I_s/E [A/(V/m)]
	Avg	Top	
Human (standing)	2.7	18.0	1.6×10^{-9}
Swine	1.4	6.7	7.0×10^{-10}
Rat (resting)	0.7	3.7	1.2×10^{-11}
Rat (rearing)	1.5	*	2.4×10^{-11}
Horse	1.5	*	2.7×10^{-9}
Cow	1.5	*	2.4×10^{-9}

*Data not available.
Note: "Avg" refers to field enhancement averaged over body. "Top" refers to field enhancement at top of body. I_s/E is normalized short-circuit current at 60 Hz.
Source: Data from Kaune (1981).

As suggested by the figure, the electric field on the surface of the body will be enhanced in certain regions, particularly on the upper body, and reduced in other regions. The average field, E_a, over the entire body has been shown by Kaune (1981) to be related to the short-circuit current by

$$E_a = \frac{I_s}{2\pi f \varepsilon_0 A_b} \tag{9.10}$$

where A_b is the body surface area. Notice the similarity of Eqs. (9.5) and (9.10).

The average represented by Eq. (9.10) will be exceeded at particular regions of the body, particularly at elongated surfaces on the uppermost body. Table 9.2 expresses field enhancement factors applying to the average, or to the topmost region of the body. The table also includes normalized values of short-circuit current, applicable to 60-Hz fields. The enhancement factor indicates the value of the incident field as a ratio of the undisturbed field. For example, the field at the top of a person's head will be 18 times the undisturbed field measured in the absence of the person. If the person were to raise one hand above the head, the electric field on the raised hand would be enhanced even further.

Table 9.2 applies to the situation where the source of the electric field is several body dimensions distant from the subjects. If, instead, the source were close to the subject, there would exist a further enhancement of the incident field due to the interaction of the charges on the body surface, and the charges on the conducting object.

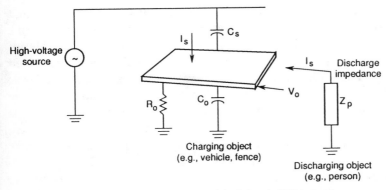

Figure 9.3   Equivalent circuit model of electric field induction.

Induced Electric Shock Principles[†]

Figure 9.3 illustrates a conceptual model of ELF electric field induction. An AC high-voltage source, such as an electric power transmission line, is capacitively coupled to a charging object through C_s. The object is also capacitively coupled to ground through C_0, and may include leakage resistance, indicated by R_0. The maximum sinusoidal current that would flow in a low-impedance connection to ground is the short-circuit current I_s discussed above.

The voltage and charge developed across C_0 can be discharged through a conductive path to ground, such as through a grounded person, represented by the impedance Z_p. The capacitive discharge may be conducted through direct contact with the skin, or, with sufficient voltage, through a spark. The critical field strength required to support an electrical discharge in air depends on the material of the two conductors, and, in accordance with Paschen's law, on the product of electrode separation and atmospheric pressure (Howatson, 1965). The voltage needed to conduct a spark discharge to dry, intact human skin is in the vicinity of 500 V (see Sec. 2.5). After direct contact is made with a charged object, a sinusoidal current I_p will flow into the body. For many cases of practical interests, $I_p \approx I_s$ (Reilly, 1979b), where I_s is the short-circuit current.

Figure 9.4 shows examples of voltage waveforms measured on an object energized by a 60-Hz source, and brought in contact with a subject's finger. These examples, were produced using equipment that simulates waveforms observed in field conditions [see Reilly and Larkin (1987) for further details]. Each record shows a combination of capacitive spark

[†]Portions of this section have been adapted from Reilly and Larkin (1987).

discharges, indicated by abrupt discontinuities in the voltage trace, and a period of continuous sinusoidal current while the finger rests on the electrode. Current spikes can be seen in coincidence with the abrupt changes in voltage, but the magnitude of these spikes is not represented accurately, because the sampling rate in these records ($100\,\mu s$) was relatively slow. In Fig. 9.4a, V_0 is 1000 V rms. A capacitive discharge occurs upon contact with the electrode at $t \approx 20\,ms$. In Fig. 9.4b, V_0 is 2000 V rms. Again, there is a single prominent contact discharge, but there are also several discharges just as the finger is withdrawn ($t \approx 57\,ms$). Expanded waveforms of capacitive discharges are shown in Figs. 2.28–2.32. If the voltage at the moment of contact is high enough (about 500 V), the discharge occurs through a spark; below the critical voltage, the discharge occurs only at the moment of physical contact with the electrode.

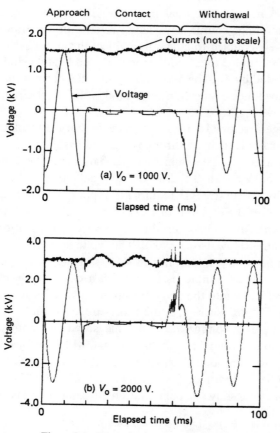

Figure 9.4   Example waveforms for AC field-induced stimuli, $C_0 = 200\,pF$.

The foregoing examples show that both the capacitive discharge voltage and the sinusoidal contact current are important components of the AC field-induced stimulus. These quantities are related in a well-defined way:

$$\frac{V_0}{I_s} = Z_0 \tag{9.11}$$

where Z_0 is the impedance of the charging object. For many cases of interest, $C_s \ll C_0$ and Eq. (9.11) can be written

$$\left|\frac{V_0}{I_s}\right| = \left[\frac{1}{1/R_0^2 + (2\pi f C_0)^2}\right]^{1/2} \tag{9.12}$$

where f is the sinusoidal oscillation frequency of the electric field. Absolute values are indicated because Eq. (9.12) applies to the magnitude of V_0 and I_s without regard to their relative phases.

The condition of greatest induced voltage on the charging object occurs when leakage resistance is large, that is, $2\pi R_0 C_0 \gg 1$. In this case, Eq. (9.12) reduces to

$$\left|\frac{V_0}{I_s}\right| \cong \frac{1}{2\pi f C_0} \tag{9.13}$$

In many practical cases, however, R_0 may be small enough that the induced voltage is substantially below its theoretical maximum (Reilly, 1979b). In such cases, the ratio V_0/I_s, given in Eq. (9.12), becomes smaller than its theoretical maximum; i.e., the open-circuit voltage becomes less prominent. From a sensory point of view, the sinusoidal current in the contact phase of the stimulus may take on greater significance than the capacitive discharge component.

In this chapter, the effects of leakage resistance are expressed in terms of the factor $k = |V_0/I_s|(2\pi f C_0)$. A functional expression for k can be obtained by multiplying Eq. (9.12) by $2\pi f C_0$:

$$k = \left[1 + \frac{1}{(2\pi f R_0 C_0)^2}\right]^{-1/2} \tag{9.14}$$

for a well-insulated object, $k = 1$, which is the value attained as $R_0 \to \infty$. The value $k = \frac{1}{2}$, for example, implies that leakage resistance reduces the induced voltage to one-half that of a well-insulated object.

A very slow approach to an AC-charged source provides conditions in which a sequence of capacitive spark discharges can occur. Both the average number of discharges and the average voltage at which they occur will increase as the approach velocity is reduced and as the peak sinusoidal voltage is increased. Figure 9.5 illustrates some typical discharge patterns associated with very slow approaches using a device that could move an electrode toward a stationary fingertip with a fixed, controllable velocity. In each example, the charging voltage is 1900 V rms, and the capacitance is 100 pF. Each discontinuity in the waveform represents a single spark discharge. Notice that the discharges tend to occur at lower voltages in successive cycles of the sinusoidal waveform, but that nearly all of the discharges extinguish at a minimum voltage of about 600 V associated with the spark plateau described in Sec. 2.5. The number of discharges is greatest at the slowest approach. In each trace, discharges begin to occur singly on the individual half-cycles of the 60-Hz waveform. Then, as electrode distance is shortened, discharges occur in clusters of increasing number and decreasing peak amplitude.

To gain some perspective about the factors that influence multiple discharges, the author and colleagues recorded a large number of waveforms while subjects tapped electrodes of different sizes, and at different AC charging voltages. The number of discharges occurring at finger approach and withdrawal were counted, and each discharge was measured. These counts and measurements were also made with an electrode moving at a controlled low velocity toward the skin. The results of these measurements were compared with a computational model, in which a grounded electrode was assumed to approach an energized electrode in

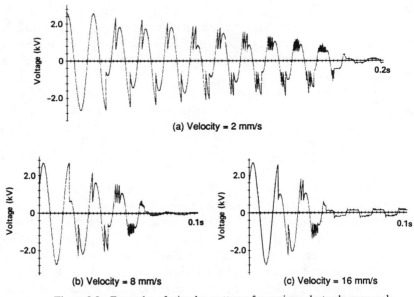

Figure 9.5   Examples of stimulus patterns for various electrode approach velocities. Energizing voltage is 1.9 kV rms. Horizontal axis is elapsed time; $C_0 = 100$ pF. (*a*) Velocity = 2 mm/s. (*b*) Velocity = 8 mm/s. (*c*) Velocity = 16 mm/s. (From Reilly and Larkin, 1987.)

random phase with respect to the 60-Hz charging voltage (Reilly et al., 1983).

Table 9.3 illustrates averages of withdrawal and approach data for two subjects, each involving 1200 tapped electrode trials. The important

Table 9.3. *Statistical summary of discharge number for AC field stimuli*

	Mean number of discharges per tap			
	Ball electrode		1-mm electrode	
Voltage (V rms)	Approach	Withdraw	Approach	Withdraw
600	0.95	0.23	0.95	0.69
900	0.97	0.37	0.94	0.93
1200	1.05	1.04	1.01	1.38
1500	1.14	1.39	1.12	1.54
1800	1.23	1.75	1.15	1.90
2400	1.57	2.71	1.36	2.98

Note: Electrode tapped with the fingertip, using light tap force.
Source: Reilly and Larkin (1987).

Table 9.4. *Stimulus characteristics for very slow approaches to electrode*

Voltage (kV rms)	Velocity (mm/s)	n_d	n_c
1.09	2	20.8	7.6
	4	11.7	3.9
	8	4.4	2.0
	16	1.7	1.3
1.31	2	27.2	8.8
	4	18.0	5.2
	8	10.1	3.1
	16	4.0	1.9
1.57	2	49.5	16.7
	4	24.8	7.6
	8	16.4	4.7
	16	6.2	2.2
1.89	2	60.6	18.2
	4	33.8	9.8
	8	20.9	5.5
	16	8.3	3.4

Note: n_d = average number of discharges during stimulus pattern; n_c = average number of waveform half-cycles during which discharges were present.
Source: Reilly and Larkin (1987).

findings for a tapped electrode can be summarized as follows. (1) Below 900 V, multiple discharges are rare. Typically, only one discharge is observed, in the approach stage. (2) As the AC voltage is raised above 900 V, the number of discharges increases. The increase is greater in finger withdrawal that in approach. (3) There are generally more withdrawal discharges for a small electrode than for a large one. There is little dependence on electrode size in the approach stage. (4) When multiple discharges occur, they cluster within individual quarter-cycles of the sinusoidal waveform.

For very slow approach velocities (less than 20 mm/s), and adequate voltage (> 500 V peak), a series of spark discharges may be obtained on individual half-cycles of the alternating induced voltage function. Table 9.4 illustrates statistical data on the number of discharges for voltages from 1.09 to 1.89 kV rms. In Chapter 7 we show how the sensory system integrates these repeated stimuli.

9.3   Direct Perception of ELF Electric Fields
Mechanisms for Human Detection

Internal body currents from electric field induction have been considered for their possible biological effects in chronic exposure to power-frequency environmental fields. However, such currents can be easily dismissed as having any ability to stimulate excitable tissue. This statement can be easily demonstrated by considering the electric field necessary to excite peripheral nerves. Consider, for example, a person of 1.8 m height standing within a 60-Hz field, and grounded through both feet. If we consider the combined cross section of the two lower legs (less bone) as 35 cm^2, and the minimum current density threshold as 0.1 mA/cm^2 (see Sec. 7.6), then we calculate from Eq. (9.9) that an undisturbed field strength of 200 kV/m would be required to stimulate peripheral nerves in the ankles. Considering that the undisturbed field is enhanced at body surfaces (18 times, for example, at the head, as noted in Table 9.2), the body would be in a state of severe corona at the stated field level. Nevertheless, low-frequency electric fields can be perceived directly by humans and animals through other means. The mechanism for human detection is thought to be due primarily to the vibration of body hair or clothing. Hair has been observed to vibrate at both 60- and 120-Hz rates in a 60-Hz electric field; these rates suggest two different mechanisms of hair vibration (Deno and Zafanella, 1982; Cabanes and Gary, 1981). When the hair is hydrated, it appears to be sufficiently conductive that charges induced on the surface of the skin are free to move along the hair shaft. This phenomenon would result in a mutual repulsion between individual hair follicles, with a maximum force occurring on each one-half cycle of the electric field. The result would be a 120-Hz vibration rate in a 60-Hz field. Dry hair, in contrast, is a very poor conductor, on which environmental ions may be held relatively immobile. The force on dry hair would consequently respond directly to the electric field, resulting in a 60-Hz vibration within a 60-Hz field. As a general observation, low temperature and humidity decrease human sensitivity to ELF field detection, whereas moisture caused by humidity or perspiration increases sensitivity. At low stimulation levels, subjects typically perceive a 60-Hz electric field as a gentle breeze on exposed body surfaces (Reilly, 1978; Kato et al., 1989). A tingling sensation between body and clothes is also observed. At higher levels, subjects describe a distributed tingling or crawling sensation on the skin.

Tests on human perception of 60-Hz electric fields were carried out at the high-voltage research facility known as "Project UHV." During two months of testing, approximately 122 men and 8 women participated. Participants walked within an electric field produced by an overhead transmission line, stopping at selected points corresponding to a gradually

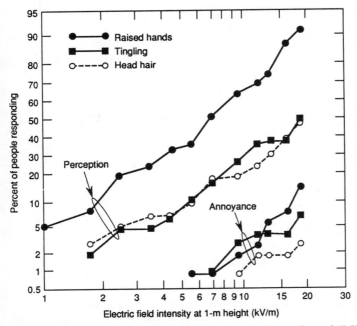

Figure 9.6   Response thresholds of men to the direct effects of 60-Hz electric field. Temperature = 50–65°F, both dry and wet conditions included. (From Reilly, 1978.)

increasing electric field. At each stopping point, participants considered the sensation on their raised hands, head, head hair, or tingling between body and clothes (hands at sides). The participants recorded one of three responses: no feeling, perception, or annoyance. Figure 9.6 (from Reilly, 1978b) summarizes the responses of the men. A few participants reported perception at remarkably low field levels, such as 2 kV/m or less. It would require more carefully controlled tests to be sure that participants are actually responding to low field levels rather than confusing other stimuli such as a breeze. For the group of 8 women, the median threshold of stimulation on the raised hands was 17.5 kV/m, versus 6.7 kV/m for a group of 60 men tested on the same day. The higher threshold for women is presumed to be due to the fact that their body hair is typically shorter than men's. The fact that E-field detection thresholds are indistinguishable in male and female mice (Stern and Laties, 1985) tends to make other gender-related explanations less plausible.

Other tests conducted with 50-Hz fields at an indoor test facility (Cabanes and Gary, 1981) have found higher thresholds for field percep-

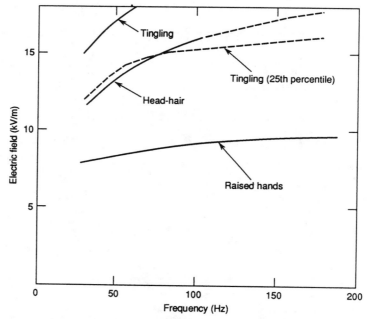

Figure 9.7   Median values of threshold of perception of hair sensations of people as a function of the frequency of the electric field. For "tingling" the 25th percentile also is given. (From Deno and Zafanella, 1982. Copyright 1987. Electric Power Research Institute. EL-2500. *Transmission Line Reference Book: 345 kV and Above*, 2nd ed. Reprinted with permission.)

tion than shown in Fig. 9.6. In tests with 75 subjects (65 male, 10 female), median thresholds were found to be 25 kV/m with the hands at the sides, and 12.5 kV/m with one raised hand. The fact that these thresholds are higher than those for outdoor tests (Fig. 9.6) might possibly be due to a less humid climate in the indoor facility. This speculation cannot be verified, since relevant climatological data were not published for the indoor tests.

Figure 9.7 illustrates the connection between the electric field frequency and perception levels (from Deno and Zaffanella, 1982). Perception thresholds were least at the lowest tested frequency (30 Hz), with a gradual loss of sensitivity as the frequency was increased. For the data represented in Fig. 9.7, the rate of hair vibration was not reported; as noted earlier, it could be either at the frequency of the applied field, or twice the frequency, depending on the degree of hair hydration.

Cutaneous vibration sensitivity has a frequency response counter to that indicated for E-field sensitivity. Hair receptors are examples of *rapidly adapting* mechanoreceptors that are responsible for detection of vibration on hairy skin. These receptors have a U-shaped threshold curve when plotted against the frequency of excitation, as shown in Fig. 3.20. A complete model for electric field detection may have to include the mechanical compliance and electric impedance properties of the hair follicles.

Human perception is greatly enhanced when a strong DC field is combined with an AC field, as shown in tests by Clairmont and colleagues (1989). Test subjects recorded reactions within electric field environments produced by conductors energized at various AC or DC potentials. Individuals recorded sensation levels at measurement locations where the AC field, DC field, and ion current densities were monitored. Sensation levels were reported on a six-point scale defined as follows:

0: not perceptible
1: just perceptible
2: definitely perceptible, but not annoying
3: slightly annoying
4: very annoying
5: intolerable

Figure 9.8, illustrating head hair perception, shows that AC and DC fields have a synergistic effect—the combined fields can produce a much stronger reaction than either field taken separately. Such synergism might be explained by a mechanism where ions are produced by the DC conductors and collected on body surfaces; the alternating component of the fields would provide an oscillatory force for hair vibration. While such an explanation appears plausible, note that even a pure DC field is perceptible. Detection of a DC field may also be due to hair vibration, which has been reportedly observed beneath a DC high-voltage transmission line. Such hair movement is not clearly understood but may be associated with corona on the tips of the hair shafts. Laboratory studies by Kato and colleagues (1986) failed to detect hair movement within a steady DC field. Differences between laboratory and environmental observations might be accounted for on the basis of the density of ambient air ions, and relative humidity.

In other laboratory experiments, Kato and colleagues (1989) exposed the hairy arm and the palm of the human hand to 50-Hz electric fields. The median perception threshold was 50 kV/m on days that the relative humidity ranged from 65% to 78%. On dry days with humidity ranging from 22% to 33%, nearly all subjects failed to perceive fields up to 115 kV/m, although one exceptional subject detected a field at 115 kV/m. In all cases, perception was associated with hair movement. When the hairless hand was exposed, the field could not be detected up to the maximum level used in the experiments (115 kV/m).

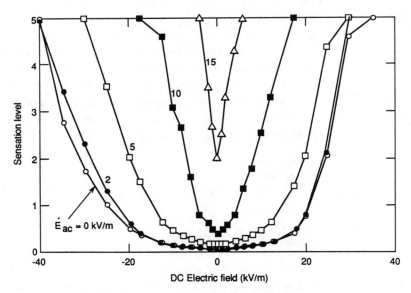

Figure 9.8   Reaction levels for exposure to mixed DC and 60-Hz AC electric fields; stimulation of head hair while standing in field. Indicated field levels apply to undisturbed field. Horizontal axis applies to DC field strength; curve parameters apply to AC field strength. (From Clairmont et al., © 1989, IEEE.)

The field levels cited in these experiments may seem large in comparison with those indicated in Figs. 9.6–9.8, but may be accounted for by differences in the definition of exposure levels. In Sato's laboratory experiments, the reported fields are those on the surface of the skin; in the environmental measurements (Figs. 9.6–9.8), the reported levels apply to the *undisturbed* field, i.e., the rms value of the field in the absence of a person. The difference between these two definitions is significant, as suggested by Table 9.2.

A variety of studies have examined detection of ELF fields by laboratory mammals (Stern et al., 1983; Stern and Laties, 1985; Sagan et al., 1987; Weigel et al., 1987). Weigel's experiments with cats have provided significant insight into mechanisms of transduction in electric field perception. In these experiments, both normal and hairless cats' paws were exposed to strong 60-Hz electric fields. Thresholds as low as 160 kV/m were noted, where the field value applies to a measurement directly on the surface of the exposed body part. That threshold value is similar to the perception threshold noted in Fig. 9.5 for human detection, if one accounts for the field enhancement on local body parts.

Weigel and colleagues isolated receptor types and their receptive fields in the cat's hind limb by monitoring action potential activity from the spinal dorsal roots that supply those regions. They concluded that the most sensitive mode of transduction was via rapidly adapting receptors. It was thought that hair vibration could activate hair follicle receptors directly, or could vibrate the skin and activate other types of nearby receptors, such as pacinian corpuscles. In some cases, the later mode appeared to be the more sensitive mechanism for transduction. But even in the hairless paw, electric field stimulation was still noted, although at somewhat increased threshold levels. Stimulation in the hairless paw was thought to be from vibrational displacement of the skin due to dielectric forces across the corneum. This postulated mechanism is very similar to the one discussed in Sec. 7.2, in connection with the detection of microampere currents.

One should exercise great caution in comparing electric field detection thresholds of humans and laboratory mammals because of differences in body hair, as well as body shape, size, and orientation. Comparisons with nonmammalian experimental species require even more caution, because of the possibility of differences in detection mechanisms. Certain highly sensitive fish, for example, are able to detect E-fields within the marine environment as low as 5 nV/cm (Kalmijn, 1990). The existence of a unique receptor and signal-processing mechanisms are said to be responsible for this remarkable capability.

9.4   Human Reactions to AC Field-Induced Shock[†]

Stimuli produced through AC electric field induction are more complex than the single capacitive discharges discussed in Sec. 7.3. We have seen in Sec. 9.2 that AC stimuli may contain one or more capacitive discharges when physical contact with the induction object is made and broken, as well as sinusoidal current during the period of maintained physical contact.

The individual discharges of the AC field-induced stimulus are identical to the capacitive discharges discussed in Chapter 7. Thus, the factors that affect sensitivity to individual discharges are also important with AC stimuli. With AC field-induced stimuli, however, additional factors are important, including multiple discharges and sinusoidal contact current. The strength of the individual discharge, relative to the contact current, will depend on both the capacitance and leakage resistance of the AC field-charged object, as suggested by Eq. (9.12). It is therefore important to evaluate AC field stimulus sensitivity throughout a range of capacitance and leakage-resistance values.

Perception Thresholds

Chapter 7 provides a treatment of the measurement and interpretation of sensory responses to various types of electrical stimulation, including capacitive discharges. Unlike tapped-electrode, single-capacitive discharges, AC thresholds depend on electrode size; thresholds with a large electrode (3.44-cm diameter) are about 40% higher than those with a small electrode (1-mm diameter). This effect can be accounted for by differences in stimuli for the two electrode sizes. Capacitive spark discharges at the point of withdrawal from the small electrode occur more frequently than with the large electrode (see Table 9.3). Furthermore, the sinusoidal current density during the contact phase of the stimulus is much higher with the small electrode. In this section, data are given only for the larger electrode, as this represents the more practical case of interest.

The sinusoidal contact component of the AC stimulus has a peak amplitude that is considerably below the peak of the capacitive discharge current. Nevertheless, the contact component can be perceptually potent—experimentally, we find that 60-Hz sinusoidal currents can be detected with large-area fingertip contact at an average current of about 0.25 mA (Reilly et al., 1983). This value is consistent with 60-Hz touched-contact thresholds noted in other studies (see Chapter 7).

To further compare sensitivity for AC field stimuli with single-discharge stimuli, the author and colleagues tested the two kinds of stimuli randomly

[†]Parts of this section have been adapted from Reilly and Larkin (1987).

interleaved during the same testing session. Capacitance was also varied randomly, among values of 100, 800, and 6400 pF. At each capacitance value, we determined the individual's threshold ratio, defined as the voltage threshold for a single capacitive discharge divided by threshold rms voltage for the AC field stimulus. For the six subjects tested, there were wide differences (about 2:1) in absolute levels of sensitivity. Nevertheless, the threshold ratios were highly similar at each capacitance. The implication of this result is that, by knowing an individual's sensitivity to one type of waveform, it is possible to determine his or her sensitivity to another waveform using a simple multiplicative constant. The constant will, of course, depend on the details of the two waveforms being compared.

Figure 9.9 plots AC threshold sensitivity contours. AC thresholds decrease with increasing capacitance, and with decreasing leakage resistance. [See Eq. (9.14) for a definition of the leakage-resistance parameter k.] The resistance effect stems from the greater sensory contribution of the sinusoidal contact current component as leakage resistance is reduced. The fact that thresholds are lower with reduced leakage resistance does not mean that reduced resistance makes the shock exposure worse. If the induction electric field is held constant, decreased leakage resistance

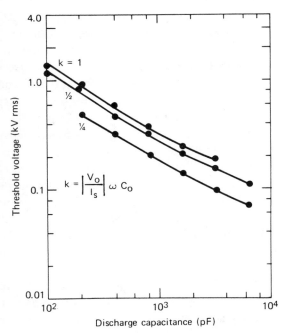

Figure 9.9   Perception threshold for 60-Hz AC field-induced stimuli. Tapped electrode. Curves for $k = \frac{1}{2}$ and $\frac{1}{4}$ indicate effect of leakage resistance. (From Reilly and Larkin, 1987.)

Table 9.5. *Multiples for suprathreshold cate-gories with 60-Hz AC-field-induced stimuli*

Response category	Capacitance		
	200 pF	800 pF	6400 pF
Unpleasant	2.1	2.2	1.8
Pain	2.9	2.9	2.4
Tolerance	4.3	6.2	3.5

Note: Tabulated values indicate multiples relative to perception thresholds for given response categories. Data apply to fingertip stimulation.
Source: Reilly and Larkin, 1987.

reduces the induced voltage even more than it does the threshold. For example, comparing $k = \frac{1}{4}$ with $k = 1$, the induced voltage is reduced by the multiple 0.25, but the perception threshold at $C = 3200$ pF is reduced by the multiple 0.52 (see Fig. 9.9).

Suprathreshold Effects[†]

Suprathreshold reactions on the fingertip were tested for AC E-field induction stimuli in a manner similar to that described in Chapter 7 for single capacitive discharges. In the magnitude-estimation procedure, perceived magnitude for AC stimuli on the fingertip was found to grow at the 2.7 power of induced stimulus voltage at 6400 pF, and at the 2.2 and 2.8 powers at 800 and 200 pF, respectively. These exponents are not very different from those observed for single-discharge stimuli (Chapter 7).

Table 9.5 lists, for six subjects, the mean affective response category thresholds as multipliers of individual perception thresholds. The multiples are for AC field-induced stimulation of the fingertip in active electrode touching. Compared with the corresponding multiples for single discharges (Table 7.3), AC stimuli reach the "unpleasant" category at approximately the same multiple; however, "pain" and "tolerance" categories are reached with generally lower multiples.

Multiple Discharges

When a person grazes a metallic object that is energized by an AC electric field, there is a possibility for multiple discrete discharges. As demonstrated in Sec. 9.2, even a seemingly slow approach produces

[†]See Chapter 7 for a discussion of the measurement and interpretation of suprathreshold sensory responses.

stimulus patterns not unlike a deliberate direct touch with the fingertip. It is very difficult to generate more than two to four discharges with a deliberate grazing motion of the limb. However, a continuous stream of discharges can be easily produced by pressing a shirt sleeve against an energized electrode. In the author's tests, continuous sparking through a 0.18-mm-thick fabric was easily produced at 1.3 kV and above. Multiple discharges occur because the fabric prevents electrical contact with the skin, which would preclude further capacitive discharges. The duration of discharges in this manner will depend largely on the reaction delay for the person to pull the arm away. A variety of studies on human reaction time show that about 0.25 s is needed to move the hand in response to a stimulus (Woodworth and Schlosberg, 1954). This period corresponds to 30 half-cycles of a 60-Hz stimulus.

To produce multiple discharges reliably in the laboratory, a device was used to move an electrode toward the skin in a repeatable controlled motion. Figure 9.5 gives examples of the discharge patterns. Sensory magnitude estimates for repeated discharges were obtained in a manner similar to that described for single-discharge stimuli (Chapter 7). On each test trial, electrode approach velocity was randomly set at either 2, 4, 8, or 16 mm/s.

Figure 7.11 illustrates perceived sensation magnitude (geometric means for three subjects), plotted against n_c, the number of 60-Hz half-cycles during which capacitive discharges were produced. The magnitude estimates were related to the affective categories "neutral" and "pain" in a separate procedure. Figure 7.11 suggests that a perceptible but neutral stimulus can become highly unpleasant if presented as a repetitive train. A rationale for representing n_c as the independent variable is that there is a trade-off between the amplitude and number of pulses in each half-cycle packet (Fig. 9.5), and that it approximately characterizes the overall stimulus. The values of n_c are the means for each velocity that was used in the procedure. The straight lines in this format indicate power functions with an exponent of 0.85, indicating nearly a one-to-one correspondence between n_c and the mean perceived magnitude; i.e., sensory enhancement was determined in this experiment almost completely by the number of half-cycles during which perceptible discharges occurred.

It is significant that the dynamic range from the threshold of perception to the limit of suprathreshold tolerance is rather narrow (Table 9.5). The limited dynamic range is especially striking when one considers that a stimulus of low sensory potency can become quite unpleasant when presented as a repetitive train. Consequently, conditions giving rise to perceptible AC field-induced shocks may, under some circumstances, also present conditions for creating unpleasant effects.

Extrapolation to Other Frequencies

If field strength is held constant (resulting in constant V_0), the induced current will be proportional to frequency [refer to Eq. (9.13)]. At the 50-Hz power frequency common in Europe, the contact current would be reduced by a factor of 50/60 (8.3%) relative to that at 60 Hz. The maximum effect on AC thresholds (Fig. 9.9) can be predicted from this observation. As capacitance is increased and leakage resistance is reduced, the contact component becomes a more salient part of AC field stimuli. This would result in an upward shift of 50-Hz thresholds, but the shift should not exceed 8.3% of the values plotted in Fig. 9.9.

If the induction frequency is increased beyond 60 Hz, the sinusoidal contact current takes on increasingly greater sensory importance in comparison with the capacitive discharge component, and, for sufficiently high frequencies, will become the dominant sensory factor. Above 1 kHz, thresholds for AC sinusoidal currents rise sharply with frequency, as noted in Chapter 7. Eventually, as frequency is increased to about 100 kHz, cutaneous electrical thresholds become so high that tissue heating determines perception, and sensory thresholds become independent of frequency.

The opportunity for multiple discharges can increase with the induction frequency. When individual capacitive discharges can be perceived, multiple discharges can greatly increase sensory potency, as noted in Fig. 7.11. Sensory effects with multiple discharges may be intensified at frequencies somewhat above 60 Hz because, at a neural level, sensory magnitude is encoded in part through the action potential repetition frequency (see Chapter 3).

9.5   Time-Varying Magnetic Field Induction

Magnetic fields are produced by all current-carrying devices. The time-varying magnetic fields produced by alternating-current devices induce electric currents in conductive material, including the human body. Under most conditions of environmental exposure, the induced currents are far too small to stimulate excitable tissue. Nevertheless, fields capable of stimulation may be associated with certain industrial or research facilities, or with medical diagnostic devices.

In this chapter, magnetic fields will be characterized in terms of *magnetic flux density*, B, having units of tesla (T). That quantity is related to *magnetic field strength*, H, by $B = \mu H$, where μ is the permeability constant of the medium, and H is the field strength in units of amperes per meter (A/m). The permeability of air (μ_0) in this system of units is $4\pi \times 10^{-7}$. The permeability of biological tissue differs little from that of air (see Chapter 2).

Consequently, the value of *B* within the biological medium may ordinarily be taken as that measured in air in the absence of the biological specimen. The system of units used above is the *International System* (SI). Flux density, *B*, is sometimes reported in the cgs system in units of *gauss*. To convert between the two units, note that 1 gauss $= 10^{-4}$ tesla.

To calculate the induction in a volume conductor, such as the human body, a simple approach treats the volume as if it were made up of concentric rings normal to the direction of the field, as illustrated in Fig. 9.10*a*. In this simple picture, the field is assumed to be oriented along the long axis of the body. The magnetic field induces an electric field within the body, having roughly circular paths, as indicated in the figure.

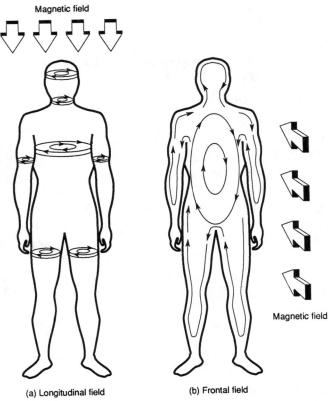

(a) Longitudinal field   (b) Frontal field

Figure 9.10   Distribution of internally induced electric fields from whole-body exposure to time-varying magnetic field. Magnetic field direction is parallel to long axis of body in part (*a*), and perpendicular to front of body in (*b*). Direction of internal E-field and induced current reverses every half-cycle of alternating magnetic field.

The induced electric field, in turn, produces circulating *eddy currents*, which, in a medium of homogeneous conductivity, would follow the path of the electric field.

In accordance with Faraday's law, the internally induced electric field **E** is related to the time rate of change of flux density **B** by

$$\oint \mathbf{E} \cdot \mathbf{dl} = -\frac{\partial}{\partial t} \int\int \mathbf{B} \cdot \mathbf{ds} \tag{9.15}$$

The first integral is taken over a closed path, and **ds** is the element of area normal to the direction of **B**. Each term in the integrals is a vector quantity. Elsewhere in this chapter the symbols B and E are used to represent scalar quantities. If B is uniform over the region inside a closed path of radius r, the induced electric field strength calculated from Eq. (9.15) is

$$E = -\frac{r}{2}\frac{dB}{dt} \tag{9.16}$$

where the direction of the induced E-field is along the circumference of the circle. In some applications the magnetic field varies sinusoidally as $B = B_0 \sin 2\pi f t$, and Eq. (9.16) becomes

$$E = (r\pi f B_0)\cos 2\pi f t \tag{9.17}$$

where the term in parentheses is the peak induced electric field during the sinusoidal cycle, and B_0 is the peak magnetic field. Frequently, calculations for sinusoidal magnetic fields express only the magnitude term in Eq. (9.17), leaving off the cosine term. For induction in a concentric ring model, such as that shown in Fig. 9.10a, Eq. (9.16) suggests that the outermost rings would have the greatest electric field strengths. According to this simple model, the maximum electric field for whole-body exposure would be computed from Eq. (9.16) with r being the maximum circle radius that can be drawn on the body in a plane perpendicular to B.

When the magnetic field is perpendicular to the long axis of the body, the pattern of induced fields and eddy currents is somewhat more complex, as suggested in Fig. 9.10b. An improved model treats the human body as a prolate spheroid, illustrated in Fig. 9.11. A long-wavelength solution for the spheriod model has been published by Durney and colleagues (1975), and expressed in applied form by Spiegel (1976). The long-wavelength assumption is that the field wavelength is much greater than the dimensions of the biological subject. The complete solution for the induced E-field due to electromagnetic exposure involves one component due to the incident electric field, and another due to the incident magnetic field. The internally induced electric field for three orientations of a sinusoidal magnetic field is given by

$$E_1 = \frac{-j2\pi f B_x(a^2\, y\mathbf{a}_z - b^2 z\mathbf{a}_y)}{(a^2 + b^2)} \tag{9.18}$$

$$E_2 = \frac{j2\pi f B_y(a^2\, x\mathbf{a}_z - b^2 z\mathbf{a}_x)}{(a^2 + b^2)} \tag{9.19}$$

$$E_3 = j\pi f B_z(y\mathbf{a}_x - x\mathbf{a}_y) \tag{9.20}$$

where E_1, E_2, and E_3 are the magnitudes of the sinusoidal electric fields
that are induced by magnetic field exposure along the x, y, and z axes,
respectively; B_x, B_y, and B_z are orthogonal magnetic field magnitudes; a
and b are semimajor and semiminor axes of the spheroid; $\mathbf{a}_x, \mathbf{a}_y$, and $\mathbf{a}_z$
are unit vectors along the x, y, and z axes; f is the frequency of the field;
and j is the phasor operator (indicating $90°$ phase shift with respect to
the sinusoidal phase of the B field). Equations (9.18)–(9.20) define spatial
vector components of the induced electric field in terms of the magnitude
of the sinusoidal variation [the term in parentheses in Eq. (9.17)]. The
vector magnitude in Eq. (9.20) is identical to the amplitude term of
Eq. (9.17) because $r = (x^2 + y^2)^{1/2}$ along a circular path.

A general expression for pulsed fields of arbitrary waveshape can be
derived from Eqs. (9.18)–(9.20) using arguments based on Fourier syn-
thesis. Designating A_n as the magnitude of the nth Fourier component
of flux density, its derivative is $j2\pi f_n A_n$. It follows that the electric field

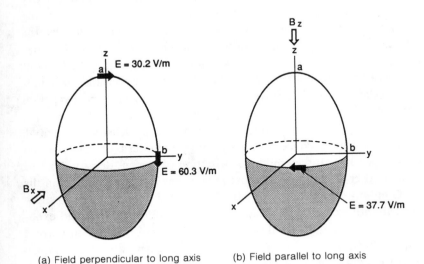

(a) Field perpendicular to long axis   (b) Field parallel to long axis

Figure 9.11   Prolate spheroid model of uniform magnetic field induction.
Bold arrows indicate direction of induced electric field at outer boundaries of
spheroid. Indicated values apply to 60-Hz magnetic field of 1 T strength, for
semimajor axis $a = 0.4\,\text{m}$, and semiminor axis $b = 0.2\,\text{m}$.

induced in an elliptical cross-sectional area is

$$E = -\frac{dB_w}{dt}\left[\frac{a^2 u \mathbf{a}_v - b^2 v \mathbf{a}_u}{a^2 + b^2}\right] \tag{9.21}$$

where $\mathbf{a}_u$ and $\mathbf{a}_v$ are unit vectors along the minor and major axes respectively, (u, v) is the location within the area, and B_w is the magnetic flux density in a direction perpendicular to the elliptical cross section. For a circular area, $a = b = r$, and Eq. (9.21) is equivalent to Eq. (9.16) on the perimeter of the circle.

Consider, for example, a spheroidal representation of the torso of a large person with $a = 0.4$ m and $b = 0.2$ m, and a uniform 60-Hz magnetic field of strength 1 T. If the field is oriented along the z axis, then Eq. (9.20) specifies that the induced electric field at the equator of the ellipsoid is 37.7 V/m. For orientation along the x axis, Eq. (9.18) specifies that the field is 30.2 V/m at the upper extreme of the spheroid, and 60.3 V/m at the lateral extreme. The induced E-field components for this example are illustrated by bold arrows in Fig. 9.11. In this example, the maximum internally induced E-field is greater by a factor 1.6 when the external magnetic field is perpendicular to the long ellipsoid axis, as compared with a parallel orientation.

The internally induced electric fields result in circulating eddy currents. The current density, J, is related to the induced E-field by the conductivity of the medium, σ, in accordance with $J = E\sigma$. For induction in living subjects, calculation of the current distribution is complicated by the widely differing conductivities of various body components (muscle, bone, blood vessels, fat, etc.). The model used in this chapter treats the body as if it were of homogeneous conductivity. The force responsible for stimulation is taken to be the induced E-field within the biological medium (Chapter 4), and the actual current density is of secondary interest.

In some applications, a knowledge of the current density is essential. One such application concerns the distribution of internal heating caused by E-M field exposure. Here, the distribution of current density must be known in order to determine the patterns of power deposition within the body. Gandhi and colleagues (1984) implemented one approach to study the distribution of power deposition resulting from magnetic field exposure. Their approach modeled the body as a series of rectangular current loops, as illustrated in Fig. 9.12. Each loop incorporates the complex impedance appropriate to the tissue in a particular location of the body. The induced electric field in each loop can be calculated in accordance with Eq. (9.15); the current value is thereby inferred using the complex conductivity (including anisotropic properties) of the elements in the loop. By simultaneously solving the loop equations for the entire matrix, one can determine the overall current distribution.

Figure 9.13 illustrates the power deposition, calculated with the above method, through a cross section containing the liver cut in a plane perpendicular to the long axis of the body. The magnetic field is assumed to be perpendicular to the cross section (as in Fig. 9.10a), and to have a sinusoidal variation at either 27.1 or 13.6 MHz. The figure indicates that power deposition generally rises when the measurement location moves from the center of the body to its outer perimeter. This overall behaviour is a consequence of the radial dependency of the induced field, indicated in Eq. (9.16). For a perfectly homogeneous medium, the power absorption is expected to vary as r^2 (since power is proportional to the square of current). In Fig. 9.13, deviations from an r^2 dependency arise from variations in conductivity. The sudden drop near the edge of the body ($r = 12$ cm) results from a sudden decrease in conductivity at the outer layers comprising fat and skin.

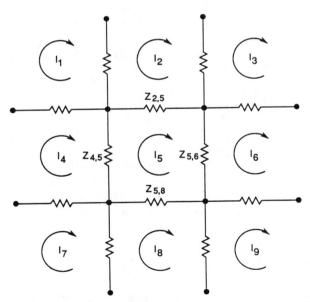

Figure 9.12 Impedance mesh model for calculating magnetically induced currents in nonhomogeneous medium. For simplicity of notation, only nine magnetic loops are indicated. Induced electromotive force in each loop is calculated according to Eq. (9.15). (Adapted from Gandhi et al., 1984.)

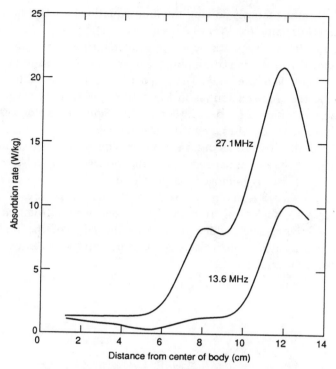

Figure 9.13   Average radial dependence of power deposition in cross section of human liver; cross-sectional plane perpendicular to long axis of body; distance measured from center of body. Magnetic field perpendicular to cross section; magnetic flux density is 56 μT; frequency of field indicated on curves. (From Gandhi et al., 1984, © 1989 IEEE.)

9.6   Excitation by Time-Varying Magnetic Fields[†]
Stimulation by Pulsatile Fields

Because the magnetically induced electric field is proportional to the time rate of change of flux density, dB/dt, it follows that pulsatile magnetic waveforms can result in large induction fields. Figure 9.14 shows examples of gradient field magnetic waveforms currently used in magnetic resonance imaging (MRI). These waveforms typically consist of gated trapezoidal and sinusoidal functions. The figure also shows associated dB/dt waveforms. Typical MRI gradient field pulse durations may be tenths of milliseconds to several milliseconds long. The field patterns may include sequences of pulses separated by intervals as short as several tenths of milliseconds and longer repetition patterns having periods some tens

[†]Portions of this section have been adapted from Reilly (1989).

of milliseconds long. Considering the individual pulse waveshapes and repetition patterns, MRI-induced currents can be quite complex.

When a large area of the body is uniformly exposed to a time-varying magnetic field, the strongest induced internal electric field occurs along the outer perimeter of the body. Accordingly, we are interested in stimulation thresholds of peripheral nerves. Neural excitation is most efficient when the electric field in the biological medium is aligned with the long axis of the excitable fiber (see Secs. 4.3 and 4.4). For calculation of magnetic thresholds, the induced E-field must be determined. Current density is, however, a more frequently cited stimulation parameter. Consequently, when considering published data, it may be necessary to refer to current density thresholds as well.

It is preferable to consider the magnetic stimulus in terms of the dB/dt waveform, as in Fig. 9.14b. With magnetic induction, a biphasic charge-balanced waveform is always produced. However, the biphasic current return can be gradual or delayed, such that a monophasic stimulus effect is approximated. The simplest dB/dt waveform to analyze is a monophasic rectangular pulse. Stimulation thresholds by single rectangular pulses can be expressed by well-defined strength-duration relationships for both nerve (Chapter 4) and cardiac (Chapter 6) tissue. Recall that when the stimulus duration is long in comparison with the excitation time constant τ_e of the excitable tissue, the minimum threshold is expressed in terms of the magnitude of internal electric field (or, alternatively, the current density). Consequently, minimum dB/dt thresholds are obtained for relatively long stimulus durations. If the stimulus is very short compared with τ_e, then the product of electric field and duration is a minimum; consequently, the threshold value of B is a minimum for very short magnetic pulses.

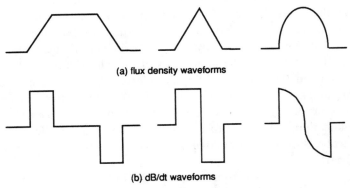

(a) flux density waveforms

(b) dB/dt waveforms

Figure 9.14   Example of flux density and dB/dt waveforms.

Peripheral Nerve and Cardiac Thresholds

Peripheral nerve and cardiac excitation thresholds have been used to specify acceptability criteria for short-term exposure to time-varying magnetic fields. Regulators of MRI diagnostic procedures, for example, have attempted to choose criteria that would preclude nerve or cardiac excitation. Magnetic stimulation may also be an intentional outcome in some diagnostic applications (Barker et al., 1987). Additional information concerning small coil systems is given in Sec. 9.7. Whether intentional or not, it is important to understand the principles of magnetic excitation. What follows is a comparison of cardiac and peripheral nerve excitation by magnetic exposure to a large area of the body. To make the comparison, we need to pay careful attention to the geometric arrangement of anatomical features, excitation thresholds, and stimulus waveform dependence of cardiac and nerve tissue.

Anatomical Considerations. The size and location of anatomical features will vary considerably from one person to another. Furthermore, the location of the heart within the torso will vary as the body position is changed from prone to erect, and also will move during the cardiac pumping cycle. The importance of body geometry is illustrated in Fig. 9.15 for an example of a large man. Body dimensions in applied situations may differ from this example, and calculations would vary accordingly. Calculated induced E-fields have been carried out by representing body cross sections as ellipses, as indicated in the drawing; semimajor and -minor axes are also shown. Numbered points in the drawing indicate positions where the E-field has been calculated according to Eq. (9.21).

Each numbered point in the figure is associated with a calculated induced E-field, in V/cm, corresponding to a magnetic field with $dB/dt = 1$ T/s. Points 3, 9, and 14 represent loci where the induced field is greatest within the exposed cross section. The other points are at various locations on the surface of the heart. Calculated E-field values generally increase as the location is moved toward the perimeter of the body, and are especially great along the minor axis of the ellipse. Points of maximum induction on the surface of the heart occur at points 1, 6, and 10 for longitudinal, frontal, and sagittal exposures, respectively. The heart's outline has been projected on the cross-sectional plane in each view. This simplification would lead to a somewhat conservative calculation of the applicable E-field as compared with a cross-sectional representation of the heart.

A somewhat more conservative calculation can be made by assuming that the entire body is uniformly exposed to the B field by using an equiva-

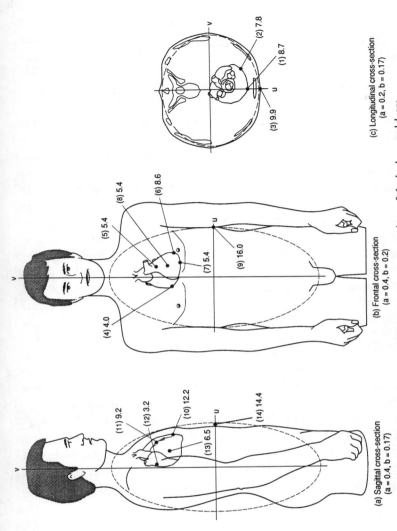

(a) Sagittal cross-section
(a = 0.4, b = 0.17)

(11) 9.2
(12) 3.2
(10) 12.2
(13) 6.5
(14) 14.4

(b) Frontal cross-section
(a = 0.4, b = 0.2)

(8) 5.4
(6) 8.6
(5) 5.4
(7) 5.4
(9) 16.0
(4) 4.0

(c) Longitudinal cross-section
(a = 0.2, b = 0.17)

(2) 7.8
(1) 8.7
(3) 9.9

Figure 9.15   Example of position of heart within three cross sections of the body; a and b are semimajor and minor axes of equivalent cross-section ellipsoids. Numbered locations are identified with calculated electric field intensity (V/m) for magnetic exposure within the elipse of $dB/dt = 1$ T/s. Direction of field is perpendicular to cross section.

lent body ellipse with a major axis equal to the height of the person. A calculation for sagittal exposure was made, using semimajor and semiminor axes of 0.90 and 0.17 m, respectively. In this case the calculated E-field was somewhat elevated relative to the case where only the torso was exposed. The increase was 11% at point (10) and 14% at point (14).

Excitation Thresholds. Theoretical excitation thresholds are described in Chapter 4 using a myelinated nerve model. With a stimulus field (or current) oriented along the long fiber axis, thresholds are calculated as 6.2, 12.3, and 24.6 V/m for fiber diameters of 20, 10, and 5 μm, respectively (Table 4.2). Corresponding current densities are 0.12, 0.25, and 0.49 mA/cm^2 in a medium of conductivity 0.2 S/m. The threshold applicable to a 20-μm fiber will be used here for a conservative calculation.

Cardiac excitation thresholds are discussed in Sec. 6.10 in terms of current density. Recall that the excitation threshold is a minimum during the unexcited (diastolic) interval of the cardiac cycle; the fibrillation threshold is a minimum during the vulnerable (partially refractory) period. The relatively lower excitation threshold will be used here to evaluate magnetic stimulation effects. For prolonged monophasic stimuli, the threshold at the fifty percentile rank is estimated to be at 0.2 mA/cm^2, and the one percentile rank is estimated at 0.1 mA/cm^2 for healthy animal hearts, although a higher percentile rank may apply to the diseased human heart[†] (Sec. 6.10). The later value is close to the theoretical threshold of 0.12 mA/cm^2 mentioned above for a 20-μm nerve fiber. Considering uncertainties in the experimental thresholds and the conductivity of the biological medium, it is reasonable to assume that the minimum E-field thresholds of nerve and cardiac tissue do not differ greatly. To facilitate the calculations in this section, minimum excitation thresholds for both nerve and cardiac tissue will be assumed equal to the theoretical value for a 20-μm nerve, namely, 6.2 V/m.

Strength-duration time constant. The S-D time constants of cardiac and nerve tissue differ widely. As noted in Chapters 4 and 7, a theoretical model for myelinated nerve indicates $\tau_e = 0.12$ ms, a value consistent with in-vivo nerve stimulation in animal experiments. With electrocutaneous stimulation in humans, an average of τ_e is about 0.27 ms. In contrast, the experimental average of τ_e for cardiac stimulation is around 3 ms—a factor at least 10 times greater than that for nerve stimulation.

[†]Statistical distribution determined from a variety of animal experiments. Measurements on humans undergoing valve replacement indicate a broader distribution (see Sec. 6.10).

Table 9.6. *Magnetic field induction: dB/dt necessary to induce 6.2-V/m electric field at selected body loci (calculation by elliptical cross-section model)*

B-field orientation	Ellipse b, a (cm, cm)	Body locus	dB/dt (T/s)
Longitudinal	17, 20	1	71.3
		3*	
Frontal (torso)	20, 40	6	68.1
		9*	38.8
Sagittal (torso)	17, 40	10	48.8
		14*	43.1
Sagittal (whole body)	17, 90	10	44.0
		14	37.8

*Points 3, 9, 14 apply to peripheral nerve loci; other points to heart. Calculations are for human geometry, indicated in Fig. 9.19.

Comparison of cardiac and neural thresholds. In accordance with the preceding, a threshold field of 6.2 V/m will be assumed as a lower bound for both peripheral nerve and cardiac stimulation by square-wave pulses of long duration. Using this criterion, Table 9.6 lists dB/dt thresholds for selected points on the body model of Fig. 9.15. Calculations are shown for the three orthogonal orientations of the field. In the case of sagittal exposure, an additional conservative calculation is shown for exposure to an elliptical representation of the entire body. Figure 9.16 illustrates dB/dt thresholds as strength-duration curves, in which the minima correspond to the values listed in Table 9.6.

The foregoing has been based on several assumptions that are generally conservative, but not the most extreme ones that can be made. Some assumptions worthy of emphasis are as follows:

1. *Exposure model.* An elliptical model of the torso has been used, with dimensions applicable to a relatively large man. Since the induced internal field is generally proportional to body size, the calculated thresholds would vary with the size of the individual, or with the size of the exposed area of the body.

2. *Conductivity.* A homogeneous conductivity model has been used. A more involved model would account for nonhomogeneous and anisotropic properties of the biological medium.

3. *Body geometry.* Calculated thresholds vary with the assumptions of overall body size, and the relative position of the heart within the torso. Calculations presented here are for a specific example of a large man. Actual geometric relationships will vary with individual body structure, body position, and the relative time within the cardiac cycle.

4. *Threshold.* The minimum threshold value of the electric field is assumed to be 6.2 V/m for both cardiac and nerve tissue. This assumption corresponds to the theoretical value for a 20-μm nerve fiber, and to the estimated 1st percentile rank for cardiac excitation of healthy animals. The corresponding one-percentile rank threshold for the human heart, especially that in a pathological con-

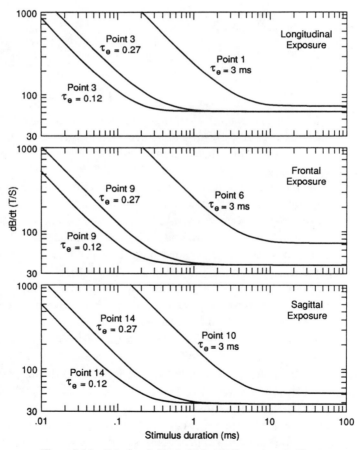

Figure 9.16   Calculated threshold for dB/dt exposure of human torso, with induced E-field of 6.2 V/m as a criterion; reference points identified in Fig. 9.15.

dition, is not well-defined, but may be lower than indicated by animal data (see Sec. 6.10).

5. *Satial uniformity.* The peak magnetic field has been assumed to exist uniformly within the exposure area. This may represent a pessimistic assumption for some applications. With MRI exposure, for example, the *B* field is not uniform along all axes, and may undergo a phase change within the exposure area.

As a general rule, the more central position of the heart within the torso in comparison with the more eccentric position of peripheral nerves results in higher cardiac thresholds. The relative geometric advantage of the heart depends on the direction of the incident field with respect to the body. Nerve and cardiac thresholds are predicted to be most disparate when the incident field is perpendicular to the frontal cross section of the body, and least disparate when the field is perpendicular to the sagittal cross section. The lowest absolute thresholds are predicted for peripheral nerves with frontal exposure, and for the heart with sagittal exposure.

A further separation of cardiac and nerve thresholds can be realized by taking advantage of the relatively longer excitation time constants of cardiac tissue. For pulsed *dB/dt* waveforms, cardiac and nerve tissue become more disparate as the stimulus duration is made shorter than the stimulation time constant of the heart (about 3 ms). For sinusoidal stimulation, a frequency above about 120 Hz results in a greater margin between cardiac and nerve thresholds.

A number of experimenters have studied the reaction of the human or animal heart to wide-area *dB/dt* exposure (McRobbie and Foster, 1985; Silny, 1986; Willis and Brooks, 1984; Cohen et al., 1990). In a review by the author (Reilly, 1991), it was concluded that, in all cases, experimental findings are consistent with the theoretical predictions discussed above. Although it can be stated that the experimental observations are not inconsistent with the theoretical models, that does not imply experimental verification. Verification will require an experimental study that takes into account the spatial distribution of the magnetic field, its waveform, the positions of anatomical features, the excitability characteristics of nerve and cardiac tissue, and a more complete model for determining the field induced within the body. Such an experimental study is recommended to more confidently determine safe operating limits for time-varying magnetic field exposure.

Sinusoidal Fields

The excitation threshold, when plotted versus sinusoidal frequency, describes a *strength-frequency* (S-F) curve. Excitation thresholds of sinusoidal and rectangular pulse stimuli have been compared in Chapter 4 using a theoretical model for myelinated nerve fibers. At low frequencies (e.g., 100 Hz), thresholds expressed as the sinusoidal peak exceed the corresponding values for a rectangular wave by 17%. Thresholds rise at both higher and lower sinusoidal frequencies. An empirical S-F relationship has been described in Chapter 4 [Eqs. (4.35)–(4.37)]. These equations have the asymptotic forms

$$E_t \approx E_0 \left(\frac{f}{f_e} \right)^a \qquad \text{for } f \gg f_e \tag{9.22}$$

$$E_t \approx E_0 \left(\frac{f_0}{f} \right)^b \qquad \text{for } f \ll f_0 \tag{9.23}$$

where E_t is the threshold electric field, E_0 is the minimum threshold, and a, b, f_e, and f_0 are empirically determined parameters describing the power functions and corner frequencies of the S-F curve. As noted in Chapters 4 and 7, $a = 0.9$ for continuous sinusoidal stimulation, and $a = 1.35$ for single-cycle stimulation; $b = 0.8$ in either case. The low-frequency corner was found to be $f_0 = 10$ Hz. As for the high-frequency corner, available data suggest a spread of values. Results with the myelinated nerve model indicate $f_e = 5400$ Hz for axonal stimulation (Chapter 4). Electrocutaneous sensory stimulation indicates a range of values, having a median of about 500 Hz (Chapter 7). With cardiac excitation, the median experimental value of f_e is 120 Hz (Chapter 6).

The upper transition frequency f_e is significantly lower for electrocutaneous stimulation as compared with the myelinated nerve model. That observation is consistent with the finding that experimental values of the strength-duration time constant τ_e are higher with electrocutaneous data. In developing criteria for magnetic stimulation, the relevance of electrocutaneous data is not clear. It is possible that a model based on axonal stimulation may be preferable to results derived from electrocutaneous sensory experiments. This speculation is suggested by the good agreement of model and experimental data when an axon is excited subcutaneously (Reilly et al., 1985). Unfortunately, available frequency sensitivity data are inadequate to provide clear guidance on this question. Lacking appropriate experimental data, it may be appropriate to assume two transition frequencies, one corresponding to the model, and one to electrocutaneous experiments.

Figure 9.17 represents an asymptotic threshold model for continuous sinusoidal stimulation. The high-frequency behavior is indicated for f_e at the three frequencies mentioned above, namely, 120, 500, and 5400 Hz. Magnetic flux density thresholds can be derived from the criteria of

Figs. 9.18 and 9.20. For calculation purposes, assume the body dimensions represented in Fig. 9.15. Results are shown in Fig. 9.18 for excitation of cardiac tissue with $f_e = 120$ Hz, and peripheral nerve with either $f_e = 500$ or 5400 Hz. Separate curves are identified for the three orientations of magnetic exposure defined in connection with Fig. 9.15. Figure 9.19 illustrates the same criteria expressed as dB/dt thresholds.

The high-frequency behavior of Fig. 9.17 might seem to indicate that internal E-field thresholds rise indefinitely as the stimulus frequency is increased; similarly, the high-frequency magnetic thresholds in Figs. 9.18 and 9.19 might appear to be extendable to arbitrarily high frequencies. However, a practical limit is reached where tissue heating becomes significant, and thermal excitation of warm receptors would dominate over electrical thresholds. For cutaneously applied currents, the crossover between electrically elicited sensory reponses and thermal thresholds occurs around 200 kHz (see Chapter 7).

The thresholds indicated in Fig. 9.18 are quite large in comparison with exposures that might ordinarily be encountered. At 60 Hz, for example, the figure indicates a field of 0.12 T as a lower bound for excitation of sensitive neural structures. That value may be compared with fields measured in various environments. In residential environments, for example, ambient 60-Hz fields are generally below 1 μT (Kaune et al., 1987; Silva et al., 1989). Highly localized flux densities around 100 μT may be

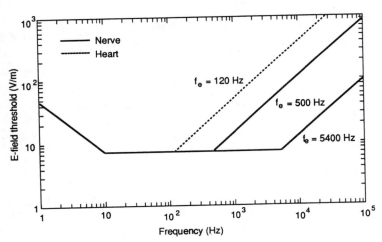

Figure 9.17 Asymptotic strength-frequency model for excitation by uniform electric field within biological medium; continued sinusoidal stimulus.
High-frequency corner of 120 Hz applies to cardiac tissue; corners at 500 and 5400 Hz apply to nerve tissue.

found within a few centimeters of household appliances (Gauger, 1985). Beneath high-voltage transmission lines, fields near $10\,\mu T$ are possible (Project UHV, 1982). These commonly encountered exposures are at least 1000 times below the levels necessary for excitation of peripheral nerves. Considerably higher magnetic exposures may be encountered by workers in certain industrial settings. Flux densities in the range 10^{-4} to $10^{-2}\,T$ have been measured at 50 and 60 Hz in both electrical welding and electrosteel industries; levels as great as 0.06 T have been observed near induction heaters (Lövsund and Ölberg, 1982; Stuchly and Lecuyer, 1989).

Direct Perception of Time-Varying Magnetic Fields

In certain animal species, specialized mechanisms for magnetic field detection can be demonstrated (Tenforde, 1989). If the dominant mechanism of human perception is due to peripheral nerve excitation, then magnetic thresholds can be determined as suggested in the previous section. At 60 Hz, for example, magnetic thresholds for the most sensitive

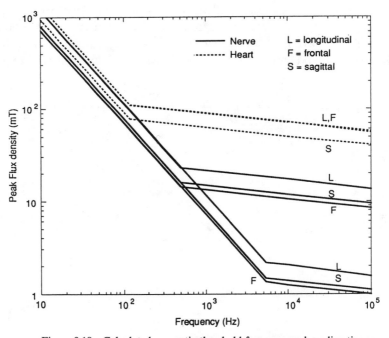

Figure 9.18   Calculated magnetic threshold for nerve and cardiac tissue; continued sinusoidal stimulus. Separate curves shown for three orthogonal orientations of magnetic exposure to the torso. Transition frequency f_e is shown at 120 Hz for the heart, and at 500 and 5400 Hz for nerve.

neural structures are estimated to be 120 mT (peak) for frontal exposure of the torso and 160 mT for longitudinal exposure.

Magnetic perception thresholds were tested in now-classic experiments by Tucker and Schmitt (1978). These experiments established that humans cannot detect 60-Hz fields as high as 0.75 mT when applied longitudinally to the whole body, or as high as 1.5 mT when applied locally to the head. An interesting finding was the extreme care that must be exercized to avoid artifactual perception due to extraneous cues in the experimental procedure, such as vibration, acoustic cues, or dimming of lights. The longitudinal fields in these experiments were below the theoretical thresholds by a factor of about 200, and therefore would not be expected to result in peripheral nerve stimulation. Apparently, if some detection mechanism exists in humans other than peripheral nerve stimulation, it requires whole-body exposure above 0.75 mT at 60 Hz to be effective.

A particularly sensitive mode of detecting magnetic fields occurs through the perception of *phosphenes*. These are perceived light patterns induced by nonphotic stimuli, such as pressure on the eyeballs, or electric energy. A unique feature of electric phosphenes is their low excitation threshold

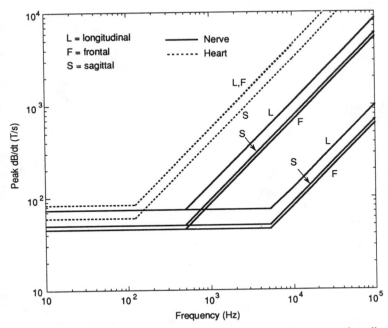

Figure 9.19   Calculated dB/dt excitation thresholds for nerve and cardiac tissue; continued sinusoidal stimulus.

and sharply defined frequency sensitivity as compared with neural stimulation. These features can be seen in the experiments of Lövsund and colleagues (1980a, 1980b). Figure 9.20 compares thresholds of magneto- and electrophosphenes. The former were elicited in individuals with the head placed between the poles of a large electromagnet (temple to temple). The latter were elicited via electric currents from electrodes placed on the temples. Magnetophosphene thresholds are shown in Fig. 9.20 on an absolute scale (Lövsund et al., 1980b); electrical thresholds are shown on a relative scale (Lövsund, 1980a). Somewhat lower magnetophosphene thresholds have been reported by Silny (1986).

A third curve in Fig. 9.20 shows electrical thresholds divided by the frequency; this display facilitates a comparison of the curve shapes of magneto- and electrical thresholds. A rationale for this representation is that the induced electric field (and current density) is directly proportional to the frequency of the magnetic field, as suggested by Eq. (9.17). Electrical thresholds, modified in this manner, ought to conform to the curve shape

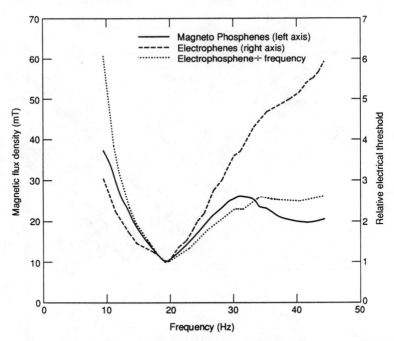

Figure 9.20 Comparison of phosphene thresholds for magnetic and electric excitation. Magnetic threshold on absolute scale (left vertical axis); electric threshold on relative scale (right vertical axis). Background illumination at 3 cd/m². (Adapted from Lövsund et al., 1980b.)

for magnetic thresholds. The two curves do, in fact, appear quite similar, suggesting that the magnetic effect is due to induced currents, rather than some other unspecified action of the magnetic field. The differences that do exist could be explained by the fact that the current pathways are not the same in the two cases. This explanation is plausible considering that the shape of the threshold-frequency curve was found to vary somewhat with the locations of the cutaneous electrodes.

Using the observed magnetophosphene thresholds, we can estimate corresponding induced electric fields using an ellipsoidal model of the head. Such a treatment is only approximate because the experimental fields in Lövsund's experiments were not uniform over the entire cross section of the head, as is assumed in the theoretical treatment. Consider sagittal magnetic exposure of an ellipsoidal model of the head, with semimajor and semiminor axes of 0.13 and 0.10 m, respectively, and a minimum threshold of 10 mT at 20-Hz frequency. With these parameters, the maximum E-field induced within the head is calculated to be 0.079 V/m. At the location of the retina and cornea, the calculated fields are 0.053 and 0.069 V/m, respectively. These calculated values are consistent with the current density threshold of 0.01 A/m^2 at the retina determined for electrophosphenes (Lövsund et al., 1980b), assuming that the conductivity of the brain is 0.15 S/m (see Chapter 2).

Phosphene thresholds are quite unlike those for peripheral nerve excitation. For one thing, the internal E-fields corresponding to phosphene perception at the optimum frequency are a factor of 100 or more below minimum thresholds for neural stimulation (Table 4.2). Furthermore, the frequency sensitivity shown in Fig. 9.20 differs greatly from that for neural stimulation (Fig. 9.17). These observations suggest that some mechanism other than afferent neural stimulation is responsible for the production of phosphenes. Indeed, some researchers have concluded that phosphenes are generated through modification of synaptic activity in the receptors of the retina, rather than stimulation of visual pathways or the visual cortex (Lövsund et al., 1980a, 1980b).

Experimental Excitation Thresholds

When comparing theoretical thresholds with experimental findings, it is preferable to examine experimental data applying to conditions similar to the assumptions used in the theoretical models. Unfortunately, available publications do not always provide a clear description of all the relevant details, such as a complete description of the stimulus waveform, its spatial orientation, the method of exposure, the biological preparation or subject, or the criterion for judging the presence of excitation.

Nevertheless, available experimental data can be useful with some judgment and interpretation.

Table 9.7 summarizes available experimental data from the listed references. The two columns labeled "Stimulus peak at threshold" are of particular interest for our comparison. In converting from field strength to current density, the value of conductivity cited in the reference was used. For items 8, 9, and 12, conductivity was not cited, so a value of 0.2 was S/m assumed. Except for item 11, stimulation current was induced by pulsed magnetic fields. For item 11, an electric field was created between metallic electrodes within the conducting medium.

The thresholds listed the Table 9.7 are not necessarily the minimum possible values, because the stimulus waveforms or field orientations were not necessarily optimum. As a result, a number of the listed thresholds should be reduced to obtain minimum possible values. Some specific comments on this point are as follows:

Items 1, 4, and 5. The dB/dt waveforms were damped cosine waves at 3.0 kHz. The sinusoidal stimulus threshold at 3 kHz would be about a factor of 2 above that for the frequency region of greatest sensitivity between 50 and 500 Hz if the upper frequency corner, f_e, is 5 kHz (Fig. 9.17).

Items 2 and 6. The dB/dt waveforms approximated a linear decay from a peak. The threshold level of a rectangular wave of the same duration can be as much as a factor of 2 below that for a triangular wave.

Items 3 and 7. The half-cycle cosine wave is biphasic. A rectangular biphasic wave with 75-μs phases would be reduced in sensitivity by about 20% relative to a monophasic wave of 75-μs duration (Fig. 4.16). Additionally, the peak threshold of a cosine wave is greater than that for a rectangular wave of the same phase duration (Fig. 4.15).

Items 8 and 9. The threshold for a 180-μs pulse will exceed the minimum threshold for a long (1-ms) pulse by about 50% (Fig. 9.16).

Items 10 and 11. Details of the dB/dt waveshape were not provided. The direction of the field in item 10 was transverse to the nerve/muscle orientation. As noted in Chapter 4, this is an inefficient orientation for stimulation. In item 11, the preparation was placed in a saline bath, and the stimulus current was applied to electrodes within the bath. The orientation of the preparation relative to the field was not stated. Judging from the fact that the threshold was about one-tenth that in item 10, a transverse field orientation is suspected in item 11.

Table 9.7. *Experimental thresholds for neural stimulation*

Item	Waveform of dB/dt or stimulus current	Duration (ms)	Subject/ prep.	Body locus	Response	Stimulus peak at threshold E (V m^{-1})	J (A m^{-2})	Reference
1	Damped cosine, 3 kHz	2.0	Rat	Chest	Twitch	34	5.1	McRobbie and Foster, 1984
2	Quarter-cycle cosine	0.083	Rat	Chest	Twitch	45	6.8	McRobbie and Foster, 1984
3	Half-cycle cosine	0.150	Rat	Chest	Twitch	36	5.4	McRobbie and Foster, 1984
4	Damped cosine, 3 kHz	2.0	Human	Forearm	Sensation	25–50	3.8–7.5	McRobbie and Foster, 1984
5	Damped cosine, 2.1 kHz	*	Human	Forearm	EMG	78	11.8	McRobbie and Foster, 1984
6	Quarter-cycle cosine	0.12	Human	Forearm	EMG	113	16.7	McRobbie and Foster, 1984
7	Half-cycle cosine	0.24	Human	Forearm	EMG	87	13.1	McRobbie and Foster, 1984
8	Monophasic pulse	0.18	Rat	Whole body	Muscle contraction	44–73	9–16*	Polson et al., 1982a
9	Monophasic pulse	0.18	Human	Wrist	EMG	70	14*	Polson et al., 1982b
10	Pulse*	2–3	Frog nerve/ muscle		*	42–96	14–32	Ueno et al., 1984
11	Pulse*	2–3	Frog nerve/ muscle		*	4.4	1–47	Ueno et al., 1984
12	Half-cycle cosine	4.0	Frog nerve/ muscle		*	10	2.0*	Irwin et al., 1970

*Information sketchy or not available.

Considering the qualifications discussed above, the thresholds for items 1–10 might reasonably be reduced by a factor of 2 to obtain minimum thresholds for comparison with rectangular monophasic stimuli. This adjustment would bring the experimental values within the range of minimum theoretical thresholds listed in Table 4.2.

The data summarized in Table 9.7 exclude references to experiments in which current was introduced through electrodes held against the skin. One may be tempted to calculate electrocutaneous current density thresholds by dividing threshold current by contact area. Thresholds determined this way, however, may provide very conservative estimates of actual current density because current density beneath a contact electrode is concentrated at the edges of the electrode. Furthermore, with cutaneous electrodes, current is preferentially conducted in discrete cutaneous channels (see Chapter 2).

Cohen and colleagues (1990) exposed two human subjects to pulsed sinusoidal fields of frequency of 1.4 kHz, repetition frequency of 14 Hz, and peak dB/dt of 86 T/s. The subjects reported sensations in the lower back and facial regions. Simultaneous ECG recordings failed to show any effects of the applied field. Dogs exposed to slightly higher fields also failed to show cardiac abnormalities. From Fig. 9.19 we except that peripheral nerves may be excited at the cited experimental levels when humans are exposed to 1.4-kHz magnetic fields but cardiac excitation would require significantly higher fields. A full evaluation of these experimental findings would require careful attention to the spatial distribution of the applied field in both magnitude and phase.

Budinger and associates (1990) tested perception thresholds in human subjects exposed with a large coil system. At a frequency of 1270 Hz, subjects reported mild sensations in the lower back when exposed to a longitudinal field of 60 T/s peak. As the frequency was raised from 600 to 2000 Hz, thresholds for longitudinal exposure were nearly proportional to frequency. Thresholds decreased as the number of sinusoidal cycles of stimulation increased; the relationship was approximately in accordance with theoretical results of Fig. 4.18. Painful stimulation occurred at about twice the perception threshold. The small dynamic range from perception to pain is similar to that noted for electrocutaneous stimulation (Sec. 7.4).

Bourland and associates (1990) exposed 20 human subjects to a longitudinal field, with the buttocks centered in the region of maximum field. Perception thresholds followed a classical strength-duration curve (Sec. 4.2). At the maximum pulse width tested (1.25 ms), the median and minimum thresholds were 75 and 45 T/s, respectively. More recently, the investigators reported the strength-duration time constant to be 148 ±

39 μs using magnetic stimulation of dogs (Bourland et al., 1991). That result may be compared to the theoretical value of 120 μs obtained with the myelinated nerve model under uniform field excitation (Sec. 4.5).

Cardiac excitation by magnetic stimuli has also been examined experimentally. At a pulse duration of 530 μs, the longitudinal magnetic threshold for an ectopic heart beat in dogs was reported to be a factor of 13.5 ± 3.7 above the threshold of skeletal muscle stimulation (Bourland et al., 1991). In comparison, the theoretical model for human exposure predicts that ectopic stimulation would occur at a factor of 6 above the one-percentile cardiac threshold (Fig. 9.16), or a factor of 12 above the median threshold. A more careful analysis of the relative human and canine anatomy would be required before these data could be confidently used for human exposure applications.

McRobbie and Foster (1985) examined the ECG of rats subjected to repetitively pulsed magnetic stimuli having maximum dB/dt in the range 3 to 30 T/s. The authors were unable to produce cardiac abnormalities, even for stimuli that caused violent muscle contractions. The magnetic waveforms used for these experiments were damped sinusoids, with frequencies in the range 670 to 2000 Hz, and decay time constants of about 1.5 s. For sinusoidal stimulation in this frequency range, one expects cardiac thresholds to be significantly above neuromuscular thresholds (Figs. 9.18 and 9.19). Consequently, the lack of cardiac stimulation in these experiments is not surprising.

Silny (1986) placed coils around dogs, such that the magnetic field was either along or perpendicular to the long axis of the body. With a single 50-Hz stimulus, Silny observed salvos of ectopic beats at applied fields above 2 T. That field is about 2.5 times the estimated median threshold for continuous sinusoidal exposure of animal hearts; it is about 5 times greater than the one-percentile threshold indicated in Fig. 9.18. In order to compare the calculated human thresholds with the observed canine data, we need to account for the relative sizes of subjects, the geometric configuration of the body and heart, and the distribution of the magnetic field strength within the exposure areas—information that is lacking in the Silny report. Nevertheless, one can conclude that the observed canine thresholds ought to be greater than human thresholds because the canine subject is smaller than a human and the coil placement in Silny's tests, with the heart roughly in the center of the coil, may not have been optimum for cardiac stimulation.

Willis and Brooks (1984) studied rats and guinea pigs under magnetic exposure. The authors were unable to observe cardiac abnormalities with dB/dt pulses of 2 T/s maintained for about 25 ms. The applied exposure was well below the predicted cardiac threshold (Fig. 9.16), especially considering the small size of the animal subjects. Therefore, the lack of observed effect is to be expected.

9.7   Local Magnetic Stimulation[†]

In recent years, considerable attention has been devoted to using localized time-varying magnetic fields for stimulating excitable tissue without the use of electrodes. The main advantage of this technique is that it provides a noninvasive means of stimulation without causing pain (Barker et al., 1985). One potential application is magnetic stimulation of the brain, although it has not yet been approved for clinical use by the FDA. Recent results from magnetic brain stimulation include measurement of central motor conduction time (Barker et al., 1987; Cracco, 1987; Hess et al., 1987; Mills and Murray, 1985), monitoring of spinal cord function during spinal surgery (Levy, 1987; Shields et al., 1988), and the ability to produce motor twitches for a single digit (Amassian et al., 1988). Nonmotor brain areas have been stimulated as well. Some examples of stimulating cognitive areas are the suppression of visual perception by stimulating the visual cortex (Amassian et al., 1989a) and eliciting a sense of movement in a paralyzed limb (Amassian et al., 1989b). Stimulation of peripheral nerves is also possible, but has not received the same attention as brain stimulation, except in a few special cases such as facial nerves (Evans et al., 1988; Maccabee et al., 1988b; Schriefer et al., 1988). Magnetic stimulation of lumbosacral roots has also shown utility (Tsuji et al., 1988; Chokroverty and DiLullo, 1989).

The circuitry used to produce a magnetic stimulus is straightforward, but there are many variations (Davey et al., 1988; Hallgren 1973; Merton and Morton, 1986; Polson et al., 1982a). The localized time-varying magnetic field used for magnetic stimulation is usually produced by rapidly discharging a capacitor into a small coil placed near the excitable tissue. Figure 9.21 shows a common circuit for generating the magnetic pulse. The figure omits the charging circuit necessary to produce the initial charge on the capacitor and the inherent series resistances of the coil, thyristor, capacitor, and cabling. These resistances are of considerable practical concern and cannot be neglected. Resistor R_s is used only to limit the current flowing into the gate of the thyristor during firing. Figure 9.21 shows the simplest possible trigger control. Many other trigger methods are possible and may be preferable.

In the circuit configuration shown, after the switch is pressed, the

[†]Written by H. A. C. Eaton, The Johns Hopkins University Applied Physics Laboratory.

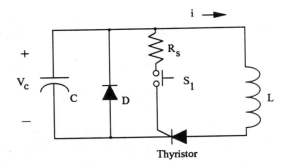

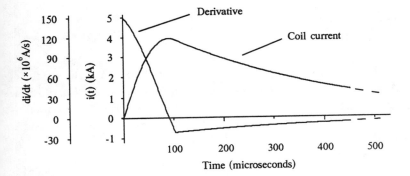

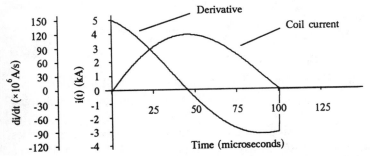

Figure 9.21 Upper: a simple magnetic stimulator circuit employing a thyristor switch. The coil, *L*, is the stimulus coil located near the excitable tissue. Middle: The coil current waveform from the circuit, and its derivative. Lower: The coil current and derivative waveforms that result when the diode is removed from the circuit.

thyristor is triggered, allowing the capacitor to discharge through the coil. Current will flow through the capacitor until the voltage at the capacitor terminals reaches zero. During this time, the coil current can be described by a series RLC circuit equation:

$$i(t) = \frac{V_c \sin(t\sqrt{1/LC - (r_1/2L)^2})}{\sqrt{L/C - r_1^2/4}} \exp(-tr_1/2L) \qquad (9.24)$$

where r_1 is the sum of the series resistances of the coil, capacitor, thyristor, and cabling. This equation is valid for the time interval from $t = 0$ (when the thyristor is triggered) until time $t = t_s$ when the diode begins to conduct, where t_s is given by

$$t_s = \frac{\tan^{-1}\left\{\frac{[2L\sqrt{1/LC - (r_1/2L)^2}]}{(r_1 - 2r_2)}\right\}}{\sqrt{1/LC - (r_1/2L)^2}} \qquad (9.25)$$

where r_2 is the sum of the parasitic resistances to the right of the diode. A four-quadrant arctangent function must be used for Eq. (9.25). The time t_s is slightly later than the time at which the current is maximal. After t_s, the current circulates through the diode, thyristor, and coil and is governed by a series RL circuit equation:

$$i(t > t_s) = I_s \exp\left[-\frac{(t - t_s)r_2}{L}\right] \qquad (9.26)$$

Where I_s is the coil current at the time that the diode begins to conduct and is given by

$$I_s = \frac{V_c \exp(-t_s r_1/2L)}{\sqrt{L/C + r_2^2 - r_1 r_2}} \qquad (9.27)$$

Figure 9.21 also shows the shape of the coil current as well as its derivative. The scales are for a hypothetical circuit having a 200-μF capacitor charged to 750 V, a 5-μH inductor, $r_1 = 50$ mΩ. These are realistic circuit parameters, although far different values are possible. The coil current rises rapidly, followed by a long decay period. The induced electric field in the tissue is proportional to the time derivative of the coil current. The induced electric field and resulting current are bipolar; such is always the case with magnetic stimulation, because the coil current always returns to zero. With the circuit of Fig. 9.21, the stimulus is a short-duration, high-amplitude pulse followed by a long-duration, low-amplitude repolarization pulse. The energy stored in the capacitor is dissipated primarily as heat in the coil resistance. The diode prevents reverse charging of the capacitor, which allows it to be either an electrolytic or a nonpolarized type in this circuit.

If the diode in Fig. 9.21 is removed, the coil current will be governed by Eq. (9.24) alone. The series *RLC* equation is valid as long as the current is positive, since the thyristor will allow current to flow in only one direction. If the diode is removed from the circuit in Fig. 9.21, the thyritor switch discharges the capacitor through the coil, which then recharges the capacitor in the opposite polarity. Using this technique, much of the energy originally stored in the capacitor is returned to the capacitor, and less heat is dissipated in the coil. Figure 9.21 also depicts the shape of the coil current and its derivative when the diode is deleted from the circuit. The stimulus and repolarization pulses are of roughly equal amplitude and duration. If the diode is removed, then the capacitor must not be a polarized type.

For a practical circuit, the capacitor may be several hundred micro-farads, and may be charged to over 1 kV. The inductor may range from a few microhenries to hundreds of microhenries, depending on the coil arrangement. Peak currents of several kiloamperes are typical. Special care must be taken to keep the inherent series resistances of all of the components to a minimum (usually well under 0.1 Ω), so that the *LC* circuit is as underdamped as possible. Because the induced current is proportional to the derivative of the current in the coil, stray inductance in the cabling connecting the circuit must be kept as small as possible, especially in the (typically long) cable between the stimulator electronics and the coil. The ringing frequency of the *RLC* circuit is usually in the 1- to 10-kHz range. Considerations of the strength-duration curve of the target tissue aid in choosing the optimal ringing frequency (see Sec. 9.6). Because this fre-quency is governed only by the values of r, L, and C, it is difficult to adjust the stimulus duration while maintaining a constant stimulus amplitude. Usually, a single fixed duration is used, in which case the amplitude of the stimulus is easily adjusted by changing the initial voltage on the capacitor.

A large amount of energy is stored in the capacitor prior to each discharge. Most circuits used to recharge the capacitor build up this charge slowly, resulting in a long interpulse period (typically a few seconds). Considerable heating of the stimulating coil results after delivery of many pulses, and coil temperature is usually monitored on commercial magnetic stimulators. Multiple action potentials are difficult to produce. One or more seconds are typically required to recharge the energy storage capac-itor, which prevents stimulation at higher rates. Allowing an *LC* circuit to "ring" continuously is not an effective means for producing multiple action potentials. If a low ringing frequency is used, the magnitude of dI/dt will be low unless very high peak currents are used. Such large currents are difficult to handle, and resistive losses will dampen the oscillations quickly, making this scheme impractical. If a high ringing frequency is chosen, the refractory period of the nerve will reduce the excitability, and will reduce it to zero if the frequency is sufficiently high. More sophisticated circuits for charging and discharging of the capacitor coupled with special coil designs can produce magnetic pulse trains capable of producing multiple action potentials (Davey et al., 1988).

It is not possible to design a coil that would allow for focal stimulation with any desired location, depth, and orientation of the induced current. The inherent shape of various body structures places limitations on the possible orientations of the induced currents. For example, in the near-spherical regions of the head, it is impossible to induce significant radially oriented (perpendicular to the skull) currents with a magnetic stimulator. This is a natural consequence of the shape of the conductor, and the necessary boundary conditions (Eaton, 1992). Continuity of current places inherent limitations on how localized the induced currents can be. Stimulus strength decreases with distance from the coil, making deeper structures more difficult to stimulate than superficial ones. The inductance and resistance of the coil will affect the necessary driving circuit, and this places practical limits on the coil design as well.

Coil current, geometry, and the tissue boundary conditions determine the pattern of the induced currents. For the case of a magnetic stimulator, the coil size and its distance from the point where the field is being computed are much smaller than one wavelength; therefore, quasi-static conditions prevail. This allows the approximation that the phase shifts due to propagation delays can be ignored. Tissue conductivities are small enough that the skin effect can also be ignored. With these approximations, the magnetic vector potential can be found from:

$$A = \mu_0 \int_{source} \frac{J\,dv}{4\pi R} \qquad (9.28)$$

where R is the distance between the differential element of current density J flowing in the coil, and the point in the medium where A is being computed. If the current in the stimulus coil is assumed to be flowing in negligibly small filaments, A can be found from

$$A = \mu_0 \int_{coil} \frac{i\,dl}{4\pi R} \qquad (9.29)$$

where dl is directed along the path of the current. Either expression for A will have zero divergence because the coil current forms a closed loop (Scott, 1966). The path of the current flowing in the cable between the coil and stimulator electronics can be neglected and the coil can be treated as a closed loop if the cable has low inductance. It can be shown that the induced electric field can be found from the negative time derivative of the magnetic vector potential minus the gradient of a scaler potential V:

$$\mathbf{E} = -\mu_0 \frac{di}{dt} \int_{coil} \frac{d\mathbf{l}}{4\pi R} - \nabla V \tag{9.30}$$

The ∇V term arises from the charge density appearing on the boundaries, at interfaces between different tissues, or other inhomogeneities. In this analysis, induced currents in the tissue are assumed to have a negligible contribution to the total magnetic field, a quite realistic assumption for biological materials. In a homogeneous, isotropic medium, either of infinite extent or having appropriately symmetric boundaries for the coil under consideration, the charge density will be zero and, consequently, only the integral term above contributes to the electric field. Several investigators have examined the induced electric fields for the special case of zero charge density for different coil arrangements. Much work in this area concerns producing a focal region of high electric field (Maccabee et al., 1988a; Rösler et al., 1989).

The boundary conditions are found from the considerations of the currents at the interfaces and boundaries. A complete expression for the boundary conditions is

$$\sigma \nabla \cdot \mathbf{E} + \mathbf{E} \cdot \nabla \sigma = -\varepsilon \nabla \cdot \frac{\partial \mathbf{E}}{\partial t} - \frac{\partial \mathbf{E}}{\partial t} \cdot \nabla \varepsilon \tag{9.31}$$

In a region where ε and σ do not vary spatially, this leads to an exponential decay of $\nabla \cdot \mathbf{E}$ and charge density. Because the integral term in the expression for $\mathbf{E}$ already has zero divergence, $\nabla^2 V$ must equal zero in such a region. Thus, Laplace's equation is solved to find V in a homogeneous region. On the boundaries of the homogeneous region, $\mathbf{E}$ must satisfy the above relation. For most purposes, the boundary conditions used to determine V can be computed for the stationary case of $di/dt = 1$, and the resultant V field is then multiplied by di/dt. This neglects the $d\mathbf{E}/dt$ terms above. Omitting these terms neglects the time lag between the application of the field and the resultant rearrangement of charge; however, this approximation is good, even for highly capacitive tissues (Eaton, 1992). The result of this simplification is that the spatial distribution of the field does not vary with time.

Equation (9.30) can be used to predict the induced electric field inside the body for nearly any practical coil arrangement. The closed line integral is performed along the path of the coil winding. For the case of a circular coil, there is no closed-form solution to the integral, but it can be represented in terms of the complete elliptic integrals (Jackson, 1962). If simple body models are used, the determination of the V field is not difficult. The finite-element approach has been used for simple-body models (Ueno et al., 1988), and can be applied to more complicated problems as well.

A simple model for the brain cavity is that of a uniform, spherically shaped conductor bounded by an insulator, representing the skull. If a circular coil is located symmetrically about a radius of the sphere model, then the boundary conditions will be satisfied by symmetry; V will be identically zero. Figure 9.22 shows the electric fields inside a sphere 11 cm in diameter, induced from a circular coil 5 cm in diameter located 1.5 cm above the conducting sphere surface. The induced electric field is computed in several

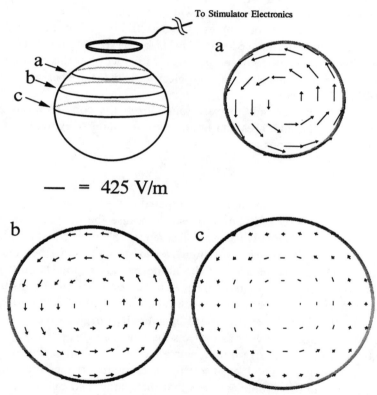

Figure 9.22   The electric field inside an 11-cm-diameter conducting sphere excited by a 5-cm-diameter circular coil located 1.5 cm above the sphere. Vectors shown are the total electric field because there is no out-of-plane component. Slices are located 3 cm, 5 cm, and 7 cm below the coil plane. The field is computed for the point in time when di/dt in the coil equals 550 A/μs. Arrowheads are omitted from very short vectors.

planar slices through the sphere. The vectors show the magnitude (length) and direction of the electric field projected in the plane shown. For the sake of clarity, if the magnitude of a vector is too small, then its arrowhead is omitted. The tail of the vector is located at the point where the field computation is made. For the case shown in Fig. 9.22, the total induced

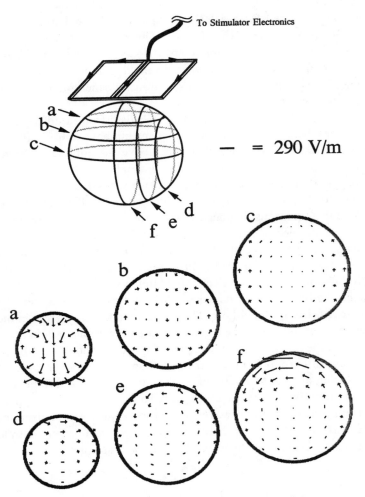

Figure 9.23   The electric field distribution inside an 11-cm-diameter conducting sphere excited by a double square-shaped coil 5 cm on an edge located 1.5 cm above the sphere. Vectors shown are the component of the field projected into the plane. The field is computed for the point in time when di/dt equals 230 A/μs. Arrowheads are omitted from very short vectors. Slices are spaced 2 cm apart.

electric field vector always lies in a plane parallel to the coil plane. Since the slices where the field is computed are all parallel to the coil plane, the field shown is the total electric field; there is no out-of-plane component. This figure illustrates that the current (which is everywhere parallel to the electric field) flows in circular loops that are strongest near the top of the sphere (closest to where the coil is located) and flow in a direction opposite that of the coil current. The scale is computed from the assumption that the coil is excited from a 500-μF capacitor charged to 500 V, coupled through a 5:1 lossless transformer. The peak di/dt in the one-turn is 5.5×10^8 A/s, and goes to zero in abaut 74 μs. The field shown is computed at the point in time when di/dt is maximal. The maximum electric field strength inside the sphere is 760 V/m directly beneath the coil winding at the sphere boundary.

Figure 9.23 shows the induced electric field from a different coil. This figure has the same 11-cm-diameter sphere, but has a double square-shaped coil 5 cm on an edge located 1.5 cm above the conducting sphere. The currents in the coil flow in the directions indicated by the arrows on the coil. The center conductor is made of two wires, each carrying a current equal to that of the outer conductors. For this arrangement, the boundary conditions significantly affect the total field. Unlike the previous example, the electric field is not always parallel to the coil plane. The field shown in Fig. 9.23 is computed assuming the same driving circuit as the previous example. The peak di/dt in the coil is 2.3×10^8 A/s, and goes to zero in about 115 μs. The electric field in the sphere is strongest (665 V/m) at the top boundary of the sphere directly beneath the center of the coil where the two squares adjoin. The return path for the current in the center region is complicated. The current divides in three main directions: to the right, to the left, and down. Because of this dispersion of the return path, the central current is much stronger than the current in other regions, producing a more focal stimulus. The strong central current magnitude decays rapidly with depth, much more so than with the circular coil. The current near the boundary away from the strong central current does not decay as rapidly with depth, because it is the return current path for both the central current at the given depth as well as currents above it.

10

High-Voltage and High-Current Injuries

MICHAEL A. CHILBERT

10.1   Introduction

Several modes of injury can result from electrical accidents, including tissue destruction, cellular excitation, and trauma secondary to the passage of current. Thermal injuries in the extremities can lead to amputation because of the deep nature of the burn. Cell lysis can also destroy tissue if there is a sufficient electric potential across the cell membrane. The effects of lysis are sometimes delayed. Cellular excitation of muscle and nerve can lead to fibrillation or transient neural dysfunction. The leading cause of death from electricity is attributed to fibrillation. Secondary injuries result from flash burns, falling, or gross contraction of the voluntary muscles.

Electrical accidents result in nearly 1100 deaths per year in the United States, 10% of which are caused by lightning. One-third of these deaths are caused by voltages below 1000 V occurring in the home and workplace. Low voltages account for more than half of the industrial deaths (Dalziel, 1978), High-voltage deaths are typically from industrial accidents. About 2% of low-voltage accidents and 10% of high-voltage accidents are fatal.

The number of accidents resulting in survivable injury are not consistently documented, because victims are not admitted to the same critical-care areas within a given facility and those not surviving at the scene are not admitted. Victims of severe burns are, of course, admitted to burn-care units, while those afflicted only with cardiac maladies are not. Typical injury reports include only the burn victims. One reason for this is that the management of electric burns is complicated and the evaluation of the burn's extent is very difficult and may require multiple procedures.

This Research was supported in part by NIH research grant GM34856 and the Department of Veterans Affairs Medical Center Research Funds, Milwaukee, Wisconsin.

Electric burn injuries account for 4–6% of all admissions to burn-care facilities (Hammond and Ward, 1988; Hunt et al., 1980; Rosenberg and Nelson, 1988). These burns cover an average of 12% of the body surface area (BSA) but result in limb amputation for 50–70% of the cases. The average BSA resulting in subsequent death is only 34%, compared to a much higher BSA percentage for thermal burns.

This chapter emphasizes electrical burn injury and its theoretical basis. The physical parameters affecting the character of injury and including impedance considerations and current distribution throughout the body are also included. Discussions of thermal and nonthermal trauma follow. Lightning injuries often have unique characteristics and traumatic sequelae; these are to be discussed as well. Important to the patho-physiology of electric trauma are the clinical observations and treatment of the accident victim.

10.2   Modes of Injury

The following sections provide background for the forms of trauma seen clinically. Thermal trauma has the greatest consequence to survivors of electric accidents; loss of limb is often the result. Nonthermal trauma has the form of lesions in neural and connective tissues. Lightning injuries often involve lesions exclusive of burns. Other trauma affects the heart rhythm and can result in fibrillation.

Thermal Injury

Thermal injury is caused by heating of tissue from the passage of current. The amount of heat generated in the tissue depends on spatial and temporal patterns of current density and tissue resistivity. The current density and resistivity are, in turn, affected by the heat generated in the tissue. Consequently, the process of thermal injury involves a feedback process.

The calculation of the heat generated in the tissue is given in terms of energy density:

$$Q_j = J^2 \rho t \tag{10.1}$$

where Q_j is the thermal energy density (J/cm^3), J is current density (A/cm^2), ρ is the resistivity (Ω/cm), and t is the duration of application(s). From this equation the temperature increase in the tissue can be determined. In its simplest form the thermal energy density can be related to the change in temperature in the following manner:

$$Q_j = pc\Delta T \tag{10.2}$$

where p is the tissue density (g/cm^3), c is the specific heat of tissue ($J/g\,°C$), and ΔT is the temperature change (°C). Equation (10.2) assumes that there are no thermal losses by conduction to adjacent media, by convection into nonstationary media, or by convection or radiation into air. In other words, all boundaries of the specific volume are adiabatic, and there are no internal heat sinks or sources in the volume. By solving for the change in temperature using Eqs. (10.1) and (10.2),

$$\Delta T = \frac{Q_j}{pc} = \frac{J^2 \rho t}{pc} \tag{10.3}$$

the result gives a greatly simplified equation for the temperature change in tissue. Equation (10.3) illustrates the primary relationship among the current density, resistivity, the duration of current, and the resulting change of temperature. Thermal injury depends on the duration of exposure and the temperature (Henriques, 1947; Henriquez and Moritz, 1947). The temperature threshold for cutaneous thermal damage is about 44°C, indicating that thermal injury will not occur below this temperature. Cutaneous burns occur when the temperature is elevated for a sufficient length of time: 45°C requires more than 3.h, 51°C requires less than 4 min, and 70°C requires less than 1 s for injury (Henriques, 1947; Henriques and Moritz, 1947). Temperature levels that cause injury in other tissues are similar, as is the injury rate for a given temperature. Electrically induced thermal injury of muscle begins at 43°C with a 1-A current in the limb for 15 min, and at 46°C the tissue damage is much greater (Chilbert et al., 1985b). Electrically induced thermal damage to peripheral nerve has been noted to occur at 48°C after several seconds.

One important aspect of electrical burning is the rate at which heat is deposited and removed. The deposition of heat is dependent on the current density and resistivity. Resistivity decreases temperature, causing further increases in current density and temperature. This aspect of resistivity change is rarely considered in burn models because of the resulting nonlinear equations. Heat is removed from the tissue by conduction into adjacent tissue, by convection and radiation into the air at the surface, and by blood flow. To account for conduction and blood flow effects, one can expand Eq. (10.2) into:

$$pc \frac{\partial^2 T}{\partial t^2} = k \frac{\partial^2 T}{\partial x^2} + k \frac{\partial^2 T}{\partial y^2} + k \frac{\partial^2 T}{\partial z^2} + w p_b c_b (T_b - T) + q_m + q_j \tag{10.4}$$

where T is the tissue temperature, k is the thermal conductivity of tissue, w is the blood perfusion rate, p_b is the blood density, c_b is the specific heat of blood, T_b is the initial temperature of blood, q_m is the volumetric rate of metabolic heat generation, and q_j is the volumetric rate of electrical heating. Note that the first three terms of the equation describe the tissue conduction and the fourth term describes the effects of vascular convection. Equation (10.4) is termed the *bioheat equation*, which was

Table 10.1. *Thermal conductivity values (k) for skin, fat, and muscle*

Tissue or medium	Conductivity, $k \times 10^3$ (W/cm °C)
Amorphous	2.09
Skin	3.40
Fat	3.23
Muscle	5.00
Water (40°C)	6.32

Source: Data from Pennes (1948), Olsen et al. (1985), Song et al. (1988).

expressed in its initial form by Pennes (1948) and has been modified extensively for various applications (Song et al., 1988). Table 10.1 lists measured values of thermal conductivity for skin, fat, and muscle of human and hog tissues. These values have been used in thermal models as well. Values for the thermal conductivity of water are included for comparison. The thermal conductivity of blood used by Song et al. (1988) was 5.0×10^3 W/cm °C. Other parameter values used were $p_b = 1.05 \, g/cm^3$ and $c_b = 3.8 \, J/g$ °C. The value for w is 2–12 (ml blood)/(ml tissue)/s for dermal perfusion.

Lee and Kolodney (1987b) developed a unidimensional, axisymmetric model for the heating resulting from a high-voltage electrical contact in the upper limb. The limb was modeled as a coaxial cylinder having a 10-cm diameter and included layers to represent bone, muscle, fat, and skin of 1.0-, 3.5-, 0.3-, and 0.2-cm thicknesses, respectively. The bioheat equation used for this model included the term for electrical heating while neglecting metabolic heat. Lee and Kolodney show that tissue perfusion is critical for the removal of heat in the tissue, but does not significantly modify the heating process during the application of current.

Thermal injury due to the passage of current is dependent on the current path, which is determined by the contact points on the body (sometimes termed entry and exit sites). Once the current is inside the body, its distribution is determined by tissue resistivity. Early perceptions of current in the body stated that it would flow equally through all tissues or that it would flow along the path of least resistance (in the vessels and nerves). Both are wrong, but are still referred to in clinical reports. Experimental evaluation of the current path through the body has shown that the resistivity of each tissue in a given cross section determines the current distribution through the cross section (Chilbert et al., 1983, 1985b, 1988, 1989; Sances et al., 1981a, 1983).

Tissue resistivity is typically anisotropic; that is, the resistivity varies with current direction. For example, muscle has a lower resistivity along its fibers than across them, and skin has a high resistivity across its layers and a lower resistivity along its layers. Some tissues can be considered isotropic when no directional differences exist in the tissue. The implications of anisotropic resistivity in thermal burns is that the direction of current through a given tissue may explain selective tissue trauma seen away from the contact site. This selective tissue destruction is well documented in the clinical literature and is seen in muscle tissue near the bone (Artz, 1974, 1979; Hammond and Ward, 1988; Luce et al., 1978; Moncrief and Pruitt, 1971; Ponten et al., 1970; Rosenberg and Nelson, 1988; Sances et al., 1979; Wang et al., 1984, 1985, 1987; Zelt et al., 1988).

Selective thermal trauma can also be related to tissue resistivity changes with temperature and events at the contact site. As the tissue temperature increases, the resistivity decreases, altering the current distribution and subsequent temperature changes in the tissue. Contact site events that change the current distribution are desiccation and arcing. Desiccation of skin at the contact will cause the total current to decrease rapidly. This limits the contact time at voltages below 1000 V and the trauma is usually limited to the contact region. With higher voltages, arcing can occur over the desiccated tissue and alter the current path.

Electroporation

When the electric field strength across the cell membrane is sufficiently large, the cell membrane will rupture and cause cell lysis (Lee and Kolodney, 1987a). The process that causes cellular rupture is called *electroporation*, which is an increase in membrane permeability due to development of aqueous pores in the membrane (Lee et al., 1988). Electroporation of the cell membrane leads to rupture when the pore size becomes large enough or when fusion of several pores occurs. Calculation of the membrane potential uses cable theory, which is discussed in Chapter 4, and depends on the cell length, cell diameter, the internal fluid conductivity, the membrane conductivity, and the thickness of the cell membrane.

The membrane potential causes the pore size to increase proportionately until an irreversible state is reached. This state occurs when the pore size reaches a diameter of one-half the thickness of the membrane. The membrane potential causing ruture is in the range of 800 to 1000 mV. Reversible electroporation has been noted to occur below a 200-mV membrane potential. The incidence of pores has been evidenced by experimentally observed increases in membrane permeability, but the exact mechanism of electrical membrane failure has not been determined (Lee and Kolodney, 1987a). Theoretical analysis of the membrane potential uses cable theory, which is discussed in Chapter 4, and has been derived for electrical injury situations by Gaylor et al. (1988).

The role of electroporation in electrical injury is related to the electric field strength in the biological medium. Studies of muscle cells in culture by Lee et al. (1988) have shown that an electric field strength between 50 and 300 V/cm will disrupt cells of 1 mm. This is consistent with values of the membrane potential of his earlier study (Lee and Kolodney, 1987a). For smaller cells, such as fibroblasts of 10-μm diameter, the electric field strength that causes rupture exceeds 1000 V/cm. The relation of this data to accident victims is given in Sec. 10.5. The membrane potential is dependent on the cell length, cell diameter, and membrane thickness until the cell is electrically long, after which the membrane potential depends on the length constant and the electrical field strength (see Chapter 4). Muscle cells greater than 1 cm in length can be considered electrically long. For small cells the membrane potential depends mostly on length, while for longer cells it depends on the diameter and for medium-length cells the dependence is on both (Gaylor et al., 1988). The electric field needed for electroporation decreases as temperature increases. Consequently, the tissue trauma caused in a region of elevated temperature is related to both electroporation and temperature.

Fibrillation

The effects of electricity on the heart have been thoroughly discussed in Chapter 6. This chapter will focus on short-duration, high-voltage, and high-current applications. Power-line accidents typically result in severe burns, whereas household voltages result in fibrillation (Sances et al., 1979). Most deaths at power-line voltages are attributed to burns or to secondary trauma such as falling. Severe burns or complications arising from them are the causes of death in the hospital. Rarely is instantaneous death attributed to fibrillation at high voltages, yet several case reports of latent cardiac anomalies have been published (Ahrenholz et al., 1988; Dixon, 1983; Guinard et al., 1987; Hammond and Ward, 1988; Rouse and Dimick, 1978; Wilson et al., 1988). The minimum fibrillation level, as discussed in Chapter 6, is well documented; several equations to approximate its character with application time have been published. Similarly, nonfibrillating and defibrillating levels have been documented. However, fibrillation can occur at high current levels with very short application times, suggesting that these levels are not constant and may vary in a similar fashion to the minimum fibrillation level.

Reports on the experimental evaluation of fibrillation began before the beginning of this century. Ferris et al. (1936) published a comprehensive study on fibrillation from currents up to 17 A for a duration of 0.3 s. Until recently, most studies of fibrillation have used current levels below 10 A to delineate minimum electrical thresholds for the cardiovascular system. Kouwenhoven (1964) reported that fibrillation does not occur above 7.5 A when the contact lasts longer than 1 s. Since 1980 this author and co-workers have been involved with the study of electrical injuries and current pathways associated with high voltages and currents (Chilbert et al., 1983, 1985a, 1985b, 1988; Prieto et al., 1985; Sances et al., 1979, 1981a, 1981b, 1981c, 1983). It was noted that fibrillation often occurred at current levels greater than 10 A, and one study investigated high-current fibrillation in hogs (Chilbert et al., 1989).

In the high-current fibrillation studies of the authors and co-workers, the current delivery system supplied 1.5 to 200 cycles of 60-Hz currents between 1 and 51 A at voltages between 500 and 6000 V. The current was applied during the preselected cardiac cycle through a triggering circuit synchronized with the cardiac waveform (Chilbert et al., 1988). Nineteen fibrillations occurred in 67 runs performed on 20 animals. Two fibrillations were induced with currents starting in the QRS interval, lasting 25.5 cycles, and occurring in animals that had been previously defibrillated. One fibrillation associated with the P wave lasted for 200 cycles at a current of 3 A. The fibrillation associated with the T–P interval was at 51 A, six cycles long, and was a first fibrillation; however, its point of initiation was at the end of the T wave (Fig. 10.1); the actual occurrence of the T wave could only be approximated by temporal extrapolation. Results show that 18 out of 32 short-duration runs involving the T wave caused fibrillation; 9 fibrillations out of 14 runs occurred below 10 A and 9 out of 18 occurred above 10 A. As the current increases, the likelihood of defibrillation increases (see Chapter 6). This also indicates that the likelihood of fibrillation decreases with current amplitude and has been noted by others (Dalziel, 1968, 1972; Geddes et al., 1986). The short-duration applications used here were always less than one complete cardiac cycle (excluding the run of 200 cycles).

The work of Ferris et al. (1936) has shown that, for current durations of 30 ms (about 2 cycles at 60 Hz) during the sensitive intervals, fibrillation can be induced by 15-A, 60-Hz currents in sheep. They also noted that the high-voltage, short-duration shocks do not show a cumulative effect (i.e., increase fibrillation's rate of occurrence), and that normal cardiac rhythm returned within 5 min in the nonfibrillating runs. Of 913 runs in 132 sheep, only 100 resulted in fibrillation. They recorded 11 fibrillations

out of 16 runs at 4 A, 6 fibrillations out of 17 runs at 12 A, and one fibrillation out of 51 runs at 24 A. This shows a decrease in the likelihood of fibrillation as the current level increases; the occurrence at 25 A is about 10
times less than that at 5 A. This is consistent with the hog data above.
For currents of 4 to 14 A applied for 30 ms during the other intervals
of the cardiac cycle, no fibrillations were recorded; however, 4-A currents
applied for 150 ms to 260 ms did cause occasional fibrillations.

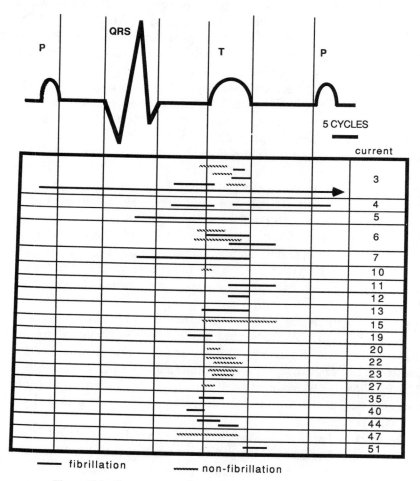

Figure 10.1   Occurrence of fibrillation during specific periods of the cardiac
cycle. Current applications causing fibrillation all are associated with the
T-wave period. Fibrillation occurs less frequently at higher current levels. All
currents involve the cardiac cycle shown if normal cardiac rhythm is
extrapolated past the onset of current. (From Chilbert et al., © 1989 IEEE.)

Table 10.2. *Heart/body ratio of animals and humans*

Species	N	Average weight Heart (g)	Average weight Body (kg)	Ratio (%) heart/body
Human	*	280	70	0.40
Hog	9	300	79	0.38
Dog	10	170	22	0.77
Sheep	25	270	56	0.48
Calf	10	420	70	0.60

Source: Data from Ferris et al. (1936), Chilbert et al. (1989).

The Ferris study tabulated body weight and heart size of sheep, hogs, dogs, and calves. One important aspect of body weight and heart size is the ratio between the two (Table 10.2). This ratio suggests that, for similar body weights, the hog heart is closest in size to the human heart, since the body weight and the ratio are very similar. The thoracic circumference of hogs and humans is also comparable for specimens of the same weight, although the cross-sectional arrangement of organs and other tissues is different. The similarities of ratio and circumference suggest that the hog is an excellent model for fibrillation in humans. Also, the results of Ferris et al. (1936) show that hogs tend to fibrillate at lower-than-average current levels, thus making the hog a conservative model for ventricular fibrillation.

The statistical inference of the current is useful for development of safe circuitry, but the occurrence of fibrillation is more accurately based on the current density in the heart. However, voltage is usually the only known electrical parameter in the accident situation, and the total current can only be estimated. Consequently, the current density in the heart is entirely unknown. If one could estimate the current density in the heart, the probability of fibrillation could be predicted more accurately.

10.3   Impedance Considerations and Current Distribution in the Body

The ensuing sections will present first the total body impedance and the effects of current and voltage on it. The total body impedance will then be analyzed in terms of the contact impedance and segmental body impedances. The segmental body impedance can be further analyzed in cross section, and the current density levels in various tissues and organs will be presented.

Total Body Impedance

The current that flows through the body in high-voltage and high-current injuries is determined by the total body impedance and its change with time. The voltage is typically constant in electrical injuries and fluctuates only when the maximum power of the source is reached (Sances et al., 1981b). The minimum total impedance typically assumed for low-voltage electrical accident analysis is $500 \, \Omega$ (Dalziel, 1978; Taylor, 1985), and variations to this value are discussed in Chapter 2. At high voltages the total impedance of the body is less in short-duration contacts, while for longer durations the values increase with time.

Figure 10.2 depicts the current–time relationship for electrical contact at three low-voltage levels. The current amplitude can be separated into three phases determined by its rate of change with time. The initial phase occurs as the contact is established (the contact impedance is rapidly reduced), sometimes termed the breakdown of the electrode–skin interface. The middle phase is the phase of maximal current flow. Increases during the middle phase can be attributed to heating of the tissue near the electrode contact and in cross sections where large current densities occur. The final phase occurs when the tissue under the electrode becomes highly resistive

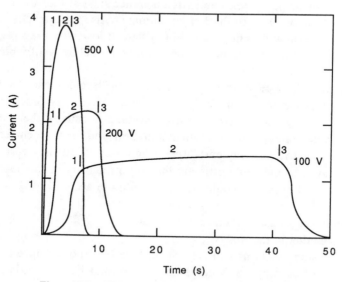

Figure 10.2   Relationship of current to time for a 5-cm disk in contact with the skin of a hog at different voltages. Note that an increasing voltage results in a decrease in the effective application time. Each trace shows the triphasic characteristics for current, which are the initial phase (1), middle phase (2), and final phase (3), as described in the text.

Table 10.3. *Measured changes with voltage in the hog*

Volts	n	Effective application time (s)	Maximum applied current (mA)	Minimum total resistance (Ω)	Total energy (J)
75	13	205	721	104	8773
100	14	92	806	124	6008
150	15	21	1087	138	2789
200	13	11	1227	163	2208
250	11	5.6	1471	170	1734

Note: Values for a 2.5-cm-diameter disk applied to the hindlimb with a large plate electrode on the other hindlimb.

and stops passing current, caused by desiccation or charring of the skin. The rate at which these three phases occur decreases as the voltage increases (Fig. 10.2). The three phases are present over a range of application times (Carter and Morely, 1969a, 1969b; Prieto et al., 1985; Sances et al., 1981a, 1981b). Application times for five voltage levels are listed in Table 10.3 and show the decrease in time with increasing voltage. Also note that the minimum total impedance increases with applied voltage. Similarly, the total energy decreases with increasing voltage. The changes in impedance and energy are caused by the rate of charring at the electrode site. The contact phenomenon of Fig. 10.2 is applicable for low voltages that do not support arcing.

High voltages and high currents are usually associated with arcing at the electrode site (Sances et al., 1979, 1981c). The initial and middle phases of the current-time plot can occur within milliseconds; the onset of arcing is the final phase. Voltages above 1000 V are sufficient to arc through air and desiccated skin. Once arcing is established, the arc length and internal impedance limit the current. Figure 10.3 shows that current falls with time when an arc is established, indicating that the arc length increases as tissue is burned away. The arcing appeared to occur over regions of uncharred tissue, taking the path of minimum resistance (Sances et al., 1981c). The tissue removal by the arc continues until the circuit is interrupted or the limb is transected. At 7200 V (Fig. 10.3), soft tissue was removed from the limb within 5 s; the remaining 11 s were required to transect the bone. At 14,400 V, limb transection time is about 3 s. Circuit interruption can occur when the nonconduction path length exceeds the voltage's ability to maintain an arc or by the standard means of a ground-fault interrupter, circuit breaker, or fuse (see Chapter 11).

Experimental studies of electrical burn injury have been performed in the hog, since this animal approximates the human in skin, anatomy, weight, and cross-sectional dimensions better than other species. The total impedance of the hog at high voltages for different electrode types and locations is given in Fig. 10.4. This figure shows that a large contact (plate electrode) on the hindlimb decreases the total impedance and that forelimb-to-hindlimb impedance is greater that hindlimb-to-hindlimb impedance. This is further exemplified in Table 10.4, which shows a decreasing impedance with increasing contact area and perimeter. As the location of the band electrodes is moved up the limbs, the impedance decreases, which is consistent with observations made previously (Prieto et al., 1985; Sances et al., 1981b). The impedance also changes with the applied voltage level.

Impedance depends on both applied voltage and duration of application (see Fig. 10.10). At low voltages applied for long time periods (those that exceed the second phase), the impedance increases with voltage while the total energy decreases (Fig. 10.5a). The decrease in total energy with increasing voltage indicates less tissue heating at higher voltages (Table 10.3). For short periods of time ($t < 1$ s), the impedance is dependent on the initial phase of the current waveform (Fig. 10.2). The short-duration impedance at low voltages will therefore decrease as the voltage increases, which is evident in Fig. 10.5b. High-voltage impedance changes are similar to those of low-voltage changes. Short-duration contacts occurring in the

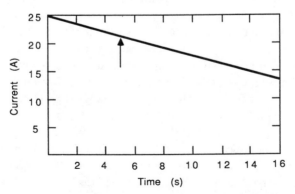

Figure 10.3   Current versus time of application for limb transection at 7.2 kV. Arrow indicates point of soft tissue removal; remaining time is needed to transect the bone. The initial value of current was 25 A and the final value was 14.8 A. The contact was between a wire contact on the hindlimb and a large plate under the other hindlimb. [From Sances et al. (1981b), © 1981 IEEE.]

first phase are typically less than 100 ms, and long-duration contacts would be denoted by the onset of arcing.

Body impedance varies with voltage, time of application, and contact parameters. Increasing the voltage increases the current and decreases the total impedance for short application times. The longer application times increased the peak current at lower voltages, since the tissue resistivity decreases with increasing temperature near the electrode sites. The electrode contact site on the limbs determined the contact area and the perimeter for the band electrodes (Table 10.4), and this affects the total current flow as well. In general, increasing the area and perimeter increases the current for a given voltage; consequently, the largest parts of the limbs had a greater contact area and perimeter and allowed greater current passage with the band electrodes. Also, the location of the contact on the body will affect the total abody impedance, since different current paths in the body have different impedances.

Contact Impedance and Segmental Body Impedance

Contact impedance will limit the level of current in the body, its duration, and its path, and the probable extent of trauma. The remaining portion of the total impedance can be divided into individual impedances for the various segments of the body. These segmental body impedances can be modified to account for tissue heating away from the contact points. Tissue heating decreases impedance, thereby increasing current. By

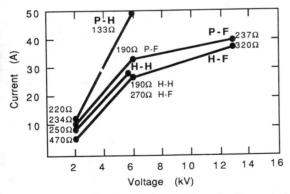

Figure 10.4 Current versus voltage for first contact in the hog. P-H is large plate electrode-to-hindlimb wire configuration, P-F is large plate electrode-to-forelimb wire, H-H is hindlimb-to-hindlimb wire contacts, and H-F is hindlimb-to-forelimb wire contacts. The plate was located under the hindlimb opposite the wire contacts. Total body impedances are shown for each measurement point and were determined from the average current for each run. [From Sances et al. (1981b), © 1981 IEEE.]

Table 10.4. *Total body resistance (Ω) for band electrodes located on the hindlimb and analogously on the forelimb (8-cm disk electrode values d included for comparison)*

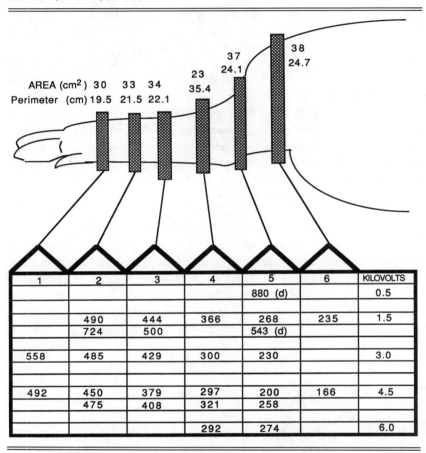

1	2	3	4	5	6	KILOVOLTS
				880 (d)		0.5
	490	444	366	268	235	1.5
	724	500		543 (d)		
558	485	429	300	230		3.0
492	450	379	297	200	166	4.5
	475	408	321	258		
			292	274		6.0

Source: Adapted from Chilbert et al., 1989.

combining information from the contact impedance and the segmental body impedances, one can better model the effects of electrical trauma.

Experimental evaluation shows that, for circular and elliptical contacts, the contact impedance is related to contact area and perimeter (Prieto et al., 1985). Figure 10.6 shows the relationship of total body impedance to the contact area and contact perimeter and gives their proportional relationships. Furthermore, the total impedance of the body can be described by the sum of the internal impedance and the contact impedance, where the contact impedance can be expressed as

$$R_c = K(A)^{-1/4}(P)^{-1/2} \tag{10.5}$$

where R_c is the contact resistance, K is a constant of resistivity, A is the electrode contact area, and P is the electrode contact perimeter. For a circular electrode $R_c = 0.3 K/r$, where r is the electrode radius. The constant of resistivity depends on the applied voltage, temperature, and skin condition (wet, dry, abraded, etc.). For electrode contact studies at 100 V and with a dry contact, Prieto et al. (1985) derived a value for K of 264 for areas between 0.3 cm^2 and 20.3 cm^2 and perimeters between 2 and

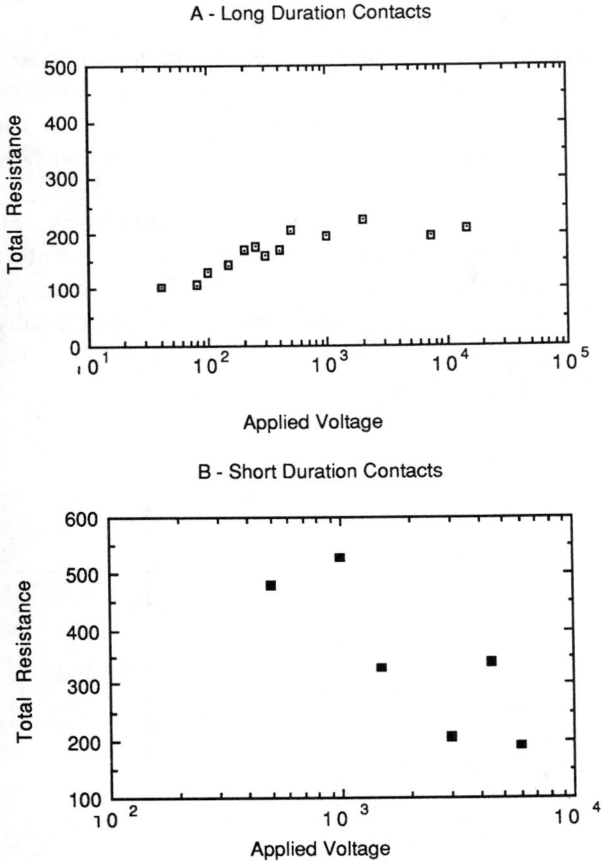

Figure 10.5   Minimum total body resistances for voltages applied between the hindlimb and a large plate on the opposite hindquarter. (a) Voltages were applied for long time periods (through the third phase of the current-time waveform). (b) Voltages were applied for short time periods (the frist phase of the current-time waveform).

28 cm. The result of a least-squares linear regression fit of this equation to the experimental data gives a correlation coefficient of 0.94. As the electrode size increases in this model, the contact resistance approaches zero, leaving only the internal impedance to limit the flow of current.

Further analysis of the contact has been performed by using finite-element analysis on an axisymmetric model in two dimensions. Figure 10.7 shows the boundary and arrangement for the analysis of a circular disk containing the skin and includes tissue layers for skin, fat, and muscle. The model is based on a 5-cm-diameter disk with a potential of 100 V contacting the tissue, which is a nonhomogeneous, anisotropic, semi-infinite medium. Model parameters are listed in Table 10.5. The model includes analysis of surface and volume heat transfer effects and temperature-dependent resistivities. The tissue layers are each homogeneous, and the skin and fat layers have isotropic resistivities. The muscle resistivity is anisotropic (Fig. 10.1), with its transverse resistivity defined in the z direction and its longitudinal resistivity defined in the r direction. The thermal properties are isotropic in all tissues.

The model determines the current density and temperature distribution in the tissue. Plots of the thermal isoclines resulting after 5 s of contact are shown in Fig. 10.8a and after 9 s in Fig.10.8b. The thermal gradient emanates from the edge of the disk and proceeds somewhat concentrically

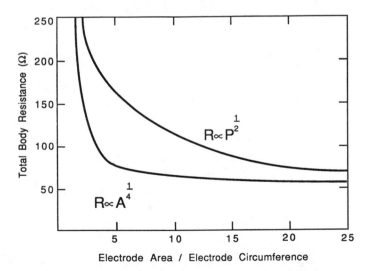

Figure 10.6   Total body resistance compared to electrode contact area and electrode contact circumference for circular and elliptical contacts. The proportional relationships indicate the best-curve fit of experimental data to the curves using a least-squares linear regression algorithm.

deeper into the tissue. Surface temperatures indicate a peak value at the edge of the disk (Fig. 10.9), showing the effect at the edge. The electrode edge effect of increased temperature has been shown experimentally by Sances et al. (1981) in saline and in the hog using a thermographic camera. The total current entering the tissue through the disk electrode is shown in Fig. 10.10. The model compensates for changes in skin that exceed 100°C by rapidly increasing the resistivity, resulting in a drop in the current after 9 s. This simulates desiccation of the skin comparable to experimental observations of Prieto et al. (1985). The analysis shows that the greatest temperature increase occurs under the edge of the disk and indicates that the current density is greatest at this region of the contact. Therefore, the majority of current entering the tissue does so near the edge of the electrode. More than 50% of the current entering the tissue enters in the outer 15% of the disk radius or in the outer 28% of the total area. These results indicate the importance of the circumference to the level of current that will flow into the body. Disk electrodes at high voltages of similar size to those used previously at household voltage levels show similar changes

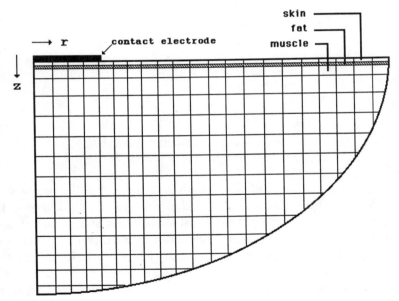

Figure 10.7   Geometric arrangement for an axisymmetric finite-element analysis of a disk on a three-layered, semi-infinite medium. The model contains three homogeneous layers for skin, fat, and muscle. The skin and fat layers have isotropic resistivity, while the muscle layer has anisotropic resistivity. The model includes surface and volume heat transfer effects and temperature-dependent resistivities. Thermal properties are isotropic in all tissues.

Table 10.5. *Finite-element analysis parameters for a disk electrode in contact with a nonhomogeneous, anisotropic, semi-infinite medium used to model electric burn injury at the contact site*

Dimensions (cm)	
Disk radius	2.5
Skin thickness	0.15
Fat thickness	0.35
Dirichlet boundary conditions	
Ambient air temperature (°C)	25
Initial tissue temperature (°C)	37
Disk potential (V)	100
Reference potential (V)	0
Material properties	
Electrical resistivity ($\Omega\,cm$)	
Skin	280
Fat	375
Transverse muscle	650
Longitudinal muscle	290
Thermal conductivity ($W\,cm^{-1}\,°C^{-1}$)	
Skin	0.037
Fat	0.020
Muscle	0.042
Density ($g\,cm^{-3}$)	
Skin	1.00
Fat	0.85
Muscle	1.05
Specific heat ($J\,g^{-1}\,°C^{-1}$)	
Skin	3.2
Fat	2.3
Muscle	3.8

Heat transfer coefficient: $13.45 \times 10^{-6}\,W\,cm^{-2}\,°C^{-1}$

due to contact area and edge length, indicating that the results of the finite-element analysis are valid at high voltages as well (Chilbert et al., 1989; Prieto et al., 1985). Theoretical studies by Caruso and colleagues (1979) showed that current is concentrated at the edge of an electrode to a degree that depends on the conductivity layers beneath the electrode (see Fig. 2.9).

The internal impedance can be represented by segmental resistances of the body, making the total impedance determination independent of contact location. Each body segment can be modeled by either a constant resistance or a temperature-varying resistance. Figure 10.11 shows segmental resistance percentages for the human body based on the internal hand-to-hand values being 100%. The calculation of segmental impedance

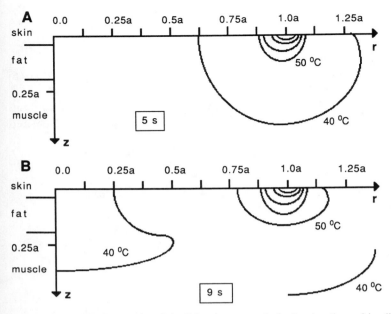

Figure 10.8   Results of the finite-element analysis showing thermal isoclines at (a) 5 s and (b) 9 s. isocline increment is 10°C per contour. The disk potential was 100 V, and the model continued until tissue temperatures exceeded 100°C. Current density levels under the disk are very similar to the thermal isoclines.

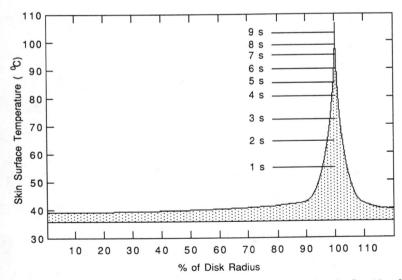

Figure 10.9 Temperature along the surface of the skin during the first 10 s of current flow. The graph approximates the transient temperature distribution for an isopotential disk of 100 V contacting a nonhomogeneous, semi-infinite medium. The peak temperature at the disk's edge is given for each second of contact.

is then

$$R_s = R_{hh} \left(\frac{S}{100} \right) \tag{10.6}$$

where R_s is the segmental resistance, R_{hh} is the internal hand-to-hand resistance, and S is the segmental percentage in Fig. 10.11 for the desired

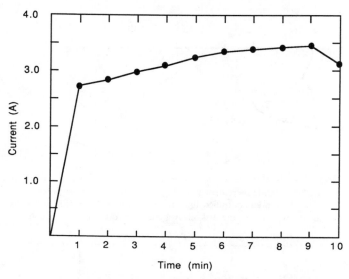

Figure 10.10   Total current flowing into the tissue from the 2.5-cm disk at 100 V for a 10-s contact. The current falls off after 9 s due to the skin temperature exceeding 100°C, when tissue desiccation would occur.

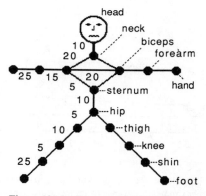

Figure 10.11   Segmental body resistances for the human body. Numbers indicate the percentage of the total internal hand-to-hand resistance between each of the designated parts of the body. Figure derived from Figs. 2.38 and 2.39 and from Biegelmeier (1985).

body segment. Chapter 2 gives some values for internal resistance in humans, but the evaluation of the contact impedance is usually quite different in those studies. The hand-to-hand resistance is used as a reference because most accidents involve a hand-to-hand contact. Experimental studies in the hog have shown that the total internal resistance is $370\,\Omega$, the forelimb is $150\,\Omega$, the hindlimb is $170\,\Omega$, and the body impedance is $45\,\Omega$. These measurements were made at voltages around $10\,V$ and at a constant 60-Hz current of $7\,mA$.

Tissue resistivity is inversely related to temperature. The slow rise in current during phase 2 (Fig. 10.2) is likely the result of decreasing resistivity with increasing temperature. Body segments in contact with the electrodes, or otherwise subject to temperature change, can be modeled with a variable resistance value that is temperature-dependent. Each segmental body resistance can be modified to change with temperature using the following equation:

$$R_{ts} = R_s(1 - 0.025\,\Delta T) \tag{10.7}$$

where R_{ts} is the resistance modified by the temperature change ΔT. Experiment shows that increasing temperature decreases resistivity at a rate of 2.5% per °C, which is about $5\,\Omega\,cm/°C$ for muscle tissue (Chilbert et al., 1983, 1985b). At low voltages, heating occurs in the vicinity of the contacts, if at all. High-voltage tissue heating occurs around limb joints, at the contact sites, and in regions of small cross-sectional area. Changes in the overall resistance will increase the total current flowing through the body and likewise increase the tissue current density. Since tissues in a given cross section have different resistivities, there will be different current densities in the different tissues; thus, specific heating in the tissues will also be different. This leads to nonuniform heating and tissue trauma.

Tissue Current Densities and Current Distribution

The extent of thermal trauma and the likelihood of fibrillation depend on the current density in tissue and current distribution in the body. Current densities have been measured in the tissues of the limbs in hogs using a constant-current source (Chilbert et al., 1985a, 1985b, 1989; Sances et al., 1981a, 1983). Figure 10.12 gives the limb current densities for given applied currents between the hindlimbs measured in the region of the hindlimb shown in Fig. 10.13. Table 10.6 lists values of current density and resistivity at specific applied current levels between the hindlimbs. The listed voltage values are initial values, since the voltage drops with time when the current is held constant. Table 10.7 gives tissue resistance, resistivity, current, current density, and energy density for

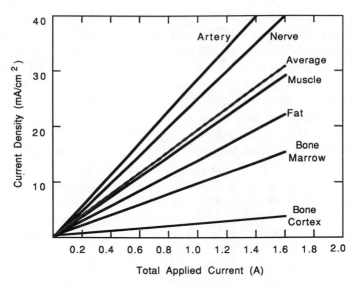

Figure 10.12   Current density versus current in the hindlimb of the hog measured about 7 cm proximal to the ankle joint; electrodes applied across the hindlimbs. The average line represents the average current density in all the tissues at that level.

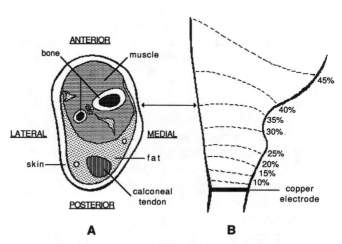

Figure 10.13   (a) Cross section of the hog hindlimb where current densities were measured. (b) Hindlimb location of measurement region shown with isopotential percentage lines of the applied voltage for electrodes placed on the hindlimbs.

Table 10.6. *Tissue current density and resistivity in the limb versus applied current from hindlimb to hindlimb in the hog*

Constant applied current (mA)	Initial applied voltage (V)	Artery		Nerve		Muscle			Fat		Bone marrow		Bone cortex	
		J	ρ	J	ρ	J_L	ρ_L	ρ_T	J	ρ	J	ρ	J	ρ
10	6	0.32	147	0.26	201	0.18	296	512	0.12	375	0.10	550	0.03	1828
30	15	1.18	145	0.86	209	0.50	282	525	0.43	360	0.27	531	0.09	1880
100	45	3.4	152	3.0	191	2.0	295	483	1.5	352	1.0	535	0.30	1832
300	135	9.8	140	8.7	197	7.0	287	501	4.9	377	3.1	547	0.85	1859
600	260	21.0	150	15.2	196	12.0	292	492	9.3	366	5.3	542	2.0	1876
1000	415	35.9	155	27.1	200	19.5	290	650	13.4	386	8.1	525	2.8	1836

Note: J in mA/cm^2, ρ in Ω cm, J_L = longitudinal current density, ρ_L = longitudinal resistivity, ρ_T = transverse resistivity of muscle.
Source: From Sances et al., 1983.

Table 10.7. *Typical values measured in the hindlimb cross section of the hog*

Tissue	Tissue resistivity (Ω cm)	Tissue resistance (Ω)	Current density (mA/cm^2)	Tissue current (mA)	Tissue area (cm^2)	Energy density (J/cm^3)
Vessel	155	911.76	51.61	8.77	0.17	412.90
Nerve	200	1666.67	40.00	4.80	0.12	320.00
Muscle (L)	290	19.24	27.59	415.72	15.07	220.69
(T)	650	—				
Fat	380	31.77	21.05	251.79	11.96	168.42
Bone marrow	550	495.50	14.55	16.15	1.11	116.36
Bone cortex	1850	898.06	4.32	8.91	2.06	34.59
Tendon	398	196.19	20.09	40.78	2.03	160.70
Dermal layers	432	50.94	18.52	157.04	8.48	148.15
Average/ total	363	8.85	22.05	903.95	41.00	176.3

Note: Applied voltage = 400 V between the hindlimbs, current = 0.9 A. Electric field strength = 8 V/cm. Resistivity and current density were measured experimentally. Cross-sectional tissue resistance = (resistivity)(tissue area)/(length = 1 cm), tissue current = (current density)(tissue area), tissue energy = (current density) (current density)(resistivity)(time = 1 s). Cross-sectional areas were determined by digitization of a limb section taken through the measurement region. Averaged values are resistivity, current density, and energy density. Total values are resistance, current, and area.

one current application of 0.9 A with an initial voltage of 400 V. This table shows how the current is distributed through the cross section of Fig. 10.13 before tissue trauma and heating alters the current path. Current density in the limbs is more likely to change at high current levels than in the body because significant heating occurs almost exclusively in the limbs and at the contacts.

Current density measured in the body structures is given in Table 10.8, along with the corresponding resistivities. Linear extrapolation of current densities (as in Fig. 10.12) in the body for higher levels of current is accurate, since the tissue temperature in the body changes minimally. The largest current densities are found in the back region. The back muscle at the level of the upper lumbar region has a current density of 0.223 mA/cm^2, and the spinal cord current density is approximately 0.280 mA/cm^2 for the same region when a current of 100 mA is applied from the forelimb to the opposite hindlimb. Various current density levels in the spinal cord are reported in Table 10.9 for three different

Table 10.8. *Current density and resistivity in the hog for 100 mA applied from left forelimb to right hindlimb*

	n	J (mA/cm²)	S.D.		ρ (Ω cm)	S.D.
Ventral intestine	4	0.071	0.077		511	43.1
Dorsal intestine	4	0.077	0.049		610	96.9
Back muscle:						
Neck	1	0.021	—		—	—
Upper thoracic	1	0.061	—		—	—
Middle thoracic	1	0.141	— Longitudinal:	270	—
Upper lumbar	4	0.223	0.079 Transverse:	1006	58.6
Lumbosacral	1	0.055	—			
Kidney	4	0.097	—		447	61.2
Liver	4	0.065	—		570	37.5
Lung	4	0.058	0.033		1605	42.8
Heart (transverse to axis)	4	0.073	0.044		800	—
Abdomen, midline (skin, fat, muscle layers)	4	0.065	0.032		547	28.6
Abdomen, side (skin, fat, muscle layers)	4	0.076	0.028		560	42.0

Note: Current density in back muscle as measured in the longitudinal direction.
Source: From Sances et al., 1983.

Table 10.9. *Spinal cord current density for 100 mA applied across the limbs*

	Transverse J (mA/cm²)	Longitudinal J (mA/cm²)
Forelimb–forelimb		
Cervical	0.0155	0.0247
Thoracic	0.0022	0.0058
Lumbar	—	0.0011
Forelimb–hindlimb		
Cervical	0.0075	0.0708
Thoracic	0.0100	0.299
Lumbar	0.0080	0.257
Hindlimb–hindlimb		
Cervical	—	—
Thoracic	0.0008	0.0020
Lumbar	0.0057	0.0040

Note: Transverse spinal cord resistivity $= 1970\,\Omega\,\mathrm{cm}$, longitudinal resistivity $= 214\,\Omega\,\mathrm{cm}$.
Source: From Sances et al., 1983.

contact orientations. High current levels in the spinal cord can lead to latent neural dysfunction (Sances et al., 1979). At a level of 10 A applied for 30 s, the spinal-cord temperature was elevated only 1.5°C, showing that temperature in the cord is not an important factor in its trauma (Sances et al., 1983). The cross-sectional area of the spinal cord is 0.07% of the body cross section, but the cord carries 0.12–0.15% of the total current.

Current distribution in the limb changes with time and temperature and is shown by the change in tissue current densities given in Table 10.10. Initial values of resistivity and current density are also shown in Table 10.7 for all tissues. Current shifts from artery, nerve, tendon, and dermal layers to muscle and fat. The increase of current in muscle and fat is due to increased temperature and tissue degradation. The decrease of current in the other tissues is due to an increase in resistivity or a slower decrease in resistivity. Nerve and vessel resistivities increase with temperature once tissue damage occurs, whereas muscle resistivity decreases with tissue damage. The ultimate distribution of tissue trauma is related to the peak temperature in the tissue. The peak temperature is determined from the energy density, which is affected by the resistivity and current distribution. However, the resistivity changes with temperature, thus altering the rate of temperature change. For muscle, where the resistivity decreases with temperature, the tissue damage occurs more selectively, as different parts of the muscle become damaged at different rates and enhance the current flow through the damaged regions. This phenomenon is the likely cause for selective muscle groups that are completely burned being next to groups having only minor trauma.

10.4   Thermal Trauma

Thermal burn injury is the most common trauma that occurs at high current and high voltage levels, although electroporation may also cause tissue damage. In this section the resultant electrical trauma will be referred to in terms of thermal burns. This section will discuss the anatomical and physiological aspects of electrical burns, which occur at the contact site, in the deeper tissues, around the joints, and in the spinal cord. Thermal trauma depends on the current density in the tissue more than the resistivity, as indicated by Eq. (10.1).

Heating at the Contact Site

Most burns are at or near the point of contact. When the injury current is high, then trauma can also be seen farther away from the con-

Table 10.10. *Simultaneous current density and temperatures 7 cm proximal to distal tibia versus time with 1-A current applied from hindlimb to hindlimb*

Time (min)	Applied voltage (V)	Artery		Nerve		Muscle		Fat		Bone marrow		Bone cortex	
		J	$T\,(°C)$	J	$T\,(°C)$	J	$T\,(°C)$	J	$T\,(°C)$	J	$T\,(°C)$	J	$T\,(°C)$
0	355	30.0	39.0	24.3	38.5	13.3	39.0	9.5	38.5	3.0	39.0	1.8	39.0
2	293	28.7	41.0	22.8	41.5	12.5	40.5	10.5	42.0	3.1	40.0	1.9	40.0
4	265	27.1	43.0	21.9	44.5	12.5	42.5	12.8	45.5	3.1	42.5	2.0	42.5
6	253	26.5	46.0	21.0	47.5	12.2	44.5	14.4	49.0	3.2	44.0	2.0	44.0
8	235	25.5	47.0	20.5	49.5	12.2	46.5	14.8	52.0	3.1	46.0	1.9	46.0
10	220	25.0	48.5	20.0	52.0	12.2	47.5	15.0	54.0	3.1	47.0	1.9	47.0

Source: From Sances et al., 1983.

tact site. The discussion here will relate the transient temperature to the severity of burn and the effects of blood flow on the extent of injury. The superficial burn is characterized by three regions of tissue destruction and charring at the electrode contact edge (Fig. 10.14). The first region of tissue destruction, which is adjacent to the electrode contact, contains desiccated and denatured tissue, indicated by erupted blisters, shrinkage of the tissue, and charring at the edge of the contact. The second region is the ischemic region, where the tissue is devoid of blood flow but retains its fluid content. The outermost region, represented by a darkened ring caused by hemorrhage and thrombosis of the microvasculature, appears several minutes after current is stopped. These burn regions extend into the deeper layers of the dermis as well (Fig. 10.15). The burns result in a full-thickness burn and will extend into underlying muscle tissue when the current is sufficiently high. Desiccation is deeper at the electrode edge than in the center. Histological evaluation of the dermis shows that the trauma from electrical burns is very similar in character to nonelectrical thermal damage. The extent of trauma can be determined by the transient temperature response of the tissue.

Henriques (1947) and Henriques and Moritz (1947) have investigated the thermal tissue response in terms of the time-temperature relationship. The general results indicate that as the peak temperature increases, the

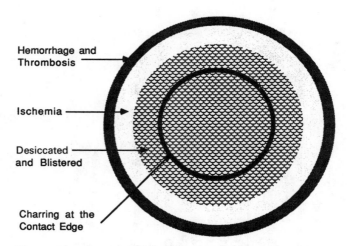

Hemorrhage and
Thrombosis

Ischemia

Desiccated
and Blistered

Charring at the
Contact Edge

Figure 10.14   Superficial burn resulting from a circular contact at 100 V. There are three distinct regions of trauma: a desiccated region adjacent to the electrode having erupted blisters, an ischemic region that maintains its fluid content, and a hemorrhagic region darkened by thrombosis. Also shown is a ring of charring that occurs at the electrode edge.

time necessary for trauma decreases exponentially. Skin will burn at 45°C after several minutes, but at 60°C only a few seconds is required. Experimentally measured temperatures are given in Fig. 10.16 for a circular electrode of 5-cm diameter having a 100-V potential and applied through the third phase of the current-time waveform (Fig. 10.2). The measured temperatures corresponds well with the finite-element model results of Fig. 10.8. In general, the peak temperature delay from the time of peak current increased primarily with distance from the edge of the electrode (Fig. 10.17). A secondary effect increasing the time delay of peak temperature is the depth from the skin surface, and regions without blood flow (ischemic regions of Figs. 10.14 and 10.15) also had longer delays. The secondary effects are due to heat radiation and convection at the surface and to slower heat conduction in ischemic tissue.

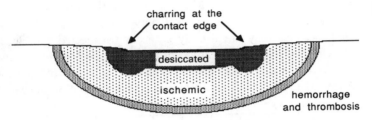

Figure 10.15   Dermal cross section beneath a circular contact at 100 V. The resultant injury is a full-thickness dermal burn. The three regions of Fig. 10.14 are present. Desiccated tissue extends deeper into the tissue at the electrode edge than at the center.

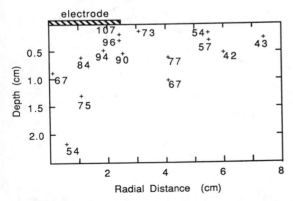

Figure 10.16   Peak tissue temperatures measured experimentally beneath a 5-cm-diameter circular contact at 100 V in five hogs. The voltage was applied until the current ceased from desiccation of the contact site (phase 3 of the current-time waveform). (Adapted from data of Prieto et al., 1985.)

Tissue perfusion by blood is an important factor in the generation of electrical burns. Comparing the above in-vitro studies to studies in situ, the results show significant variations in the total applied energy, contact time, and rate of tissue cooling. The parameters of voltage, electrode size, average power, and minimum total resistance were the same. The total energy applied was 24% less in situ than in vivo, the contact time was 21% less in situ that in vivo, and the rate of tissue cooling was 50 to 75% less in situ than in vivo. This factor is important to the evolution of electrical burn trauma, since the data indicate that blood flow helps lessen the severity of the trauma in short-duration contacts.

Heating of the Tissues

When currents are high and/or application times are long, trauma extends beyond the contact site, requiring surgical intervention and possibly amputation. The trauma affects primarily the limbs and is typically located in muscle tissue. Peripheral nerves and blood vessels are also affected, because they are usually adjacent to affected muscle. Bone is usually the least affected, but adjacent muscle is affected more than distant muscle. Deep tissue injury depends critically on the current pathway. As noted by Eq. (10.3), the distribution of current is based on the electric field

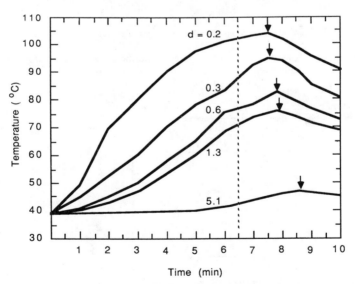

Figure 10.17   Plots of tissue temperatures beneath the electrode at different depths (cm) with time, 100 V contact. Peak temperatures do not occur simultaneously, but at times related to depth below the electrode edge following the current falloff of phase 3, as indicated by arrows. Broken line indicates current cessation.

strength of a given cross section and the tissue resistivity. Once the current distribution is established, the heating can be evaluated.

Heating of tissue and subsequent resistivity changes have been investigated in hogs and dogs (Sances et al., 1983; Chilbert et al., 1985b). Table 10.9 shows that current density in fat decreases with temperature but increases in artery, muscle, and nerve. Tendon and dermal current density decreases with temperature. The current path will shift due to changes in the resistivity. This change is due to a direct temperature dependence of resistivity and to tissue degradation. Tissue degradation causes increased resistivity in nerves and vessels while decreasing muscle resistivity. Neural tissue increases 80–100% in resistivity when thermally injured, while muscle decreases as much as 80%. The changes in muscle resistivity become more important when one considers that a large change in impedance could be indicative of burn severity.

Studies performed by the author (Chilbert et al., 1985a 1985b) indicate that tissue temperature, resistivity, and the severity of trauma can be correlated in muscle tissue. Figure 10.18 shows the measurement sites in the dog gracilis muscle for temperature and resistivity. Electric burn trauma was induced by passing 1-A currents at 60 Hz until the distal measurement site reached a temperature of 60°C. Peak temperatures along the muscle decreased proximally as the limb cross section increased

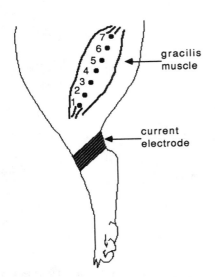

Figure 10.18   Measurement sites in the gracilis muscle of the hindlimb of a dog. Temperature and resistivity were measured at each site. Site 1 is the most distal and site 7 is the most proximal.

(Fig. 10.19). Regions 1 and 2 exceeded 50°C, while the remaining regions remained below 47°C. Severe trauma was noted in regions 1 and 2, while regions 6 and 7 had edema with minimal structural changes. Regions 3, 4, and 5 were in a transition zone between viable and nonviable tissue. The resistivity increased proximally, but all values were less than control values (Fig. 10.20). The initial resistivities at the onset of the 1-A current increased 35% because of muscular contraction; they then declined with increasing temperature. The resistivities continued to decline even as the tissue cooled, indicating further progression of tissue injury.

Thermal trauma to peripheral nerves is also related to the maximum temperature reached. The temperature in a nerve is most likely dependent on the temperature of the surrounding tissue. The peak temperature of nerve in Table 10.10 shows that it is only 2°C less than the tissue in which it was measured (i.e., fat). Neural activity (indicated by averaged evoked response) decreases with temperature as neurons are damaged. Degradation of the peripheral nerve response with a thermal injury (not electrically induced), caused by slowly increasing the temperature, shows the development of trauma (Fig. 10.21). Important features of the evoked response indicate that certain neurons are more sensitive to increased temperature, noted by a larger decrease at 45°C and at 50°C. Which neurons are affected

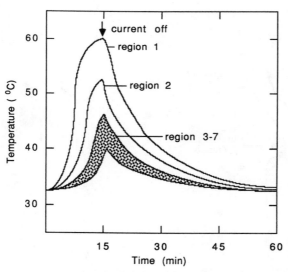

Figure 10.19   Temperature increase caused by the flow of a 1-A current with time. Current was applied for 15 min. Temperatures at regions 1 and 2 were elevated above 50°C, while the other regions were between 40 and 46°C and are grouped together.

at these temperature levels in not presently known. For rapid heating of the nerve the temperature of trauma is above 50°C, where recovery of the evoked response is usually rapid once the temperature reaches that of the body. However, heating of the nerve with electric current causes permanent damage at lower temperatures. The evoked response is always immediately affected by current, at levels above 3.5 mA/cm² along the nerve, causing blocking of the response from 5 to 20 min. Once temperatures exceed 45°C, permanent reductions in the evoked response are observed, and above 50°C the response nerve returns. This suggests that the current may affect neurons directly exclusive of temperature, possibly by electroporation. Limited studies in the spinal cord are consistent with peripheral nerve information, but the temperature threshold for electrically induced trauma is lower.

Thermal trauma to bone is seldom seen except at very high voltages (Sances et al., 1981b). The high resistivity of cortical bone prevents significant direct heating and protects the marrow as well. Table 10.10 shows that bone has the lowest temperature in cross section. Muscle tissue near the bone is often burned, while muscle tissue away from the bone is not

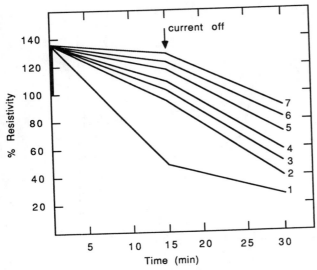

Figure 10.20   Changes in resistivity measured during and following the application of a 1-A current. Resistivities of muscle were measured in the transverse direction and perpendicular to the direction of current flow and are given in percent of control values. Initial increase of resistivity is due to the contraction of the muscle by stimulation from the current. Initial resistivity is for contracted muscle at time = 0 min. Muscle resistivity continues to decrease with decreasing temperature after current is turned off at $t = 15$ min in region 1.

affected. Some clinicians have erroneously attributed this phenomenon to heating of the bone. The cause is due to higher current densities in the muscle near the bone (Sances et al., 1983). The result is that muscle near the bone heats faster, which decreases the resistivity and increases the current density, thus propagating the injury. The reason for this is that the energy in the tissue is related to the square of the current and the resistance; consequently, the energy is more dependent on current changes than changes in resistance.

Joint regions in the limbs are more susceptible to trauma than in tissue around the longbones. In cross section the joints have a large quantity of bone compared to other more conductive tissues. This creates very high current densities in the more conductive tissue, consequently causing severe trauma at the joint while leaving tissues by the long bones viable (Sances et al., 1983; Zelt et al., 1988). Also, the joint capsules have been known to disrupt explosively from the buildup of steam (Sances et al.,

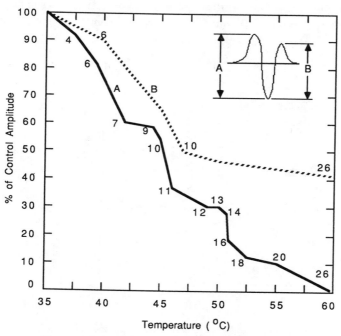

Figure 10.21 Changes in the evoked response of the tibial nerve in the hog with temperature. The reductions in the peak amplitudes were permanent. The first component of the evoked response is denoted by A, and the second is denoted by B. Numbers on curves indicate time in minutes after the temperature increase was started. Amplitude is given in percent of normal evoked response amplitude.

1981b, 1983). However, if the victim has the ability to flex the joint, the current will have an alternate path through the skin surfaces above and below the joint, thus preserving the joint. This has been noted clinically around the elbow, where burn marks appear on the skin proximally and distally.

10.5   Nonthermal Trauma

Lesions in tissue not caused thermally are likely caused by electroporation (Lee et al., 1988). Typically, the electric field strength needed to rupture muscle cells must exceed 100 V/cm. Similar levels are probably necessary for neural cell damage as well. Also, it has been noted that temperature increases the likelihood of electroporation occurrence at a given temperature. Studies in the peripheral nerve of hogs show that no significant alterations in the evoked response occur at current densities up to 167 mA/cm². This current density level results in an electric field strength of 33.2 V/cm, which is below the minimum level of 100 V/cm cited by Gaylor et al. (1989). To attain field strengths in the limb, the applied voltage could exceed 5000 V. For spinal cord injury to occur, the minimum applied voltage would need to be between 35 kV and 50 kV for a hand-to-foot path. These values were predicted from measurements in the hog given in Tables 10.4 and 10.9 and in Fig. 10.3. From Table 10.9 one notes that a current density of 0.3 mA/cm² results when a current of 100 mA is applied from the forelimb to the hindlimb and the spinal cord has a resistivity of 214 Ω cm. Since the electric field strength should be above 100 V/cm, the current density in the cord needed to cause irreversible electroporation would have to be 500 mA/cm². To obtain this current density in the spinal cord, an applied current of 167 A between forelimb and hindlimb is needed. To estimate the voltage level for 167 A, refer to Table 10.4: At electrode location 4, nearly 20 A will flow at 6 kV. A simple extrapolation of the voltage-current ratio (6 kV-20 A) shows that 50 kV is needed to generate 167 A in the body. However, resistance decreases with increasing voltage, therefore using the values of 7.2 kV and 25 A from Fig. 10.3 and extrapolating to the 167-A level, the applied voltage needed is 35 kV. The value for humans may be lower, since reports in the clinical literature have cited spinal cord dysfunction at 25 kV (Kanitkar and Roberts, 1988).

10.6   Lightning Injuries

Lightning accidents occur infrequently relative to other electrical incidents, but often enough to be well documented. According to Biegelmeier (1986), one-third of all lightning injuries are fatal. The immense power of lightning is shown by the effects of its millions of volts and 5,000 to 200,000 A (Sances et al., 1979; Strasser et al., 1977). Reviews of lightning injury have been given by Taussig (1968) and Silversides (1964). Persons struck by lightning who were clinically dead have been revived, and maladies seen afterward eventually disappear in most cases (Taussig, 1968).

The afflictions commonly encountered with lightning are mostly neurologic. Burns can occur at the entry points and may be accompanied by internal lesions. The burns often have a splashlike or arborlike appearance, assumed to be caused by current tracking over the body (Artz, 1979). Skeletal fractures occur when the person falls or is thrown by muscular contractions. Effects on the nervous system can occur immediately, can be delayed for several hours, or can occur days after.

Immediate effects of lightning strike depend on the severity of the incident. The victims usually undergo a total arrest of all body functions, which slowly return over a given length of time (Taussig, 1968). Fatalities can occur when the victim has a long respiratory and cardiac recovery period and no assistance, such as CPR, is administered (Strasser et al., 1977; Taussig, 1968). Abdominal lesions are also notably fatal, especially when internal hemorrhage is present (Artz, 1979). For the less serious accidents, the heartbeat is present with respiration intact or soon returning, leaving the victim in an unconscious state. Upon arousal, the victim may act disoriented and may have paresis of some or all the limbs (Critchley, 1934; Silversides, 1964). Other, less common deficits are visual, auditory, and speech (Critchley, 1934). Often amnesia occurs with lightning insults (Strasser et al., 1977; Taussig, 1968). Hypertension is normally observed, with anxiety and, at times, neurotic behavior. These symptoms usually vanish within a week (Strasser et al., 1977).

Secondary effects appear within the first few days after the accident. Except for an occasional change in the electrocardiogram or trauma to internal organs, all secondary effects are neurologically oriented, consisting of paralysis (usually of the legs), muscle pain throughout the body, photophobia from the intense light, and various autonomic disturbances. These effects normally vanish within a week. Latent effects (seen after 5 days) usually are resolved within a month after the accident. Neurologic symptoms are thought to be associated mainly with the progression of vascular disease or dysfunction. Neurologic symptoms in this stage are also common to high-voltage power-line injuries, which will be discussed later (Silversides, 1964).

Soft tissue lesions are found more frequently with lightning accidents than in high-voltage accidents (Massello, 1988; Sharma and Smith, 1978). Neurologic lesions often occur where there is an interface between regions of different resistivities. In lightning accidents one commonly sees splitting of the cortical layers of the brain and petechial and subarachnoid hemorrhage (Critchley, 1934; Silversides, 1964). Several explanations of the lesions have been suggested, but all lack supportive evidence. One hypothesis suggests that the lesions are caused by the heating effects, which produce pockets of gas or steam. This is not generally accepted as a good explanation, since the heat evolved is assumed to be insufficient to cause boiling of the tissue. Following this line of reasoning was the suggestion of fluid electrolysis; however, the time involved and current necessary may exclude this as a possibility. The most acceptable theory is electroporation causing cellular lysis at or near the tissue interfaces. The lesions have been seen more often at tissue interfaces, where charge accumulation can occur (Critchley, 1934; Strasser et al., 1977).

10.7   Clinical Observations

Of all electrical injuries, most occur to those working for electric utility companies. In one report, 95% worked for utility companies, of which 50% or more were linemen (Butler and Gant, 1978). The voltage levels for most of these injuries exceed 1000 V and are at the standard commercial frequency. Reviews of the general aspects of electrical injury have been presented by many (Dixon, 1983; Sances, 1979; Skoog, 1970; Wu, 1979). Clinical reviews have shown common observations among patients (Haberal, 1986; Hammond and Ward, 1988; Luce et al., 1978; Sharma and Smith, 1978; Solem et al., 1977). Some authors have focused reviews on the neurologic sequelae noted in electric accidents (Critchley, 1934; Silversides, 1964; Strasser et al., 1977). Others have reviewed the aspects of electric burns (Artz, 1974; Butler and Gant, 1978; Rosenberg and Nelson, 1988; Rouse and Dimick, 1978).

The various forms of trauma seen with electrical injury are the same as those seen in thermal burns with the addition of others unique to electrical injury. Some aspects of electrical injuries are similar to crush injuries. Clinical reports and reviews present observations in the general areas of burns, lesions, neurologic effects, and cardiovascular effects. The reports and reviews usually focus on one of the areas, while the patients may present symptoms from all areas.

Electrical Burns

The pattern of electrical burns can vary greatly from those seen in thermal injuries (Artz, 1979). Electrical burns follow the current path, whereas thermal burns start at the surface from one location and radiate into the surrounding tissue (Janzekovic, 1975; Luce et al., 1978; Ponten et al., 1970; Sances et al., 1979). Deep tissue necrosis is often evident; although tissue may appear viable, it may be damaged and require secondary procedures. The transformation of visibly viable tissue to necrotic tissue has been termed *progressive necrosis*, but experimental studies indicate that the visible viable tissue is actually necrotic from onset of the trauma (Zelt et al., 1988). Artz (1979) has reviewed and presented the various aspects of electrical injury, especially those associated with electric burns. Also, reference is made to internal lesions and renal sequelae. The following is a summary of burn wound observations:

Contact-site wounds usually signify deep tissue destruction locally. The contact sites can be defined in terms of the primary site, which is the contact with the energized source, and the secondary site, which is the contact with the ground or neutral source. The primary wound is usually charred and depressed. The secondary wound is dry, depressed, and has the appearance of the current exploding outward. Massive swelling is evidence of extensive tissue damage caused by heat and electroporation.

Skin injury varies from small circular spots to large areas of charring. Adjacent to the charred tissue is a whitish-yellow ischemic region. Surrounding the ischemic skin is an area darkened by vascular hemorrhage and thrombosis. All three regions are relatively cold and without sensory perception.

Vessel damage may extend beyond the general area of injury. Thrombosis has been noted away from the burn injury. Typically, extreme vascular spasms, thrombosis, and necrosis of vessel walls are observed. In non-thrombotic, damaged vascularization, hemorrhaging may result and lead to serious complications.

Muscular trauma is caused by the direct heating of current or by occlusion of the arterioles supplying the muscle. Damage to the muscle is usually uneven, affecting groups of fibers while not affecting adjacent areas. Uneven damage is characteristic away from the contact sites and may not be noticed initially (Zelt et al., 1988).

Burns caused by arcing imply high temperatures at the contact site and extensive deep tissue destruction. Traumatic limb amputation has been documented (Sances et al., 1979). Cursory burns are caused when arcing ignites the victim's clothing.

Renal failure is more common in electrical injuries than in thermal burns. Renal damage is caused by direct electric involvement of the kidneys and/or renal vessels, or more often by the abnormal breakdown of protein from other injured tissues. Devitalized muscle in electrical injury causes renal complications similar to those seen in severe crush injury (Artz, 1974). There is also a greater occurrence of hemoglobinurea and hematuria in the electrically injured patient. One critical complication is acute tubular necrosis, probably caused by myoglobin breakdown.

Oral burns are most often seen in children and cause severe burns to the lips, tongue, and dentition. This form of electrical burn has been reported extensively particularly in recent years, because of the special treatment needed (Barker and Chiaviello, 1989; Donly and Nowak, 1988; Palin et al., 1987; Sandove et al., 1988; Silverglade, 1983). Of particular concern is tissue contraction of the lips and cheeks, which is prevented by special splints (Sandove et al., 1988).

Most pediatric electrical injuries admitted to the hospital occur to the mouth (35–60%) and most victims of oral burns are under the age of 4 (Baker and Chiaviello, 1989; Palin et al., 1987). Oral burns account for almost all severe burns. Electrical trauma in children results from defective equipment or the momentary lapse of supervision or lack of knowledge on the part of the parent. Nearly all injuries occurring to children are preventable, especially with routine maintenance.

Tissue Lesions

Tissue lesions are evident at the tissue interfaces and are commonly associated with neural dysfunction and abdominal complications. Autopsy upon electrical accident victims often reveals submucosal hemorrhages dispersed throughout the gastrointestinal tract. Abdominal lesions are associated with a high mortality rate. They are difficult to detect and treat. Delayed fatality is usually associated with abdominal lesions of the gastrointestinal tract (Artz, 1979). Typically the patient is comatose, and vital signs degrade over several days until death occurs.

Lesions have the most significance in neural tissues of surviving victims. Once a nerve has been severed, its function seldom returns. Lesions may also be responsible for latent neural dysfunctions that appear up to 3 years after the injury (Sances et al., 1979; Silversides, 1964). When lesions are noted in the spinal cord, they are seldom complete transections. Clinical signs of these lesions are spastic paresis with little or no sensory deficit. The lesions usually are not suspected until the onset of patient ambulation (Baxter, 1970). These afflictions, like many associated with the nervous system, are not well understood and are subject to conjecture.

Neurologic Sequelae

Neurologic sequelae are manifest at all levels of the nervous system and are either permanent or transient. Permanent effects are the result of thermal trauma or lesions, while transient disorders can disappear within days or last for months (Hooshmand et al., 1989). Discussed below are the permanent and transient sequelae seen peripherally, seen in the spinal cord, and associated with cortical abnormalities.

Damage to peripheral nerve can be caused by electroporation (lesions) or excessive heating of the tissue. The ulnar, radial, and femoral nerves have the highest incidence of injury due to heating. Occasionally lesions are formed in the peripheral nerves, causing sensory or motor deficits. Some cases have been reported as progressive with time, the victim slowly losing sensation in the limbs and other clinical abnormalities developing (Kinnunen et al., 1988). Injury to the peripheral nerves, as stated above, is permanent, and recovery is minimal. Peripheral nerve damage is usually associated with adjacent tissue injury.

Transient disorders often seen are neurovascular, neuromuscular, and sensory. The neurovascular disorders include vascular constriction and spasmodic reactions, both of which reduce blood flow and vascular dilation (Christiansen et al., 1980; Hooshmand et al., 1989). Vascular spasms cause a reduction of blood flow in an area, usually associated with tingling, numbness, and, at times, paralysis. Vasodilation is believed to be the cause of fainting spells sometimes encountered after electrical injuries. Neuromuscular disorders are paresis, paralysis, hypertension, and muscular pain.

Current having a magnitude great enough to affect the spinal cord or nerve roots has a transthoracic path. The most common transthoracic path is from hand to hand, with hand to foot being nearly as common. Head-to-extremity paths, although not very common, will also affect the spinal cord. With hand-to-hand contact, the current level is greatest in the cervical region (Table 10.9). Hand-to-opposite-foot current paths mostly affect the heart and thoracic spinal cord. Head-to-extremity current paths affect the upper brain centers and cervical spinal cord (Butler and Gant, 1978; Sances et al., 1979; Solem et al., 1977).

Permanent damage to the spinal cord may be caused by vertebral displacement caused by falling or by muscular contractions or by lesions in the cord or nerve roots. Skeletal fractures in electrical injury are normally attributed to falling or being thrown by muscular contraction. Lesions in the spinal cord account for many of the permanent disabilities of motor function. Lesions are seen mainly in cases of power-line contact above 11,000 V (Butler and Gant, 1978). The magnitude of the lesions in the spinal cord is not readily apparent until the onset of ambulation (Baxter, 1970; Solem et al., 1977).

Transient effects of current on the spinal cord appear at all stages of recovery: immediate, secondary, or latent. The immediate effects are loss of consciousness, respiratory arrest, and neural effects on the heart rate. Unconsciousness may cease immediately after current flow has stopped or linger for a day or more (Silversides, 1964). Respiratory arrest occurs from the sustained contraction of intercostal muscles and interruption of the neural respiratory centers. Sustained contractions of the intercostal muscles last only while the current is applied. Spontaneous breathing has been delayed for several hours in some victims, but usually returns within 20 min. Secondary effects of the electric injury involving the spinal cord are temporary paralysis of the limbs, vascular spasms, and muscle pains (Silversides, 1964).

Immediate effects of electrical accidents involving the head may include convulsions, coma, cerebral edema, hysteria, amnesia, tinnitus, deafness, and visual disorders. Convulsions are induced in electrical injury from the passage of current through the cerebral cortex, analogous to those induced in electroconvulsion therapy. Coma is observed in patients who have been in cardiopulmonary arrest for relatively long periods of time. Clinical signs of electric coma are dilated, unresponsive pupils, along with no reflex response. Cerebral edema is seen only in the severest cases (very high voltages and lightning), typified by a softening of the tissue (Critchley, 1934). Hysteria is mentioned throughout the literature; its severity is minimal. Specific hysteric symptoms are agitation, confusion, amnesia, and transitory auditory and visual dysfunction (Critchley, 1934; Silversides, 1964). Amnesia is the most common symptom. Often the cranial nerves are directly affected; tinnitus and deafness are observed in victims, particularly when the victim does not lose consciousness. Permanent deficits in cranial nerves are associated with lesions along the cranial nerves.

Visual disorders include blindness, blurred vision, photophobia, and cataracts (Silversides, 1964). Blindness is caused as a result of current effects on the optic nerve or interference with the photoreceptors in the retina (Al-Rabiah et al., 1987). Permanent blindness is caused by lesions along the optic tract or separation of the retina from the choroid. Blurred vision results when the innervation of the lens musculature and iris have been affected. This condition is typified by dilated pupils. Photophobia occurs when the retina is exposed to intense light from an arc or lightning flash. Cataracts have been known to develop after lightning strikes and very-high-voltage contacts with the head (Moriarty and Char, 1987).

Secondary conditions associated with the cerebrum involve the prolonging and delayed development of the immediate symptoms discussed above. Secondary effects are mainly hysterical complications. In addition to those stated above are speech loss, disorientation, and narcolepsy (Critchley, 1934; Silversides, 1964).

Development of latent effects includes psychosis, hemiplegia, aphasia, epilepsy, choreoathetosis, and other hysterical conditions (Critchley, 1934; Silversides, 1964). Psychosis is rare and may be related to preaccident manifestations. Hemiplegia and aphasia (not as common) are seen in conjunction with cerebrovascular disturbances. Prior existence of vascular disease correlates well with hemiplegia and aphasia resulting from electrical accidents. Posthemiplegic Parkinson's disease has been known to develop, either the unilateral or the asymmetric bilateral form (Silversides, 1964). The initial convulsion in a developed epileptic condition may have started during the current flowing in the body. If the current path includes the head, the formation of an epileptic focus is likely. Actual lesions are rare, leaving a diffuse focus as the cause. Victims who developed epilepsy respond well to chemotherapy, which usually is withdrawn over a period of time.

Cardiovascular Effects

Electrical accidents have also produced latent cardiovascular disorders that are usually transient. Some of these affects may be associated with fibrillation and defibrillation of the victim (Wilson et al., 1988). Because these effects are usually less life-threatening than the complications of burn, they are seldom reported. However, the occurrence of cardiovascular complications is frequent (Guinard et al., 1987). High current levels flowing through the chest can cause cardiac arrest, which is then followed by a somewhat normal cardiac rhythm. Most cardiovascular effects follow high-voltage electrical injuries. Latent effects include atrial fibrillation, premature ventricular contractions, bradycardia, tachicardia, ventrical and atrial ectopic foci, conduction branch block, and nonspecific S-T interval and T-wave changes (Butler and Gant, 1977; Jones et al., 1983; Skoog, 1970; Solem et al., 1977).

The causes of latent cardiac effects and delayed arrhythmias is unknown. Some suggested causes include the residual effects of fibrillation, enzymal reactions with the cardiac tissue, neural dysfunction, or coronary artery spasms, The fibrillation of the heart or a sustained myocardial contraction by the current may alter the metabolic activity of the heart and result in altered cardiac rhythm (Solem et al., 1977). Certain enzymes are released from necrotic muscle tissue, such as various forms of creatine kinase, which can alter the function of the cardiac cells or mitochondria and again alter the cardiac rhythm (Ahrenholz et al., 1988). Since various forms of transient neural anomalies have been noted in other areas of the nervous system, neural alterations in the inervation of the heart may also occur (Skoog, 1970). Similarly, the vascular spasms noted in the periphery may also occur in the coronary arteries and lead to altered cardiac rhythm

(Luce and Gottlieb, 1984; Skoog, 1970). Fortunately, these disturbances are usually transient and disappear within a week (Jones et al., 1983; Luce and Gottlieb, 1984).

Latent vascular effects, such as spasms or peripheral coldness, have been attributed to neural disturbances (Kinnunen et al., 1988). Similarly, transient neural disturbances have been attributed to vascular spasms (Silversides, 1974; Skoog, 1970). Some latent effects are attributed to loss of endothelial cells and elasticity in the vessel wall, which leads to vascular impairment (Wang and Zoh, 1983). Also, the vasculature may be impaired by arachidonic acid metabolites, such as thromboxane, that cause vasoconstriction. The latent effects are often a result of the initial trauma, but appear as healing occurs.

10.8   Clinical Treatment

Treatment of electrical injury begins at the scene of the accident with removal of the victim from the electrical source and the observations of vital signs. A lack of pulse or respiration requires immediate cardiopulmonary resuscitation. One should note specific details of the accident, such as the contact voltage, whether the victim fell, if there are fractures, and if there are burns evident at the contact sites. Rapid transportation of the victim to an adequate trauma center is essential (Dixon, 1983).

The potential for considerable fluid loss coincident with deep tissue damage is great, therefore fluid replacement must be started as soon as possible. Intravenous isotonic fluids are administered to maintain the urine output at approximately 100 ml/h. Traditional burn formulas cannot be used to calculate fluid replacement requirements. These formulas underestimate the required fluid volume because they are based only on the skin surface area involved; fluid replacement principles for a crush injury are more appropriate (Artz, 1979; Dixon, 1983). If myoglobinuria is present, sodium bicarbonate is administered to alkalinize the urine, which inhibits pigment precipitation. Also, an osmotic diuretic, usually mannitol, can be used to flush the renal tubules (Baxter, 1970; Dixon, 1983, Hunt et al., 1980; Moncrief and Pruitt, 1971).

The electrocardiogram should be monitored, and echocardiography may be useful in assessing cardiac damage. Monitoring the hemoglobin and hematocrit will indicate the degree of hemolysis, and arterial blood-gas measurements are used to determine the acid-base state of the victim (Artz, 1974; Dixon, 1983).

The external appearance of the electrical burn wound is not indicative of the extent of underlying tissue damage. The electrical burn may be masked by flame or flash burns (Moncrief and Pruitt, 1971). Initially, the wounds are thoroughly cleansed. If the initial examination reveals a charred limb, an absent or diminished pulse in a limb, a loss of sensory or motor nerve function in a limb, or evidence of considerable limb swelling, then an escharotomy and fasciotomy should be performed. Fasciotomies have both a therapeutic and diagnostic function, as they may restore circulation and permit the surgeon to inspect the underlying musculature visually (Baxter, 1970; Wang 1985). Amputation of at least one extremity is frequently required following high-voltage injuries. If a limb is clearly necrotic, then amputation should be performed immediately (Parshley et al., 1985). Amputations are generally done in the first week post-injury, after the patient's condition has stabilized. Arteriograms may be helpful in determining the initial level of amputation (Hunt, 1979).

Although there is general agreement that obviously necrotic tissue should be debrided as soon as possible, there is still controversy over the treatment of tissue that is questionably viable (Barnard and Bostwick, 1976; Baxter, 1970; Rouse and Dimick, 1978). The first method advocates early exploaration and debridement, with repeated surgical procedures after 24 to 72 h until all nonviable tissue has been identified and removed (Artz, 1974; Hunt et al., 1980; Luce and Gottlieb, 1984; Parshley et al., 1985). Alternatively, a more conservative approach may be used, where the debridement procedure is delayed until a definite delineation between viable and nonviable tissue becomes evident, usually after 8 to 10 days. Proponents of the aggressive approach claim earlier wound closure and less incidence of infection, while proponents of the conservative approach feel that unnecessary multiple surgical procedures are avoided, the local blood supply is not altered by surgical trauma, and viable tissue is less likely to be removed. Zelt et al. (1988) showed experimentally that all tissue injury is histologically evident at the onset of injury; progressive necrosis does not occur.

Several methods have been advanced to determine the viability of questionable tissue so that excision of the necrotic tissue can be conducted as early as possible. Arteriography has been used to assess vascular damage, but this method does not consistently visualize the smaller arterial and arteriolar vessels in the muscles (Hunt et al., 1980). Thus, only occlusion of the larger vessels can be reliably determined with arteriography. The technetium-99 m pyrophosphate scan has been used to identify the extent of cardiac and skeletal muscle damage, including deep areas of muscle necrosis that may be missed by visual inspection during a fasciotomy (Hunt, 1979). Other studies have suggested that electrical impedance or nuclear magnetic resonance spectroscopy can be useful in determining tissue necrosis (Chilbert et al., 1985a, 1985b).

Tissue resistivities were compared with histologic changes as well as results of ^{31}P NMR spectroscopy. Abnormalities in anatomic as well as metabolic measures were shown to correspond well to resistivity changes. Measurements of impedance at several frequencies were made to compute the phase-plane plot of muscle tissue (Fig. 10.22). Impedance is unique for each tissue as well as its condition, reflecting traumatic injury, viability, edema, etc. (Ackmann and Seitz, 1984). Likewise, changes in the phase angle are indicative of alterations in tissue, but are more proportionately related to the number of cells present. The relaxation frequency occurs at the point of maximum reactance. These boundary changes (caused by cell lysis) alter the relaxation frequency; loss of cells increases the relaxation frequency. With electrical burn injury, the amount of cell lysis is determined by the amount of heat generated in the tissue. Figure 10.20 shows how muscle resistivity changes with temperature and has been correlated to trauma (Chilbert et al., 1985b). The level of trauma, as indicated by ^{31}P NMR spectroscopy and histology, shows a 75% decrease in resistivity with severely burned tissue, whereas tissue only slightly affected by the electricity showed a decrease of 25% or less. Using impedance techniques

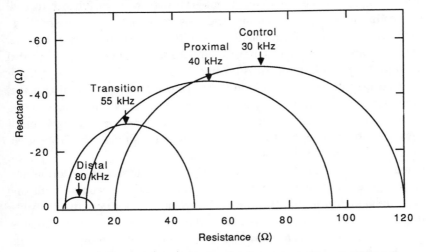

Figure 10.22   Impedance and relaxation frequency changes seen in muscle tissue with electric burn. The proximal tissue impedance curve corresponds to tissue that is only mildly edematous, the transition tissue impedance curve corresponds to tissue showing greater edema and some loss of cell structure, and the distal tissue impedance curve corresponds to tissue that shows gross destruction of cell structure and extensive damage.

to determine tissue viability allows separation of compartmental changes (edema) and boundary changes (cell lysis) so that tissue vitality is better determined. Figure 10.22 indicates both the impedance change and relaxation frequency change seen in electric burns. Once the probability of tissue survival is determined for a given change in impedance and relaxation frequency, clinical measurements can be performed and appropriate excision of nonviable tissue can be made. This will eventually lead to reduced morbidity in the electric burn patient by minimizing surgical procedures and chances for infection.

It is essential that the bacterial population of the wound be controlled to prevent infection, especially clostridial myositis, which may develop as a result of inadequate excision of necrotic tissue (Artz, 1979; Hunt et al., 1980). Topical antibacterial agents are used rather than systemic agents, because the blood flow is often compromised in the region of the wound. Temporary wound closure can be achieved with porcine xenografts, autografts, or homografts, while permanent wound closure is usually achieved with an autograft. The use of a free muscle flap, utilizing the latissimus dorsi, for immediate coverage of deep electrical injury has been advocated. However, if the debridement prior to grafting is inadequate, then muscle necrosis or infection may be masked by the vascularized muscle flap. A viable cutaneous homograft may be a useful tool to test the readiness of a wound for surgical closure or definitive autografting. Vein grafts, using autotransplantation of the saphenous vein to replace the radial and ulnar arteries, have been used to restore circulation to the hand and prevent limb necrosis (Wang et al., 1984, 1985).

Electrical injuries involving only the scalp may be managed with the same procedures used for soft-tissue electrical burns of other parts of the body (Worthen, 1982). Contrary to earlier procedures for treating an electrical burn of the scalp and skull, a full-thickness scalp flap should be used to cover the exposed skull as soon as possible, regardless of the depth of damage to the skull. The presence of a vascularized cover promotes skull regeneration, as ingrowth and revascularization occur on both the pericranial and meaningeal surfaces (Worthen, 1982). Alternatively, split-thickness skin grafts can be applied to the skull after the cortical bone is removed to expose cancellous bone, although later resurfacing with rotational scalp flaps is required to restore the hair. A graft consisting of autogenous greater omentum covered with autogenous split skin has been used to cover a large irregular area of exposed skull following a high-voltage electrical injury. Extensive destruction of the face requires a multistage reconstructive procedure, which may extend over a period of several years.

11

Standards and Protective Measures

WALTER SKUGGEVIG

Standards and codes are used to judge to risk of electric shock from electrical products offered for sale to the consumer. An appreciation of the role played by these standards and codes can be obtained by understanding the rationale for them. But who writes these standards? What do they cover? Who judges the acceptability of electrical products?

11.1 Codes and Standards

Several organizations cooperate to write safety requirements for electrical products. The National Electrical Safety Code, for example, is published by the Institute of Electrical and Electronics Engineers (IEEE, 1990) and covers supply lines and equipment used by electric power utilities. The IEEE also publishes a guide for safety in AC substation grounding (IEEE, 1986), which contains criteria for proper grounding in order to minimize the risk of electric shock at distribution, transmission, and generating plant substations. The National Electrical Code (NFPA, 1990) is published by the National Fire Protection Association (NFPA) and covers electrical installations in public and private buildings or other structures, from the point of connection to the utility line at the service drop (overhead) or service lateral (underground) to the electrical power outlet.

Safety standards are continually revised to keep them current. For example, the National Electrical Code is revised on a 3-year cycle. Each third year, the public is invited to submit proposed revisions for the Code. Twenty code-making panels consisting of representatives from equipment manufacturers, contractors, installers, large users of electrical equipment, testing laboratories, inspectors, and utilities meet to decide whether to accept or reject each of the submitted proposals. Each new edition of the National Electrical Code is published after the proposed revisions are given public review, and are presented to the membership of the NFPA for adoption.

Individual product standards cover the products that are used in residential, commercial, and industrial settings. These products include power distribution equipment such as wire and cable, panelboards, circuit breakers, fuses, etc. They also include electrical products such as toasters, refrigerators, vacuum cleaners, etc. Performance standards covering electrical products are commonly written by trade organizations such as the National Electrical Manufacturers Association (NEMA). In the United States, most of the safety standards covering these electrical products are developed and published by Underwriters Laboratories, Inc. (UL).

Some products are used in special tasks or environments. Medical equipment, for example, is used with patients who might be more sensitive to an electrical stimulus than the general population. Consequently, the requirements for these products are often more stringent than the requirements for ordinary products. A number of organizations cooperate to address these special problems. For example, the Association for the Advancement of Medical Instrumentation (AAMI), the NFPA, and UL have standards covering medical products. The Center for Devices and Radiological Health, which is part of the Food and Drug Administration of the U.S. Department of Health and Human Services, has jurisdiction over medical devices, and has standards for their safety. This governmental body also administers the Radiation Control for Health and Safety Act, which addresses the safety of products that radiate energy in various forms. The government standards that cover radiation do not contain requirements for electric shock. Concerns about electric shock are addressed in other standards, mostly those of UL. Examples of products involved with hazardous radiation might be microwave ovens, television receivers, and ultraviolet lamps.

The development of safety standards is a complex process that involves the cooperation and contributions of many individuals and organizations. UL often uses Industry Advisory Conferences (IACs) to help develop UL standards. An IAC is a meeting among UL engineers and a number of representatives from an involved industry. The IAC provides the opportunity for proposals to be considered and evaluated by knowledgeable people. Not all proposals become requirements—only those with merit survive. An IAC meeting can occur a couple of times per year or once every few years, depending on the activity in the industry with respect to changing technology.

A report of each IAC meeting is issued by UL, and is sent for comment to all members of the involved industry and to the appropriate Engineering

Council of UL. Each Engineering Council consists mainly of engineers associated with the authorities having jurisdiction and interest in the product area under consideration (such as the electrical inspection authorities), any government officials having responsibility for establishing safety standards in this field, and the appropriate insurance interests. In addition, if the product is one that is used directly by consumers, drafts of proposed requirements are sent to UL's Consumer Advisory Council, which has wide representation. The input from all these sources is used to formulate a UL safety standard.

UL submits its standards for recognition as American National Standards by the American National Standards Institute (ANSI). ANSI is a private, nonprofit, nongovernmental organization that does not write standards itself. ANSI's objectives include providing a means of obtaining national acceptance of standards on many subjects, including product safety, by consensus of groups of interested parties. ANSI coordinates standards activities in the United States and represents the interests of the United States in international standards activities.

Standards such as the National Electrical Code and the product standards produced by UL are voluntary standards. They do not, by themselves, carry the force of law. However, many municipalities and local inspection authorities adopt these voluntary standards—perhaps with some modifications. This is the process by which the standards are enforced.

The Consumer Product Safety Commission (CPSC) is an agency of the U.S. government that was established in 1972 under the Consumer Product Safety Act. The CPSC was given the responsibility for public safety, including the safety of household electric products. One of the functions of the CPSC is to collect product-safety statistics. The CPSC investigates incidents and collects data containing descriptions of product-related accidents that happened in the United States. Their files contain information that can identify hazardous electric products and can reveal hazard patterns. Although the CPSC has written a few mandatory standards to address hazards associated with certain products, the main thrust of the agency has been to influence the already established voluntary standards process to resolve safety problems that they perceive by their data-gathering system.

There are other standards-making organizations that also have a role in determining product safety. For example, the American Society for Testing and Materials (ASTM), of Philadelphia, Pennsylvania, develops test procedures to evaluate the properties of materials that can be used in the manufacture of electric products and many other applications.

ASTM does not test materials itself; the testing is done by independent laboratories.

Manufacturers make products for worldwide use. This puts pressure on the standards-writing organizations to unify the standards of different countries so that products will not need to be specially built or modified in order to meet the requirements of an importing country. However, differences among nations in economic, cultural, and legal matters can make harmonization difficult or inappropriate. Some variations in requirements among countries are necessary because of differences in the normal practices or habits of the populace, or because of differences among the electrical systems in the various countries. For example, different standard line voltages, different electrical system grounding techniques, etc., exist in different countries. Other variations among safety standards might not be necessary, such as the arbitrarily chosen values of the safety factors included in limits. The trend to harmonize requirements is particularly evident among the "high-tech" industries such as electronic data-processing equipment, medical equipment, and electrical instrumentation.

In order to promote communication among researchers of various nations, and to help keep safety standards up to date with respect to new and significant knowledge in the field of electric shock, two international symposia were held. One was held in Toronto, Canada, in 1983 (Bridges et al., 1985), the other in Freiburg, West Germany, in 1986 (Hauf, 1986). Many European countries use the standards produced by the International Electrotechnical Commission (IEC), which is based in Geneva, Switzerland. Bridges et al. (1987) elaborate on issues associated with safety criteria for ventricular fibrillation that were discussed at the first symposium in 1983, and explore some of the differences among existing standards.

Testing laboratories must interpret the standards and evaluate the many products that are manufactured throughout the world for sale in the United States. These laboratories are private, independent laboratories, unlike their counterparts in other countries, where the laboratories are generally affiliated with, or are part of, the government. UL is one of a number of laboratories that evaluate products for consumption in the United States.

Products made in other countries for consumption in the United States are tested by laboratories of the United States. To expedite the process, UL has made reciprocal agreements with testing organizations of other countries, such as Verband Deutscher Elektrotechniker (VDE), the Canadian Standards Association (CSA), the Japan Machinery and Metals Inspection Institute (JMI), and the Japan Electrical Testing Laboratory

(JET). Where the volume of products imported into the United States from a particular country is great, the testing laboratory might have permanent facilities to test the products in a laboratory in the exporting country. For example, UL has testing laboratories in the Far East.

If a new type of product is developed and no published standard exists that covers it, a testing laboratory must depend on its experience with related products to evaluate it. If the product turns out to be the first of many of its type, a new standard would have to be written to cover it.

UL and other testing laboratories maintain periodic inspection procedures to ensure that the production of all listed products continues to comply with the requirements in the years that follow the initial investigation.

11.2   Limits for Electric Current Through the Human Body

Safety standards cover many aspects of product safety, including electric shock, fire, and mechanical injuries. The concern regarding electric shock includes limiting the current inadvertently conducted by a person under both normal and abnormal conditions of product use. As discussed in the previous chapters of this book, electric current through the body can produce various physiological effects, including perception, pain, startle reaction, muscle tetanization, respiratory interference, cardiac arrhythmias, ventricular fibrillation, electrical burns, etc. From the viewpoint of UL's product safety standards, four physiological effects are considered critical—involuntary startle reaction, muscle tetanization or inability to let go, ventricular fibrillation, and electrical burns.

The threshold value to produce an effect may vary widely among individuals. It is prudent to select a limit that protects most of the population. IEC Publications 479–1 and 479–2 (IEC, 1984, 1987) contain descriptions of thresholds for these physiological effects. The subject is discussed by Biegelmeier (1987) in a guide to IEC Publication 479.

The term "threshold" means a current level at which a certain physiological effect begins to appear. A "limit," on the other hand, is based on the threshold, but usually includes a safety factor. In general, the more severe or life-threatening the physiological effect, the higher the safety factor should be.

Alternating Current at Power Frequencies

Table 11.1 shows threshold current levels for continuous 60-Hz sinusoidal current as described in IEC Publication 479, and, for comparison, shows limits used by UL to address each of the physiological effects.

Table 11.1. *Threshold levels and UL limits for continuous 60-Hz current*

Physiological effect	Threshold for continuous 15- to 100-Hz current, Described in IEC Publication 479 (mA)	Limit used by UL for continuous 60-Hz sinusoidal current (mA)
Involuntary muscular reaction	0.5	0.5[a]
Inability to let go (Tetanized muscle)	10	5.0
Ventricular fibrillation	35	20

[a] Ordinarily, a limit of 0.75 mA is applied to stationary or fixed, cord-connected products provided with an equipment grounding conductor.

Perception and Startle Reaction

Electric current of only à few microamperes can be perceived through the fingers if the contact with the electrode is made with a light rubbing action, or with a tapping motion (see Sec. 7.2). However, this is not considered hazardous and is not generally of concern. A startle reaction can result when the current level exceeds the perception threshold. At 0.5 mA rms, a 60-Hz sinusoidal current can be perceived by most people who contact surfaces with bare hands. Work with volunteer subjects done at UL during the late 1960s and early 1970s by an ANSI C101 committee (Stevenson, 1971) has shown that 0.5 mA is not likely to produce a potentially hazardous startle reaction. Examples of accidents that can be precipitated by a strong enough startle reaction are falls from high places, spills of hot liquids, and injuries from cutting blades. A further discussion of startle reactions is given in Sec. 7.11.

Under some circumstances it is not feasible to design a product so that all parts in all areas exposed to the user will meet the 0.5-mA limit. An example might be within the enclosure of a product where energized electrical components are present, but where the user is expected to perform a routine servicing operation. Warning markings are often required under these conditions to advise the user of the possibility of an electric shock during the servicing operation, but the consequences of the user making a mistake and touching a live part inside this portion of the enclosure should be no more severe than to experience an involuntary muscular reaction. Specifically, the more severe physiological effects of concern are inability to let go ("freezing" onto the equipment), ventricular fibrillation, and electric burns.

Inability to Let Go

The principal source of information concerning the so-called let-go threshold is the work done by Dalziel (Dalziel and Lee, 1968). This work is discussed in detail in Chapter 8. The limit at 60 Hz for let-go in many UL standards is 7.07 mA peak, which is 5.0 mA rms for a sinusoidal current. Dalziel states that for 60 Hz sinusoidal current, the let-go threshold for the 99.5 statistical percentile was 9 mA rms for men and 6 mA rms for women. The let-go threshold for a child was assumed by Dalziel to be about half that of an adult male, although very little work was done with children. This might suggest that the limit for let-go should be 4.5 mA rms; however, UL has developed considerable experience using the 5-mA rms limit, beginning with radio receivers in the 1940s (UL, 1945) and then with many other similar products after that. The field safety record did not indicate that any reduction in the limit was warranted.

UL limits addressing inability to let go are often written in terms of peak rather than root mean square current because the Dalziel data are presented in those terms (Dalziel, 1943), and these data suggest a useful method for determining a suitable limit for commonly encountered waveforms. The waveforms tested by Dalziel included sinusoids with various levels of DC added, and half- and full-wave rectified sinusoidal current.

Muscular tetanus can last as long as current flows through the body. When the current stops, the effect stops. The concern about the ability to let go is therefore with continous body currents that are not interrupted. If the current flows through the torso, the possibility exists that breathing can be stopped by the current, and suffocation can result. However, it would take more than a few seconds to cause serious injury or death by suffocation.

More current is necessary to cause ventricular fibrillation than inability to let go when the current pathway is from hand to feet. One might be tempted to conclude that if let-go concerns are satisfied, ventricular fibrillation concerns are also satisfied automatically. However, this is not entirely true, for the reason stated below.

Ventricular Fibrillation

Unlike muscular tetanus, ventricular fibrillation can be "triggered." A current that is too short in duration to be of concern with regard to inability to let go can still be hazardous since it can initiate ventricular fibrillation if the pathway involves the torso. Even though the current might be rapidly interrupted, ventricular fibrillation, once triggered, is not spontaneously reversible in humans. A person in ventricular fibrillation will die within a few minutes from lack of blood circulation unless special measures are taken with defibrillating equipment.

Effect of Body Weight on the Ventricular Fibrillation Threshold

There is a question whether it is justifiable to assume that the threshold of ventricular fibrillation for humans varies as a function of body weight. It is known that the threshold for large animals is greater than the threshold for small animals. However, it has not been proven whether this arises from an interspecies sensitivity, or represents a body-weight dependence within species. See Sec. 6.7 for a more detailed discussion of this issue. UL has taken a conservative approach, and has scaled down the limit for ventricular fibrillation to cover small children on the basis of a presumed body-weight scaling formula. The IEC, on the other hand, uses the assumption that experimental data from dogs provides a margin of safety when applied to human exposures. The IEC does not use a body-weight correction.

The weight of a small child of the age where it begins to be ambulatory is approximately 8.4 kg (18.5 lb), as derived from anthropomorphic data (HSRI, 1977). According to this data, 5% of the sample of 12- to 15-month-old male and female children weighed 8.4 kg or less.

A simple body-weight model can be formulated as follows: Consider two cylinders of the same density (weight per unit volume) and the same proportion of height and diameter, but of different overall size, representing the adult and the child. The ratio of the two currents flowing parallel to the axes of the cylinders that produce equal current density in the cylinders is equal to the ratio of the respective cylinder weights (or volumes) raised to the 2/3 power.

$$\frac{I_1}{I_2} = \left(\frac{W_1}{W_2}\right)^{2/3} \tag{11.1}$$

By assuming that $I_2 = 20$ mA and $W_2 = 8.4$ kg, the following relationship between fibrillating current and body weight is obtained:

$$I = 4.84 \, W^{2/3} \tag{11.2}$$

Bridges (1985) explored the same model and its relationship to data on minimum fibrillating current versus body weight obtained from Dalziel (1972) and Geddes (1975). He observed good correlation among the fibrillating current data and the "2/3 power model," but mentioned that scaling according to weight is valid only for animals of the same general proportions. There are many other factors that should be considered when evaluating body-weight relationships, as noted in Sec. 6.7. However, it is interesting that Eq. (11.2) is nearly coincident with the curve representing the 0.5% minimum nonfibrillating current as a function of body weight shown in Fig. 6.15.

Electrical Burns

The subject of electrical burns is treated in detail in Chapter 10. The source of burns can be from a heat source external to the body or from current flowing through body tissue. As a general rule, body cells die when their temperature reaches approximately 45°C for prolonged periods. To protect against electrical burns, it is desirable to have a simple rule that can be applied in the same way that a leakage current limit is applied. At low frequencies, the current limits for reaction, let-go, and ventricular fibrillation are lower than the current levels that can produce a burn, and so burns do not present a hazard. However, at high frequencies this is not true.

IEC committees and UL have both applied a simple rule covering electrical burns that places a limit on leakage current regardless of the frequency. UL recommended 25 mA to the Consumer Product Safety Commission as a limit beyond which there is a risk of electrical burns (UL, 1981). Hart (1985) reported that burns occur when current densities exceed 300 to 400 mA rms per square centimeter, and small-area touch burns of 0.2 cm² can be expected at 60 to 80 mA rms. In later years, a limit of 70 mA rms was considered by UL to address electric burns. The IEC Safety Requirements for Electrical Equipment for Measurement, Control, and Laboratory Use (IEC, 1990a) limit the current that can flow through a person's body to 70 mA rms to avoid possible burns at higher frequencies. A more conservative limit of 70 mA peak is used in the IEC requirements for safety of information technology equipment, including electrical business equipment (IEC, 1986). Electronic equipment often uses switching power supplies that produce frequencies that range up to a megahertz. Addressing electrical burns by limiting the leakage current is a simplistic approach, but the requirement is easy to apply with uniformity and repeatability.

Alternating Current in Combination with Direct Current

Occasionally, it is necessary to evaluate a combined alternating and direct current. A telephone circuit, for example, can have an AC ringing signal superimposed on an always-present DC level. Another example might be the poorly filtered or unfiltered output of a DC power supply.

The work done to form the basis of let-go limits for AC + DC current was by Dalziel (1943). The work done to form the basis of ventricular fibrillation limits for AC + DC current was by Knickerbocker (1973). Figure 11.1 illustrates the limits used by UL for combinations of AC and DC current, both for let-go and for ventricular fibrillation. The limit for the peak value of the composite waveform is expressed as a function of the magnitude of the DC component.

For let-go considerations, the limit for a pure 60-Hz AC waveform is 7.07 mA peak (5 mA rms for sinusoidal current). As a DC component is introduced and increased, the composite wave is limited to 7.07 mA peak for all values of the DC component between zero and 4.21 mA. As the DC component is increased beyond 4.21 mA, the peak of the composite waveform is limited according to Eq. (11.3), which is illustrated in the figure. From the standpoint of let-go, the DC leakage current limit is 30 mA. The let-go limit for AC + DC used by UL follows the shape of the curve suggested by Dalziel's data for adults, but is scaled down

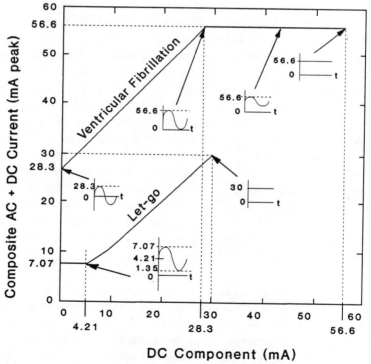

Figure 11.1   Limits for combinations of AC and DC current. (From UL, 1988.)

according to body weight to protect children as well.

$$I_{\text{AC}+\text{DC peak}} = \frac{-\ln(I_{\text{DC}}/30)}{0.68} + I_{\text{DC}} \qquad 4.21 < I_{\text{DC}} < 30\,\text{mA} \qquad (11.3)$$

Where inability to let go is not of concern, such as with interrupted currents, the risk of ventricular fibrillation becomes important. An example of frequent and periodic interruptions of current is the pauses between telephone ring signals. In the Standard for Safety for Telephone Equipment (UL, 1987), the UL limit is 20 mA rms for a pure AC 60-Hz sinusoidal current lasting more than a second (but not more than a few seconds). The limits discussed here assume that the current pathway through the body is hand (or hands) to feet, and that the person can be a child of small body size, including a small child weighing 8.4 kg.

Knickerbocker (1973) found that for shocks of more than 0.5 s applied to anesthetized dogs, the ventricular fibrillation hazard is approximately the same for sinusoidal AC current of constant peak-to-peak amplitude, regardless of the presence of a DC component, provided that the composite current is bidirectional. If the ratio of DC to AC is high enough to cause the composite current to be unidirectional, the ventricular fibrillation hazard is approximately the same if the peak value of the composite current is the same.

UL's ventricular fibrillation limit for continuous sinusoidal AC current in combination with DC is in accordance with these findings. According to Fig. 11.1, the limit for a pure AC sinusoidal current is 28.3 mA peak, which is equivalent to 20 mA rms. If a DC component less than the peak of the AC component is added, the limit for the peak value of the composite current increases so that the peak-to-peak value of the AC component remains unchanged. In this range, the DC component can be ignored. At the point where the peak value of the composite waveform equals the peak-to-peak value of the AC component, the limit is 2 times 28.3, or 56.6 mA peak. As the DC component increases so that the instantaneous current no longer reverses, the limit for the peak of the composite current is 56.6 mA peak. The peak-to-peak value of the AC component must then be reduced as the DC component increases. Finally, the AC component disappears, and the limit is 56.6 mA DC. From the standpoint of ventricular fibrillation, DC current lasting more than a second shall not exceed 56.6 mA.

Knickerbocker's work was done with the polarity of the DC component positive at the hind end of the torso of the animal, and negative at the head end. The DC current flow was therefore up through the torso of the animal, the direction for which there is greatest sensitivity of the heart to the current (see Sec. 6.10). The limit derived from Knickerbocker's work therefore covers the worst case with respect to the direction of DC current flow.

Effect of Current Duration on the Ventricular Fibrillation Threshold

Questions often arise in product evaluations regarding how fast an interrupting protective device has to operate in order to prevent ventricular fibrillation in the event of a current fault. The speed of interruption is important when considering a product such as a ground-fault circuit interrupter (GFCI), which is intended to protect against electrocution from faulty electrical equipment. Similarly, questions arise about the maximum permitted magnitude and duration of a pulse of current that is produced intentionally. For example, the pulse from an electric fence controller must be strong enough to be effective in containing cattle, but should not be capable of electrocuting people.

In the 1960s, UL established limits for interrupted 60-Hz current. These limits appeared in the Standard for Safety for Ground-Fault Circuit Interrupters (UL, 1985). This was done on the basis of experimental work done by several investigators, including Kouwenhoven et al. (1959) and Ferris, King, Spence, and Williams of Bell Telephone Laboratories and Columbia University (Ferris et al., 1936). Much of this work has been discussed in Chapter 6. These investigators used mainly dogs and sheep as their subjects to determine the threshold fibrillating current for various current durations.

The UL limits for the duration of body current as a function of the current magnitude were chosen with a comfortable safety factor. Much fewer data were available in the 1960s than today, and the fibrillating threshold data were scattered over a band. This scatter suggested that future fibrillation threshold data might show lower threshold values than was then evident. For this reason, and the serious consequences of error, it seemed prudent to choose a conservative limit. A second reason for conservatism was the question about how animal data translates to humans. A third reason was to provide a margin to accommodate natural variations associated with manufacturing tolerances or product aging. There is much more support today for these limits. Although the uncertainty has diminished, many of the reasons for conservatism are still valid, and the same limit is applied today.

IEC Publication 479–1 (IEC, 1984) contains information about the threshold of ventricular fibrillation as a function of current duration based on experimental work by several investigators using animal subjects. These include the same investigations used by UL, but it is important to realize that the IEC document presents threshold information based on the available data, and does not suggest or imply that the threshold be used as a limit without any modification. Figure 11.2 illustrates the UL limit and the IEC threshold for durations down to 4 ms. The figure also shows a plot of some of the data that were used to determine the threshold and establish the limit for single-occurrence current pulses. The UL curve is similar to the experimental relationship in Fig. 6.12.

In recent years, UL has considered acceptability limits for products that produce high-current, single-occurrence pulses of very short duration, such as a microsecond. For example, some designs of building-wiring testing devices produce high-amplitude, short-duration current to test the integrity of newly installed wiring. Information on which to base amplitude limits for these very-short-duration impulse currents was analyzed by A. W. Smoot of UL from sources including work done by investigators studying the effects of capacitive discharges (Kouwenhoven, 1956; Kouwenhoven, 1958; Peleska, 1963, 1965; Myklebust et al., 1985). The duration of the capacitive discharge current was described as one, two, or three time constants of the exponentially decaying current waveform. Fibrillating threshold information about short-duration transient currents having other waveforms was included in the work as well (Knickerbocker,

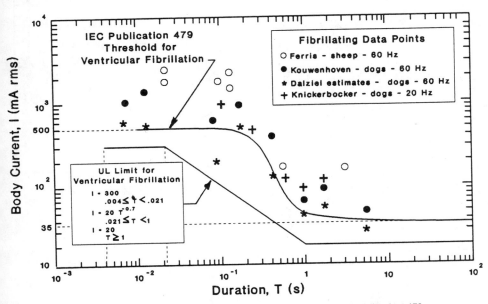

Figure 11.2 Threshold of ventricular fibrillation from IEC Publication 479 and UL's limit for ventricular fibrillation. (From UL, 1988.)

1973; Younossi et al., 1973; Green et al., 1985; Roy et al., 1985). Durations of pulses as low as $0.2\,\mu s$ were included in the data. Figure 11.3 shows the extension of the limit used by UL for certain applications involving nonrepetitive, short-duration current pulses.

IEC Publication 479-2 describes the threshold of ventricular fibrillation for single (nonrepetitive) impulse currents with durations as low as 0.1 ms. For comparison, Fig. 11.3 also shows the IEC threshold for single-impulse currents.

The UL limit is intended to be applied to bidirectional as well as unidirectional nonrepetitive pulses, but the limit might be overly conservative for bidirectional pulses. Although experiments such as the work done by O. Z. Roy on rabbit hearts (Roy et al., 1985) suggest that the heart can tolerate appreciably higher magnitudes of current in the form of a bidirectional pulse as opposed to a unidirectional pulse (see Fig. 6.17), an argument in favor of using the same limit for both bidirectional and unidirectional pulses is that changing the elements comprising the circuit impedance by the inadvertent addition of external resistance or rectification can, in some cases, cause a circuit that produces bidirectional

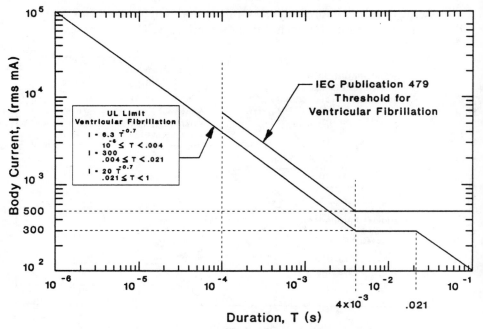

Figure 11.3 UL's limit for current pulses extended to pulses of short duration. The threshold for impulse currents in IEC Publication 479 is shown for comparison. (From UL, 1988.)

pulses under test conditions in the laboratory to produce unidirectional pulses in the field.

The testing laboratory must judge whether a product is to be accepted or rejected. The risks involved in these judgments are evident. If the product is rejected in error, the manufacturer suffers an undeserved loss of business and the public loses the use of a potentially useful product. If the product is accepted in error, the consumer might be placed at an unreasonable risk of injury or even death. If the technical information on which a good decision can be made is sparse, the decision must be based on whatever information might be available, but with an appropriate amount of conservatism to compensate for the uncertainty. Until more data with more impulse waveshapes become available, it is prudent to apply one limit for short-duration pulses regardless of whether the pulses are unidirectional or bidirectional.

Determining the Value of an Impulse Current

The determination of whether the product meets the requirements must be made rapidly and reproducibly by engineers and technicians with conventional laboratory instruments. Therefore, it is essential that the requirements be written in an uncomplicated and unambiguous way, leaving as little as possible to individual judgment. Scientific accuracy is sometimes sacrificed to achieve this objective, but is compensated for by conservatism.

The UL limit for short-duration current pulses is expressed in terms of a root-mean-square (rms) measure of the pulse. Although the rms measure does not correlate perfectly with the potential for excitability of the heart, and it does not have a fibrillation threshold by itself (the limit is a function of the current duration), it is used in a specified way to express a useful limit for testing laboratories.

Consider the equation for calculating the rms value:

$$I_{rms} = \sqrt{\frac{1}{T}\int_0^T i^2(t)\,dt} \qquad (11.4)$$

where T is the pulse duration in seconds, $i(t)$ is the instantaneous current magnitude in amperes as a function of time, and t is time in seconds.

If the current pulse is a simple rectangular waveshape, it is logical to assume that the value of T in the equation should be equal to the pulse width expressed in seconds. However, if the current pulse has no abrupt end, it is not obvious what T should be. Yet the limit for the rms current amplitude as a function of duration is not related to the energy of the impulse, and the laboratory's choice of T can therefore influence whether the product is accepted or rejected. Clearly T must be specified precisely.

Consider an example that illustrates this problem. Suppose we are to determine the acceptability of an exponentially decaying current from a capacitive discharge having a peak value of 4400 mA and a decay time constant of 0.1 ms. Suppose the duration of the pulse is described as three time constants, or 0.3 ms. Using Eq. (11.4), we would compute the rms value to be 1794 mA. The current pulse would be considered acceptable, since the limit shown in Fig. 11.3 for a pulse with a duration of 0.3 ms is 1842 mA rms. However, if the duration is taken to be 10 time constants (1 ms), the rms value of the current pulse is computed to be 984 mA rms. For this calculation, the identical pulse is rejected because 984 mA exceeds the limit value of 793 mA at 1 ms. The only difference between the two evaluation procedures is the arbitrary decision regarding the pulse duration, yet the results disagree.

IEC Publication 479 specifies three time constants as the duration of an exponentially decaying waveform of current; UL makes the same specification. Waveforms other than exponentials that are not abruptly switched off, for example, an oscillatory decaying current, must have a defined duration as well. One might define the end of such a pulse as the time when the instantaneous current falls below a specified level for the last time. The specified level could be 7.1 mA, which is the let-go limit for a continuous sinusoidal 60-Hz current. Another definition that can be used is that the current pulse is considered to end when the instantaneous current falls below 5% of the peak value of the pulse for the last time. This would be consistent with the three-time-constant definition used for exponentially decaying current, but it can be uncomfortably liberal for some waveforms that peak high and then begin to fall at a slower rate. UL has used these definitions for judging pulses with complex waveforms, but not having rules that cover all types, UL has reserved the right to analyze pulses and decide acceptability on an individual basis.

Repetitive pulses with a repetition frequency sufficiently high to have more than one pulse occur within the period of a single heartbeat are not covered by the limits described in Figs. 11.2 and 11.3. A much lower limit, perhaps lower by more than an order of magnitude, would be appropriate for these pulse trains, for reasons made clear in Sec. 6.5.

Capacitive Discharge

Current pulses due to capacitive discharges can be considered as a special case, which can be treated in a manner that saves laboratory time. The waveshape of body current is nearly exponential (see Figs. 2.33

and 2.34). The acceptability of a capacitive discharge can be stated in terms of the voltage across the capacitor prior to discharge. Figure 11.4 shows recommended limits of voltage across a capacitor prior to discharge as a function of its capacitance. The limit represented by the upper curve applies to ventricular fibrillation and is derived from the limit shown in Figs. 11.2 and 11.3 using a body impedance model consisting of a 500-Ω resistor, except for longer-duration, lower-current discharges, in which case the body impedance is likely to be higher. For large values of capacitance, the voltage limit is the same as that which would be applied to any DC voltage. At small values of capacitance, the charge is minimized, attaining a value of 40 μC at the smallest indicated capacitance (.001 μF). For comparison, the threshold values of perception and of pain that appear in IEC Publication 479 are also plotted in Fig. 11.4.

The threshold of perception depends on several variables, such as the size of the electrodes and the size of the person involved (see Sec. 7.3). The threshold is presented in Fig. 11.4 as a shaded band of values; the IEC threshold of intolerable pain is presented as a single line. The

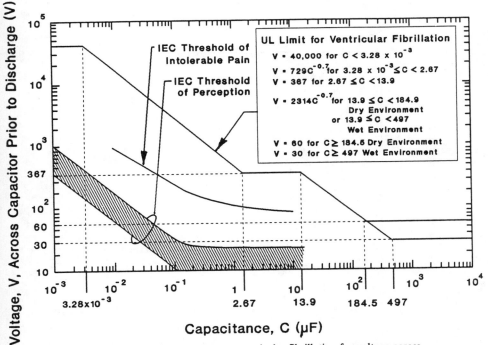

Figure 11.4   UL's limit to prevent ventricular fibrillation for voltage across a capacitor prior to discharge through the body. The IEC thresholds for intolerable pain and for perception are shown for comparison. (Limit from UL, 1987.)

threshold charge given by $Q = CV$ is minimized at small values of capacitance. The IEC thresholds are between 0.367 and 1 μC at the smallest indicated capacitance (10^{-3} μF); in Fig. 7.4, a somewhat lower range from 0.1 to 0.25 μC is shown at the same capacitance. Lower charge thresholds are expected if the capacitance is reduced further—at 100 pF, for example, Fig. 7.4 indicates a range from 0.1 to 0.25 μC, depending on the type of electrode contact. The IEC tolerance threshold represented in Fig. 11.4 was determined with a single subject. Section 7.4 should be consulted for a more thorough discussion of pain and tolerance thresholds.

Effect of Frequency on the Current Limit

Generally speaking, a low frequency, such as 50 or 60 Hz, has advantages for transmitting power. Transformers can produce whatever voltage is appropriate for the location along the power line, and the low frequency minimizes capacitive losses and series inductance along a lengthy transmission line. However, from the standpoint of electric shock, frequencies of 50 and 60 Hz produce the maximum physiological effect.

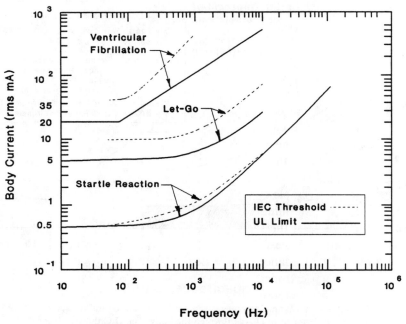

Figure 11.5   UL's limits for sinusoidal AC current as a function of frequency compared to IEC Publication 479 threshold values. (UL limits from UL, 1988.)

As the frequency of the body current rises beyond the power frequencies
of 50/60 Hz, physiological thresholds increase (cf. Figs. 6.4, 6.8, and 7.13).
Figure 11.5 shows the current limits as a function of frequency that UL
considers appropriate to safeguard against startle reaction, inability to
let go, and ventricular fibrillation. The UL limits for reaction and let-go
are based to a large extent on the data reported by Dalziel (1950b,
1956). The UL limit for ventricular fibrillation is based on data reported
by Ferris et al. (1936), Kouwenhoven et al. (1936), and Geddes and
Baker (1971). These leakage current limits for let-go and ventricular
fibrillation were recommended by UL to the Consumer Product Safety
Commission, and the development of the limits is discussed on UL's
report to the CPSC (UL, 1981). For comparison, Fig. 11.5 also shows the
threshold values that appear in IEC Publication 479. The subject of effect
of frequency is discussed further in this chapter with respect to the
measurement of leakage current from products.

11.3   Leakage Current

Leakage current paths exist in all electric products between the
electrically energized circuit parts and accessible parts. Leakage current
can be defined as all currents, including capacitively coupled currents,
that may be conveyed between exposed conductive surfaces and ground
or other exposed conductive surfaces. Since the electric circuits in a
building are normally grounded, i.e., connected to the earth via a
grounding electrode, a person who is in contact with the earth or grounded
equipment can possibly conduct leakage current when touching an electric
product. Consider a typical product connected to a grounded source of
supply as in Fig. 11.6. The electric load is represented in the figure by a
resistor. The leakage current pathway can be by way of resistance or
capacitance between the circuit parts of the product and conductive parts
that are accessible to the user. The figure shows a representation of this
inadvertent pathway by a single capacitor. If the body impedance is
represented by a simple resistor, the magnitude of the current through
the body can be determined by the quotient of the source voltage divided
by the impedance of the series combination of the body resistance and
the capacitive reactance.

Human body impedance including the limbs with wet or damp contact
with large-area electrodes at the extremities can be estimated to be
approximately 500 to 1500 Ω (see Chapter 2). These resistance values represent current pathways from one or both hands to feet. If, for example,
the body impedance is 1500 Ω and the source is 120 V, and if the leakage
current is not more than 0.5 mA (in order to meet the requirements that

limit the leakage current), the capacitive reactance must be a minimum of almost 240,000 Ω—a value much greater than the body impedance. Therefore, for most products having leakage current from the 120-V line, the magnitude of the leakage current will be determined largely by the internal product impedance (the resistance, or the reactance of the capacitive coupling, between circuit parts and accessible parts), and not by the body impedance. Under these circumstances, the product appears to the body impedance as if it were a "constant-current source," and variations in the body impedance will result in very little change in the body current. This makes the choice of an electrical model to represent the human body in a circuit under investigation less demanding with respect to accuracy. Errors in the choice of body impedance model will not measurably affect the results. In fact, the higher the internal product impedance, the less influence the body impedance will have on the magnitude of the leakage current, and the more a simple approximation of body impedance will suffice for test purposes.

If the supply voltage is less than 120 V (such as from a lower-voltage secondary power supply within the product), the product could meet the requirement for a maximum allowable leakage current limit with a lower product impedance. Consider a product with electric terminals accessible inside a cabinet that is routinely opened for servicing. The available current might be limited to 5 mA from these terminals that are not casually

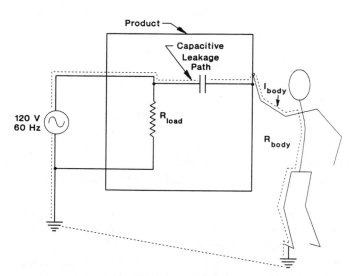

Figure 11.6   Leakage current from an electric product through capacitive coupling between live parts and accessible conductive parts.

contactable during use of the product. The minimum product impedance needed to meet the leakage current requirement is on the order of a few kilohms, which is comparable in magnitude to the expected body impedance. Under these circumstances, variations in body impedance would have a greater influence on the body current magnitude, and it would then be necessary to simulate human body impedance more accurately when the leakage current is measured from a product.

Body Impedance Models

Under contract with the Consumer Product Safety Commission from 1979 to 1981, UL reviewed the published literature on the subject of electric shock (UL, 1981). Based on this information, UL developed a versatile body impedance model for the human body for a variety of common electric injury scenarios. The model was based on information from many sources; the work done by Freiberger (1934), which included many measurements on a large sample of cadavers, provided an especially comprehensive treatment of body impedance. The body impedance model developed by UL consists of five components: two parallel combinations of a resistor and a capacitor, each in series with a resistor as shown in Fig. 11.7. The parallel combinations of capacitor and resistor represent

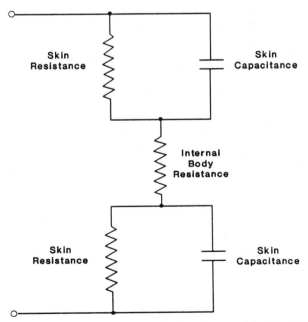

Figure 11.7   Five-component body impedance model. (UL, 1981.)

the skin contacts with the electrodes; the resistor represents the internal body. These components represent a simplified version of the body impedance model shown in Fig. 2.4.

The value of each component in the five-component model depends on various parameters, including the pathway of current through the body, body size, electrode contact area, contact pressure, voltage drop across the skin, and duration of current through the body. Despite these complexities, the model can only approximate the human body impedance in any particular situation. Some of the many variables are not known precisely, and can only be estimated from available data. For example, the model is based on the assumption that the interface between the skin and the electrodes is always moist with perspiration, but is not immersed. (Immersed bodies are another subject entirely, involving large, low-impedance contacts with the fluid and current pathways both inside and outside the body.)

Most uses of body impedance models in the field of product evaluation require simplicity. Commonly used body impedance models assume a body-current pathway of hand to feet, and a value of body impedance for this pathway that is on the low end of the range for the population. The typically used body impedance model consists of a resistor, or if more accuracy representing the human body is desired, a three-component model, rather than the more complex five-component model.

Historically, the human body with hand (or hands)-to-feet contact has been represented in product evaluations by a $1500\,\Omega$ or $2000\text{-}\Omega$ resistor for ordinary dry environments, or a $500\text{-}\Omega$ resistor for wet environments.

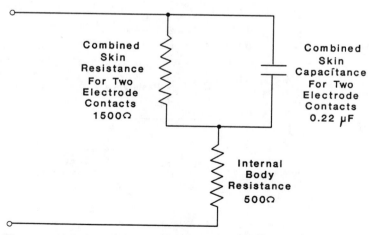

Figure 11.8   Three-component body impedance model. (IEC, 1990.)

More recently, Working Group 5 of IEC Technical Committee 74 has developed a three-component model as shown in Fig. 11.8 (IEC, 1990b). The 500-Ω resistor represents the internal body from one hand to the legs or buttocks. The resistor-capacitor parallel combination represents both the skin contact at the hand and the skin at the second, larger contact at the legs or buttocks. The hand contact represents a hand-hold such as with a metal handle; the second contact represented in the model is assumed to be much larger and would therefore have low impedance relative to the hand contact.

Measuring Leakage Current from Products

When evaluating a product, it is often important to determine its acceptability rapidly, possibly with some sacrifice in accuracy. It is also essential that test results be repeatable, with minimal variations due to chance.

In the 1960s ANSI Committee C101 developed the leakage current-measuring instrument that is shown in Fig. 11.9. This instrument is specified in the American National Standard for Leakage Current for Appliances (ANSI, 1986). It consists of a 1500-Ω resistor representing a person's body, and a 0.15-μF capacitor in parallel with the resistor. A resistor of 1500 Ω was considered representative of the body impedance, hand to feet, under the most severe conditions normally expected with dry finger contact (UL, 1945). The voltage across the resistor is used to determine acceptability. In most applications of the instrument, if this voltage is less than 0.75 V, the leakage current is considered to be acceptable from the standpoint of a startle reaction. Stationary or fixed products are not likely to be frequently disconnected and reconnected to the power supply, and therefore have higher reliability associated with the continuity of the power-supply conductors. These products with an equipment grounding conductor in the cord are more likely to be grounded, and are therefore permitted to have 50% higher leakage current than portable products. The purpose of the capacitor in the instrument is not to simulate the impedance of the human body, but to permit a

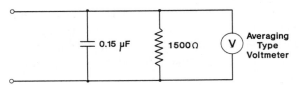

Figure 11.9   Leakage-current-measuring instrument specified by ANSI Standard C101.1. (ANSI, 1986.)

single numerical limit to be used with the instrument independent of the frequency of the source. At any frequency, when the meter reading is 0.75 V, the current through the instrument is equal to UL's leakage-current limit for startle reaction shown in Fig. 11.5. By design, at any frequency, the ratio of the total current through the instrument divided by the current through the resistor is equal to the ratio of the reaction threshold current at the higher frequency divided by the reaction threshold current at 50/60 Hz.

The measuring instrument indication coincides numerically with milliamperes only at low frequencies; consequently, a new term was introduced into the ANSI C101.1 Standard—*measurement indication units* (MIUs). New UL standards and existing UL standards being revised are starting to include MIUs to describe the units, replacing "milliampere," which is a misnomer.

In the past, nearly all 120-V consumer products had impedance between the live parts and accessible parts that was very high relative to the impedance of the leakage-current-measuring instrument. Therefore, the product impedance determined the leakage current with much greater influence than the practically negligible measuring-instrument impedance. Under these conditions, the accuracy associated with the impedance of the measuring instrument was not critical. Today, the situation has changed. Many modern products contain secondary power supplies for various sophisticated features. Some of these lower-voltage power supplies are coupled to accessible parts through impedances that are comparable to human-body impedance values. The products do not always appear to be "constant-current sources," but instead appear to be more like a poorly regulated constant-voltage source in which the human body (or its simulation for test purposes) has a much greater influence in determining the magnitude of the leakage current. For these circuits it is important to simulate the human body impedance more accurately with the input impedance of the leakage-current-measuring instrument.

At frequencies up to approximately a few kilohertz there is good correlation between the human impedance characteristic and that of the ANSI instrument. However, at higher frequencies the two impedance characteristics diverge. In the 1980s, the design of leakage-current-measuring instruments was reexamined, and new leakage-current-measuring instruments were proposed by IEC TC74 Working Group 5 (IEC, 1990b). Figures 11.10 and 11.11 show the schematic diagrams for the new instruments.

As part of the IEC TC74 Working Group 5 committee, W. F. Hart measured the voltage and current thresholds versus frequency for a sensation and interference with muscle control (IEC, 1985; Smoot, 1985). The current levels were somewhat below the threshold of muscle tetanus, and were determined by the tolerance level of each of the four adult male subjects. The current pathway was from a hand holding a conductive handle to the forearm of the same arm. The hand was dry; the forearm was wet with a soap-water solution. The frequency characteristic of the body impedance of the hand-to-forearm path can be calculated from the current and voltage data, and is shown in Fig. 11.12. The input impedance as a function of frequency of the ANSI leakage-current-measuring instrument, and the body impedance model proposed by TC74 Working Group 5, are also shown in Fig. 11.12 for comparison. If a longer pathway through the body had been involved in the experimental work of Working Group 5, the measured values of body impedance would have been higher. The data suggest a value of approximately 250 Ω for the impedance of

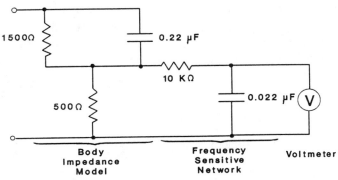

Figure 11.10   Leakage-current-measuring instrument developed by IEC TC74 WG5 for the reaction current limit. (IEC, 1990.)

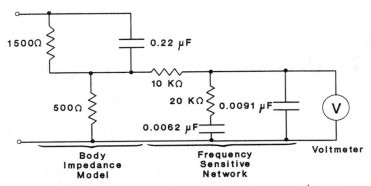

Figure 11.11   Leakage-current-measuring instrument developed by IEC TC74 WG5 for the let-go current limit. (IEC, 1990.)

the model at high frequencies; however, this represents the short current pathway used in the experimental work, which did not involve the torso. The body impedance model is designed to have 500 Ω impedance at high frequencies in order to compensate for this.

The leakage-current-measuring instruments can be viewed as having three parts: (1) the body impedance model, (2) the frequency-sensitive network, and (3) the meter. The input to the measuring instrument is the body impedance model, and it simulates the impedance of the human body representing an appropriate scenario of contact with the product under test. Both measuring instruments proposed by IEC TC74 Working Group 5, shown in Figs. 11.10 and 11.11, contain the same body impedance models. Each consists of three components—a 500-Ω resistor representing the internal body resistance, in series with a parallel combination of a 1500-Ω resistor and 0.22-µF capacitor representing the skin contact with the electrodes. The additional components simulate human frequency response to sinusoidal currents. One network is used for startle reaction (Fig. 11.10), and the other for "let-go" (Fig. 11.11). This body impedance model can be used to represent a person with a hand-hold on an electrode and another larger electrode contacting the legs or buttocks. The body impedance model produces an output voltage that is proportional to the total current through the stimulated body.

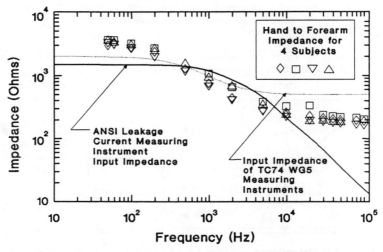

Figure 11.12   Hand-to-forearm impedance measurements taken on four adult male subjects (IEC, 1985; Smoot, 1985) compared to input impedance as a function of frequency of the leakage-current-measuring instrument specified by ANSI Standard C101.1.

As with the ANSI C101 instrument, discussed above, the output voltage from the network can be expressed in terms of measurement indication units (MIUs), and the voltage reading can be compared to a single numeric limit that is independent of frequency. For example, the limit can be expressed simply as 0.5 MIU or 5.0 MIU, depending on the physiological effect of concern.

The frequency-sensitive network makes it unnecessary to measure the frequency of the leakage current, and eliminates the need for a "look-up table" in the requirements to state the limit as a function of frequency. For example, if the reaction instrument reading is 0.25 V rms or less, the leakage current through the instrument is acceptable. The instrument designs are based on physiological data involving sinusoidal current. The limits for leakage current as a function of frequency (Fig. 11.5), on which the instrument designs are based, are derived from sinusoidal AC waveforms. However, a product often produces a waveform of current leakage that is not purely sinusoidal. The leakage current might contain harmonics of a fundamental frequency, or it might be a complex waveform resulting perhaps from a combination of several contributing sources. It might contain random noise. The question arises how to evaluate these leakage currents. Can the frequency-sensitive network be used with validity when the leakage current waveform is nonsinusoidal?

Some work has been done to explore the accuracy of this instrument when the current is not sinusoidal (Hart, 1985; UL, 1990). Hart used mixed frequencies of sinusoidal currents from 30 Hz to 100 kHz. UL used waveforms between 60 Hz and 100 kHz including rectangular waves with 10% duty cycle, and sinusoidal, square, and triangular waves. UL also used 60-Hz positive rectangular waves with 25% duty cycle, and 60-Hz phase-controlled sinusoids switched by SCRs and Triacs at 90°, 125°, and 150°. The results show that the voltage at the output of the frequency-sensitive network does not correlate precisely with the physiological levels of sensation for all waveforms; however, the errors introduced by using the measuring instruments with other than sinusoidal waveforms does not produce conditions that are unsafe. Therefore, UL has decided to use the new measuring instruments with the frequency-sensitive network, as well as the ANSI C101 leakage current instrument, for many nonsinusoidal waveforms. It is an expedient to evaluate products quickly, and does not produce an error that would be harmful to consumers of listed products.

The ANSI C101 leakage current instrument of Fig. 11.9 responds to the average of the full-wave rectified voltage across its terminals. The new reaction instrument of Fig. 11.10 has been specified in the UL Standard for Safety for Telephone Equipment (UL1459) with an rms-responding voltmeter. This subject has been further considered by ANSI C101, which now proposes to specify an RMS instrument to be used with the circuit of Fig. 11.10. The voltmeter suggested for use with the let-go instrument of Fig. 11.11 is a peak-responding meter, often an oscilloscope, in order to be consistent with the limits for "let-go," which are expressed in terms of peak current.

11.4   Implications of Voltage Criteria

From the product designer's standpoint, there is an advantage to having electric shock limits expressed in terms of voltage rather than current. For example, a designer who is selecting a voltage for a secondary power supply inside a product might choose a voltage that is inherently safe from the standpoint of electric shock. Then it becomes unnecessary to measure the leakage current or to be concerned from the point of view of electric shock about possible leakage current pathways.

The National Electrical Code (NFPA, 1990) and UL treat 60-Hz sinusoidal voltages less than 30 V AC rms, and direct current less than 60 V DC, as being nonhazardous from the standpoint of the ability to let go and the potential for ventricular fibrillation in dry locations. For situations where startle reaction cannot be tolerated, UL limits the voltage on accessible parts in dry locations to 42.4 V peak, including DC. For wet locations (immersion not included), these limits are half of the values for dry locations. Voltage limits such as these appear in many UL standards, and appear in Article 725 of the code. The conspicuous absence of field reports of incidents involving voltages considered nonhazardous by these rules tends to confirm their validity, although there might be no formal rationale written for them. Reports of electrical injuries and electrocutions are continually gathered by UL and by the CPSC. IEC standards describe safety extra-low voltage (SELV) circuits as having not more than 42.4 V peak or 60 V DC accessible under normal conditions. Under single-fault conditions,[†] the accessible voltage shall not exceed 42.4 V peak or 60 V DC for longer than 0.2 s, and shall not exceed 71 V peak (the peak of a 50-V rms sinusoidal voltage) and 120 V DC.

The voltage limits mentioned above apply to sources with frequencies less than approximately 100 Hz. For higher frequencies, the voltage limit must be adjusted. Figures 11.13 and 11.14 show the current through the pathway hand to forearm, and the voltage between the hand and forearm electrodes, both as a function of frequency, that produced an equal

---

[†]Single-fault means one defect introduced in the product, such as breakdown of a layer of basic insulation, or failure of a single component.

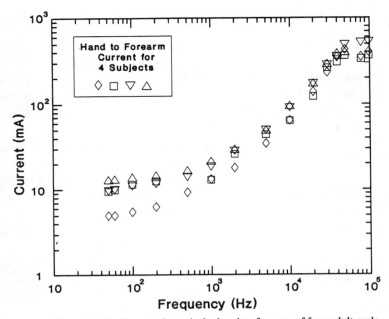

Figure 11.13   Current through the hand to forearm of four adult male subjects to produce the same sensation and muscular control throughout the frequency range 50 Hz to 100 kHz. (IEC, 1985; Smoot, 1985.)

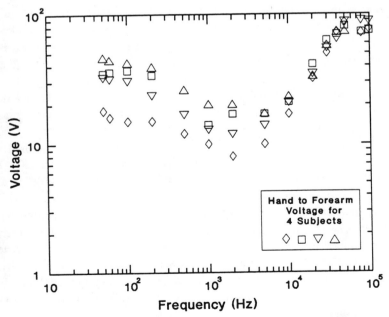

Figure 11.14   Voltage from hand to forearm as a function of frequency of four adult male subjects that correspond to the current values shown in Fig. 11.13. (IEC, 1985; Smoot, 1985.)

sensation and muscular reaction. These thresholds describe an equal subjective combination of muscular and sensory response. Stimuli were in all cases restricted to the discomfort level acceptable to the test subjects. Despite the subjectively defined threshold criteria, the shapes of the frequency characteristic curves for current and voltage are consistent for all four subjects. The current threshold for each subject rises as the frequency is increased. The voltage thresholds describe a U-shaped contour. The minimum threshold voltage is less than half of the voltage at 50/60 Hz, and occurs at approximately 1 to 2 kHz. For each subject, the frequency band over which the dip occurs is approximately two decades wide. The initial decrease in the voltage threshold occurs because body impedance begins to decrease at a lower frequency than the frequency at which the let-go (or near let-go) threshold current begins to rise. Hart reported similar findings from earlier experimental work (Hart, 1985), in which he measured impedance at several frequencies of five male subjects from hand to upper arm and found impedance characteristics for wet skin and high contact pressure.

It has troubled some that the current limits cannot be combined with the body impedance models of 500 or 1500 Ω to produce voltage limits by a simple application of Ohm's law. For example, a current limit of 5 mA and a 500 Ω body impedance would suggest that the maximum acceptable voltage should be 2.5 V, not 30 V. The apparent inconsistency arises because the resistance of the skin varies inversely with applied voltage (see Sec. 2.2). The 500-Ω value for total body impedance can be encountered in shock scenarios involving sufficiently high voltages (e.g., 120 V); below 30 V, the skin resistance is typically higher (see Fig. 2.19).

Under some circumstances, such as when contact area is large and particularly when the contact area is wet or immersed, a 30-V limit may be too high. The body impedance must be at least 6000 Ω in order for the body current to be less than 5 mA from a 30-V supply. With dry contacts, the body impedance values are likely to be higher than 6000 Ω, but with wet contacts the body impedance could be less than 6000 Ω at 30 V. The voltage limit used by UL and by the National Electrical Code is 15 V for conditions leading to wet contacts. It should be recognized that even a sweaty skin condition can significantly lower skin resistance. Where immersion is likely, such as in swimming pools, a let-go voltage limit would have to be considerably lower—perhaps only a volt or so—because of the many pathways by which current can enter the body (not just through the limbs), and because of the high likelihood of very low body impedances.

11.5   Protective Systems and Devices

Safety features are required to be part of certain electrical systems and products. These systems include grounding, double insulation, ground-fault circuit interrupters, shielding of high-voltage parts, and polarization of receptacles and certain plugs.

Grounding

Electrical grounding consists of two parts, system and equipment grounding. The National Electrical Code requires certain electrical systems to be grounded, that is, connected to the earth. Figure 11.15 shows an example of a typical 120/240-V service to a building and the system-grounding connections.

The National Electrical Code (NFPA, 1990) requires that systems and circuit conductors be grounded to limit voltages due to lightning, line surges, or unintentional contact with higher-voltage lines, and to stabilize the voltage to ground during normal operation. In addition, systems and

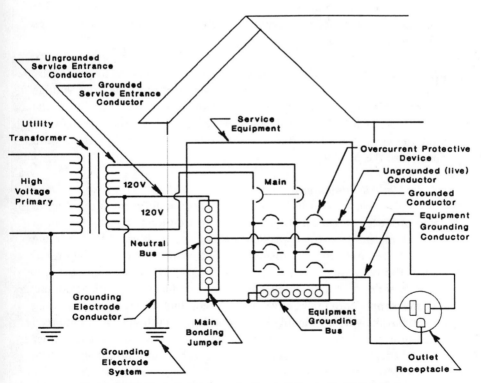

Figure 11.15   Typical electric service to a building in the United States.

circuit conductors must be grounded to facilitate overcurrent device operation in case of ground faults.

In the United States, one of the current-carrying conductors, referred to as the grounded circuit conductor, is connected to earth at the utility transformer and again where the electric supply enters the building at the service equipment. The National Electrical Code forbids making a connection between the grounded circuit conductor and ground at any location in the building on the load side of the service equipment. In our example, the means of making the connection to earth at the service equipment is by a grounding electrode conductor connected to the grounding electrode system, which may consist of a buried metal water piping system supplemented by a driven ground rod.

The National Electrical Code also states that conductive materials enclosing electric conductors or equipment, or forming part of such equipment, must be grounded to limit the voltage to ground on these materials and to facilitate overcurrent device operation in case of ground faults. An important objective of equipment grounding is to promote an equipotential environment, that is, to connect simultaneously contactable conductive surfaces together and to earth so that no harmful voltage differences exist.

In the event of a fault between a live part inside a grounded electric product and the enclosure, the low-impedance connection to ground by way of the equipment grounding conductor suppresses the fault voltage on accessible parts, thereby reducing the shock hazard. In addition, the fault condition will be short-lived because the high current through the fault will cause the overcurrent protective device to open the circuit and deenergize the product.

Double Insulation

Double insulation is an alternative to equipment grounding in some product categories. Its principle is to isolate electrically energized parts by providing two independent insulations, each capable of meeting the safety requirements for insulation. The redundant configuration enhances the reliability of a product's insulation system. In some cases, it is not feasible to achieve independence for two insulations. For example, a power-supply cord might have insulated conductors within an insulating jacket, but the whole assembly is vulnerable to a cutting edge. Reinforced insulation, which is designed to have high reliability, can be used either as a substitute for double insulation, or as a part of a double-insulation system. More information is available in UL's Standard for Safety for Double-Insulation Systems for Use in Electrical Equipment (UL, 1983).

Ground-Fault Circuit Interrupter

The ground-fault circuit interrupter (GFCI) is probably one of the most significant devices to be developed in the field of electric safety. It was introduced in the United States in the early 1960s, and has since been required by the National Electrical Code to be installed in certain branch circuits and as part of certain products. In Europe, the equivalent device is called a residual-current device (RCD). The RCD operates on the same principle as the GFCI but was developed independently and with a few differences that will be explained presently.

The two-conductor GFCI (and RCD) continuously compare the magnitude of current in the energized conductor to that in the grounded conductor. If the current in the two conductors differs by more than a specified amount, the device rapidly trips and opens the circuit. Typically, a GFCI trips in less than 25 ms. The presumption is that the current flowing through an unintended path to ground might be through a person's body as illustrated in Fig. 11.16. If the person's body were to touch parts that are connected across the line rather than between a live part and ground, the resulting body current would flow through the same pathway as normal electric load current, and no protection against electric shock would be provided. However, when a person simultaneously touches a live part and ground, the device would sense differential current, and can be effective against most electric shock scenarios.

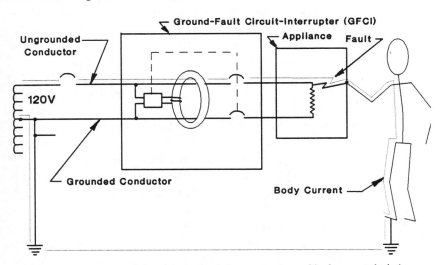

Figure 11.16   Ground-fault circuit interrupter (or residual-current device) sensing body current by comparing the magnitudes of the currents in the conductors serving the appliance. If the difference exceeds approximately 5 mA, the ground-fault circuit interrupter opens the circuit.

A GFCI acts independently from other safety mechanisms that might also be used. When it is used to protect a product that is equipped with an equipment grounding conductor or that is double-insulated, each system contributes to protect the user against electric shock—one system does not interfere with, nor depend on, the other. Being current-sensitive, it protects the user regardless of the body impedance. Most GFCIs in the United States are rated at 5 mA differential current, and are intended to prevent inability to let go as well as ventricular fibrillation. In those situations where the body current is limited in magnitude to just above let-go levels by impedance in series with the body, the GFCI protects the person from "freezing" onto the circuit. It has been argued that the probability is rather low of having just enough impedance external to the body to keep the body current within the range for a long interval that would produce the effects of let-go inability but that would not either trip the device or cause a reaction that would free the person from the circuit. In Europe, an RCD with a rating as high as 30 mA is considered suitable for protection of people against electric shock. At this rating, the protection provided by the RCD would be against ventricular fibrillation, but not against inability to let go.

An advantage of the 30-mA-rated RCD in Europe is that it can trip satisfactorily using only the power available in the fault current itself; it does not need auxiliary power to open the contacts. As such, it is inherently invulnerable to being inoperative if the grounded line conductor is inadvertently open. The GFCIs in the United States take power from the line at the input to the device to open the contacts when the device is tripped. The power available from 5 mA at 120 V is not sufficient by itself to cause the rapid tripping required in this application. If the GFCI is not a type that is likely to be located where the continuity of the grounded conductor is sufficiently reliable, the device is required by UL to protect regardless of the continuity of the grounded conductor. Many listed GFCIs satisfy this requirement by including circuitry that will not energize the output unless there is voltage across the input terminals.

If a GFCI is used to protect against electric shock from a product that itself generates high voltage, such as with an internal autotransformer, it might not trip fast enough to prevent ventricular fibrillation due to the higher body current. For example, GFCIs typically contain mechanical contacts that interrupt the circuit when the device trips. These contacts require a certain minimum time to separate—one might estimate this minimum time to be on the order of 15 to 20 ms. If the body impedance is as low as 500 Ω, and the high-voltage source is capable of applying 1000 volts across the body, then the body current would be 2 A. According

to the current limit as a function of duration in Fig. 11.3, the duration of
body current of this magnitude should be limited to 264 μs, which is
beyond the tripping-speed capability of the typical GFCI.

Shielding

The problem of reducing the risk from electric products that
have high-voltage sources within them has often been addressed by shielding
the high-voltage parts from the user. The shielded source and its load are
completely enveloped by a conductive enclosure. A transformer would
have part of the shield interposed between primary and secondary wind-
ings, usually in the form of a copper sheet. If the source is an autotransformer, the high-voltage circuitry might not be isolated from ground
because the primary is connected to a grounded supply, and so the shield
enveloping the high-voltage circuitry is made inaccessible to the user, and
is connected to one side of the line. In either type of shielding, a fault
involving the high-voltage parts would not result in high voltage on
accessible parts, and therefore the risk from electric shock is reduced.
Such a fault would probably cause an overcurrent or thermal device to
open the secondary circuit, but a GFCI that might be protecting the
product will have to protect only against electric shock from a circuit
with voltage not greater than line voltage.

Polarization

Where the physical layout of a product is such that parts
connected to one side of the line are more accessible than parts connected
to the other side of the line, it is advantageous to have the more accessible
parts connected to the grounded side. Then these parts will not be
hazardous if they were touched, and they can act as a grounded shield
enveloping the more hazardous parts. Polarization of a cord-and-plug-
connected product is accomplished by using a polarized attachment
plug. The polarized plug used in the United States has a wide blade that
fits only in the grounded-conductor side of the supply. Grounding-type
plugs fit into grounding-type receptacles in only one way because of the
third grounding prong, and that serves to make grounding-type plugs
inherently polarized. The likelihood of the polarity being wired correctly
is not perfect. Wiring mistakes are made. For example, sometimes a
receptacle will be installed with the wiring interchanged despite the color
coding that is required.

Receptacle manufacturers in the United States discontinued production
of nonpolarized designs of general-use receptacles in the 1930s. Before
the 1930s, both types of receptacles were sold. A survey of receptacles in

the homes of UL employee-volunteers in New York City and its suburbs was taken in the late 1950s to estimate the extent of nongrounding-type or miswired receptacles in homes. The results showed that 83% of the sampled receptacles were properly wired, polarized receptacles; 6% were improperly wired, polarized receptacles; and 11% were nonpolarized receptacles. In the years that followed this survey, many of the non-polarized receptacles in existence at that time have been replaced by newer polarized designs. The chances of achieving proper polarization with a polarized plug today are likely to be better.

Example

The following example illustrates several safety mechanisms discussed above. Consider a person using an electric sander to smooth the deck of a small boat while he is standing knee-deep in salt water. The sander is plugged into an extension cord, which in turn is plugged into a receptacle on the dock. As the person is working, the mated plug and extension cord connector slide off the deck of the boat and fall into the water. We will examine the effects of various protective mechanisms and see which ones will help the person in this situation—and which ones will not.

First, suppose that the sander has an equipment-grounding conductor in the cord and a grounding-type plug. When the plug and connector are immersed in the salt water, a strong electric field exists in the water between the blades of the plug. If the equipment-grounding conductor is intact between the grounding prong on the plug and earth ground at the service equipment in the building, the voltage with respect to earth of the grounding prong and the enclosure of the sander will remain low, and the person will probably not suffer any adverse effects.

However, suppose the equipment-grounding conductor is open between the grounding prong on the plug and earth ground at the service equipment in the building, perhaps because the person chose an inappropriate extension cord that has no equipment-grounding conductor. The voltage with respect to earth of the grounding prong of the plug on the sander will be approximately 60 V (half line voltage) as it samples the electric field in the salt water. The grounding prong is connected to the enclosure of the sander via the equipment-grounding conductor in the cord, and then to earth via the person's body standing in salt water. Under these conditions of large-area contact, with the person's hands on the sander and the feet in the salt water, the body impedance is likely to be low. If the body impedance were $500\,\Omega$ and the voltage across the body were 60 V, the current through the body including the torso would be

120 mA. This current is high enough to cause adverse physiological effects, which could include ventricular fibrillation and possibly death.

If a GFCI had been installed in the outlet on the dock, the current through the water directly to earth, and the body current via the sander, would have been sensed by the GFCI as differential current because this current does not flow back to the supply through the grounded circuit conductor. The GFCI would have tripped immediately after the plug and connector fell into the water. The person might have felt a "jolt" and perhaps would have reacted to the short-duration current, but no physiological effect more severe than that would have resulted. If the equipment-grounding conductor had been intact and supplemented the GFCI, the pulse of body current would have been reduced in magnitude, and the resulting sensation would have been milder.

If the GFCI had been provided in the plug of the sander instead of on the dock, it would not have sensed the body current and the current to earth in the water as differential current. The source of this current is from the interface between the plug and connector, which is on the line side of the GFCI. Even if the plug-type GFCI were to trip, the interrupter's contacts would not be in the path of the body current, which is through the equipment-grounding conductor in the cord. A plug-type GFCI on the sander would not be effective in this case. A double-insulated sander would have eliminated the equipment-grounding conductor in the cord and therefore would have eliminated the path of the body current in this scenario. The risk of electric shock would not be entirely eliminated by the double-insulated construction, since an electric field will exist in the water near the plug and connector that is accessible to the person.

No general conclusions can be drawn from this example because the various protective mechanisms would have been effective or ineffective in different ways had the scenario been changed. For example, if the sander had fallen into the water (not the plug and connector), and had the person reached into the water to retrieve it while it was energized and submerged, double insulation might not have been effective because both insulations could have been simultaneously bridged by the water entering the product through ventilating openings. Grounding probably would have been effective if the enclosure had been constructed to provide sufficient shielding and the grounding conductor had been intact between the sander enclosure and earth ground connected at the service equipment in the building. In this scenario, with the source of the body current on the load side of the plug, a GFCI located either at the dock or in the plug of the sander would have been effective.

References

Accornero, N., G. Bini, G. L. Lenzi, and M. Manfredi (1977). Selective activation of peripheral nerve fibre groups of different diameter by triangular shaped stimulus pulses. *J. Physiol.* 273: pp 539–560.

Ackmann, J. J., and M. A. Seitz (1984). Methods of complex impedance measurements in biologic tissue. *CRC Crit. Rev. Bioeng.* 11(4): 281–311.

Agnew, W. F., and D. B. McCreery, eds. (1990). *Neural Prostheses: Fundamental Studies.* Prentice-Hall, Englewood Cliffs, NJ.

Agnew, W. F., D. B. McCreery, T. G. H. Yuen, and L. A. Bullara (1989). Histologic and physiologic evaluation of electrically stimulated peripheral nerve: Considerations for the selection of parameters. *Annals Biomed. Eng.* 17: 39–60.

Ahrenholz, D. H., W. Schubert, and L. D. Solem (1988). Creatine kinase as a prognostic indicator in electrical injury. *Surgery* 104(4): 741–747.

Albuquerque, E. X., and S. Thesleff (1968). A comparative study of membrane properties of innervated and chronically denervated fast and slow skeletal muscles of the rat. *Acta. Physiol. Scand.* 73: 471–480.

Allessie, M. A., F. I. M. Bonke, and F. J. G. Schopman (1973). Circus movement in rabbit atrial muscle as a mechanism of tachycardia. *Circ. Res.*, 33: 54–62.

Allessie, M. A., F. I. M. Bonke, and F. J. G. Schopman (1976). Circus movement in rabbit atrial muscle as a mechanism of tachycardia. II. The role of nonuniform recovery of excitability in the occurrence of unidirectional block, as studied with multiple microelectrodes. *Circ. Res.* 39: 168–177.

Allessie, M. A., F. I. M. Bonke, and F. J. G. Schopman (1977). Circus movement in rabbit atrial muscle as a mechanism of tachycardia. III. The "leading circle" concept: A new model of circus movement in cardiac tissue without the involvement of an anatomical obstacle. *Circ. Res.* 41: 9–18.

Alon, G., J. Allin, and G. F. Inbar (1983). Optimization of pulse duration and pulse charge during transcutaneous electrical nerve stimulation. *Austral. J. Physiotherapy* 29(6): 195–201.

Al-Rabiah, S. M., D. B. Archer, R. Millar, A. D. Collins, and W. F. Shepherd (1987). Electrical injury of the eye. *Int. Ophthalmol.* 11(1): 31–40.

Altman, K. W., and R. Plonsey (1989). Excitation in a model for time-dependent electrical nerve bundle stimulation. *Proc. Eleventh IEEE-EMBS Conf.*, pp. 975–976.

Amassian, V. E., R. Q. Cracco, and P. J. Maccabee (1988). Basic mechanisms of magnetic coil excitation of nervous system in humans and monkeys and their applications. *Proc. 10th Ann. Conf. IEEE-EMBS*, pp, 10–17.

Amassian, V. E., J. B. Cracco. R. Q. Cracco, L. Eberle, P. J. Maccabee, and A. Rudell

(1989a). Supression of visual perception by magnetic coil stimulation of human occipital cortex. *EEG Clin, Neurophysiol*, 74: 458–462.

Amassian, V. E., R. Q. Cracco, and P. J. Maccabee (1989b). A sense of movement elicited in paralyzed distal arm by focal magnetic coil stimulation of human motor cortex. *Brain Res*. 479: 355–360.

Anderson, A. B., and W. A. Munson (1951). Electrical stimulation of nerves in the skin at audio frequencies. *J. Acoust. Soc. Am.* 23(2): 155–159.

Anderson, N. H. (1970). Functional measurement and psychophysical judgment. *Psychol. Rev.* 77: 153–170.

ANSI (1986). Leakage current for appliances. ANSI C101.1-1986, American National Standards Institute.

ANSI (1987). Procedures for the development and coordination of American National Standards. American National Standards Institute.

Antoni, H. (1979). What is measured by the so-called threshold for fibrillation? *Prog. Pharmacol.* (Stuttgart) 2/4: 5–12.

Antoni, H. (1985). Pathophysiological basis of ventricular fibrillation. In J. F. Bridges, G. L. Ford, I. A. Sherman, and M. Vainberg (eds.), *Electrical Shock Safety Criteria*, Pergamon Press, New York, pp. 33–43.

Antoni, H., J. Toppler, and H. Krause (1970). Polarization effects of sinusoidal 50-cycle alternating current on membrane potential of mammalian cardiac fibers. *Pflugers Arch.* 314: 274–291.

Antzelevitch, C., and G. K. Moe (1981). Electrotonically mediated delayed conduction and reentry in relation to "slow responses" in mammalian ventricular conducting tissue. *Circ. Res.* 49: 1129–1139.

Artz, C. P. (1974). Changing concepts of electrical injury. *Am. J. Surg.* 128(5): 600–602.

Artz, C. P. (1979). Electrical injury, In C. P. Artz, Moncrief, J. A. and Pruitt, B. A. (eds.), *Burns: A Team Approach*, W. B. Saunders, Philadelphia.

Ashton, W. D. (1972). *The Logit Transformation*. Hafner, New York.

Atkinson, W. H. (1982). A general equation for sensory magnitude. *Perception & Psychophys.* 31: 26–40.

Attneave, F. (1962). Perception and related areas. In S. Koch (ed.), *Psychology: A Study of a Science*, McGraw-Hill, New York, pp. 619–659.

Babkoff, H. (1976). Magnitude estimation of short electrocutaneous pulses. *Psychol. Res.* 39: 39–49.

Bajzek, T. J., and R. J. Jaeger (1987). Characterization and control of muscle response to electrical stimulation. *Annals Biomed. Eng.* 15: 485–501.

Baker, M. D., and C. Chiaviello (1989). Household electrical injuries in children; Epidemiology and identification of avoidable hazards. *Am. J. Dis. Child.* 143(1): 59–62.

Banks, R. S., and T. Vinh (1984). An assessment of the 5-mA 60-Hz contact current safety level. *IEEE Trans. Pwr. Sys.* PAS-103(12): 3608–3614.

Barker, A. T., I. L. Freeston, R. Jalinous, P. A. Merton, and H. B. Morton (1985). Magnetic stimulation of the human brain. *J. Physiol.* 369: 3P.

Barker, A. T., I. L. Freeston, B. Jalinous, and J. A. Jarratt (1987). Magnetic stimulation of the human brain and peripheral nervous system: An introduction and the results of an initial clinical evaluation. *Neurosurgery*, 20: 100–109.

Barnard, M. D., and J. A. Boswick, Jr. (1976). Electrical injuries of the upper extremity. *Rocky Mt. Med. J.* 73: 20–24.

Barta, E., D. Adam, E. Salant, and S. Siderman (1987). 3-D ventricular myocardial electrical excitation: A minimal orthogonal pathways model. *Annals Biomed. Eng.* 15: 443–456.

Basser, P. J., and B. J. Roth (1991). Stimulation of a myelinated nerve axon by electromagnetic induction. *Med. Biol. Eng. Comp.*, 29(3): 261–268.

Bauwens, P. (1971). Introduction to electrodiagnostic procedures. In S. Licht (ed.),

Electrodiagnosis and Electromyography (3rd ed.), E. Lichts, New Haven, CT, chap. 7.

Baxter, C. R. (1970). Present concepts in the management of major electrical injury. *Surg. Clin. North Am.* 50(6): 1401–1418.

Beck, C., and Rosner, B.S. (1968). Magnitude scales and evoked potentials to percutaneous electrical stimulation. *Physiol. Behav.* 3: 947–953.

Beck, C. S., W. H. Pritchard, and S. H. Feil (1947). Ventricular fibrillation of long duration abolished by electric shock. *J. Am. Med. Assoc.* 135: 985–995.

Beeler, G. W., and H. Reuter (1977). Reconstruction of the action potential of ventricular myocardial fibers. *J. Physiol.* 268: 177–210.

Bernhardt, J. H. (1985). Evaluation of human exposures to low frequency fields. In *The Impact of Proposed Frequency Radiation Standards on Military Operations*, Lecture Series 138, Advisory Group for Aerospace Research and Development (NATO), Surseine, France.

Biegelmeier, G. (1978). Report on the electrical impedance of the human body. Presented to the International Electrotechnical Commission, Technical Committee No. 23, Austria.

Biegelmeier, G. (1982). Report on the electrical impedance of the human body. IEC Technical Committee No. 64, *Electrical Installations of Buildings*, Austin, Texas.

Biegelmeier, G., (1985a). New knowledge of the impedance of the human body. In J. E. Bridges, G. L. Ford, I. A. Sherman, and M. Vainberg (eds.), *Electrical Shock Safety Criteria*, Pergamon Press, New York, pp. 115–132.

Biegelmeier, G. (1985b). New experiments with regard to basic safety measures for electrical equipment and installations. In J. E. Bridges, G. L. Ford, I. A. Sherman, and M. Vainberg (eds.), *Electrical Shock Safety Criteria*, Pergamon Press, New York, pp. 161–172.

Biegelmeier, G. (1985c). The impedance of the human body. *Rev. Gen. L'Electricite* 11: 817–832.

Biegelmeier, G. (1986). *Wirkunen des Elektrischen Stroms auf Menschen und Nutztiere*. Vde-Verlag, Berlin.

Biegelmeier, G. (1987). Effects of current passing through the human body and the electrical impedance of the human body: A guide to IEC-Report 469. ETZ Report 20, VDE-Verlag, Berlin.

Biegelmeier, G., and E. Homberger (1982). Über die wirkungen von unipolaren impulsatromen auf den menschlicken körper. *Bull. ASE/USE* 73, Sept. 18.

Biegelmeier, G., and W. R. Lee (1980). New considerations on the threshold of ventricular fibrillation for a.c. shocks at 50–60 Hz. *IEE Proc.* 127(2), Pt. A: 103–110.

Biegelmeier, G., and J. Miksch (1980). Effect of the skin on the body impedance of humans (translated from German). *Electrotechnik and Maschinenbau, mit Industrieller Electronik und Nachrichtentchnik* 97(9): 369–378.

Biegelmeier, G., and K. Rotter (1971). Electrical resistances and currents in the human body. *Electrotechnic and Maschinenbau*, pp. 104–114.

Bigland-Ritchie, B., R. Johansson, O. C. J. Lippold, S. Smith, and J. J. Woods (1983). Changes in motoneuron firing rate during sustained maximal voluntary contractions. *J. Physiol.* 340: 335–346.

Bigland-Ritchie, B., F. Bellemare, and J. J. Woods (1986). Excitation frequencies and sites of fatigue. In N. L. Jones, N. McCartney, and A. J. McComas (eds.), *Human Muscle Power*, Human Kinetics Publishers, Champaign, IL, pp. 197–211.

Bini, G., G. Cruccu, K. E. Hagbarth, W. Schady, and E. Torebjörk (1984). Analgesic effect of vibration and cooling on pain induced by intraneural electrical stimulation. *Pain* 18: 239–248.

Birks, R., H. E. Huxley, and Katz (1960). The fine structure of the neuromuscular junction of the frog. *J. Physiol.* 150: 134–144.

Bishop, G. H (1943). Responses to electrical stimulation of single sensory units of skin. *J. Neurophysiol.* 6: 361–382.

Bishop, G. H. (1946). Neural mechanisms of cutaneous sense. *Psychol. Rev.* 26: 77–102.

Blair, E. A., and J. Erlanger (1933). A comparison of the characteristics of axons through their individual electrical responses. *Am. J. Physiol.* 106: 524–564.

Blair, H.A. (1932a). On the intensity-time relations for stimulation by electric currents. I. *J. Gen. Physiol.* 15: 709–729.

Blair, H. A. (1932b). On the intensity-time relations for stimulation by electric currents. II. *J. Gen. Physiol.* 15: 731–755.

Bostock, H. (1983). The strength-duration relationship for excitation of myelinated nerve: computed dependence on membrane parameters. *J. Physiol.* 341: 59–74.

Bostock, H., T. A. Sears, and R. M. Sherratt (1983). The spatial distribution of excitability and membrane current in normal and demyelinated mammalian nerve fibers. *J. Physiol,* 341: 41–58.

Bouman, M. A., and H. A. van der Velden (1947). The two-quanta explanation of the dependence of the threshold values and visual acuity on the visual angle and the time of observation, *J. Opt. Soc. Am.* 37: 908–919.

Bourland, J. D., W. A. Tacker, and L. A. Geddes (1978). Strength-duration curves for trapezoidal waveforms of various tilts for transchest defibrillation in animals. *Med. Instrum.* 12: 38–41.

Bourland, J. D., J. A. Nyenhuis, G. A. Mouchawar, L. A. Geddes, and D. J. Schaefer (1990). Human peripheral nerve stimulation from z-gradients. *Society for Magnetic Resonance in Med., Proc. 9th Annual Meeting,* Aug. 18–24, NY: 1157.

Bourland, J. D., J. A. Nyenhuis, G. A. Mouchawar, L. A. Geddes, D. J. Schaefer, and M. E. Riehl (1991). Z-gradient coil and eddy-current stimulation of skeletal and cardiac muscle in the dog. *Society for Magnetic Resonance in Med., Proc. 10th Annual Meeting,* Aug. 10–16, San Francisco.

Bowman, B. R., and R. C. Erickson (1985). Acute and chronic implantation of coiled wire interaneural electrodes during cyclical electrical stimulation. *Annals. Biomed. Eng.* 13: 75–93.

Bowman, B. R., and D. R. McNeal (1986). Response of single alpha motoneurons to high-frequency pulse trains. *Appl. Neurophysiol.* 49: 121–138.

Boxtel, A. (1977). Skin resistance during square-wave electrical pulses of 1 to 10 mA. *Med. Biol. Eng. Comp.* 15: 679–687.

Bracken, T. D. (1976). Field measurements and calculations of electrostatic effects of overhead transmission lines. *IEEE Trans. Pwr. Apparat. Sys.* PAS-95: 494–502.

Brazier, M. A. (1977). *Electrical Activity of the Nervous System.* Williams & Wilkins, Baltimore.

Brengelmann, G., and A. C. Brown (1965). Temperature regulation. In T. C. Ruch and H. D. Patton (eds.), *Physiology and Biophysics,* 19th ed., W. B. Saunders, Philadelphia, pp. 1050–1068.

Bridges, J. E. (1985). Potential distributions in the vicinity of the hearts of primates arising from 60 Hz limb-to-limb body currents. In J. E. Bridges, G. L. Ford, I. A. Sherman, and M. Vainberg (eds.), *Electrical Shock Safety Criteria,* Pergamon Press, New York, pp. 61–70.

Bridges, J. E., G. L. Ford, I. A. Sherman, and M. Vainberg (eds., 1985). *Electrical Shock Safety Criteria.* Pergamon Press, New York.

Bridges, J. E., M. Vainberg, and M. C. Wills (1987). Impact of recent developments in biological electrical shock safety criteria. *IEEE Trans. Pwr. Del.* PWRD-2(1): 238–248.

Brooke, M. H., and K. K. Kaiser (1970). Muscle fiber types: How many and what kind? *Arch. Neurol.* 23: 369–379.

Brooks, C. M., B. F. Hoffman, E. E. Suckling, and O. O. Orias (1955). *Excitability of the Heart.* Grune & Stratton, New York.

Budinger, T. F., H. Fischer, D. Hentschel, H. E. Reinfelder, and F. Schmitt (1990). Neural Stimulation dB/dt thresholds for frequency and number of oscillations using sinusoidal magnetic gradient fields. *Society for Magnetic Resonance in Med., Proc. 9th Annual Meeting,* Aug, 18–24, NY: 276.

Burke, R. E. (1968). Firing patterns of gastrocnemius motor units in the decerebrate cat. *J. Physiol.* 196: 631–654.

Burke, R. E. (1981). Motor units: Anatomy, physiology and functional organization. In V. B. Brooks, (ed.), *Handbook of Physiology Section 1: The Nervous System. Vol.* III. *Motor Systems,* American Physiology Society, Bethesda, MD, pp. 345–422.

Burke, R. E. (1986). The control of muscle force: Motor unit recruitment and firing patterns. In N. L. Jones, N. McCartney, and A. J. McComas (eds.), *Human Muscle Power,* Human Kinetics Publishers, Champaign, IL, pp. 97–106.

Burke, R. E., D. N. Levine, P. Tsairis, and F. E. Zajac (1973). Physiological types of histochemical profiles in motor units of the cat gastrocnemius. *J. Physiol.* 234: 723–748.

Burton, C. E., R. M. David, W. M. Portnoy, and L. A. Akers (1974). The application of Bode analysis to skin impedance. *Psychophysiology* 11(4): 517–525.

Bütikoffer, R., and P. D. Lawrence (1978). Electrocutaneous nerve stimulation—I: Model and experiment. *IEEE Trans. Biomed Eng.* BME-25(6): 526–531.

Bütikoffer, R., and P. D. Lawrence (1979). Electrocutaneous nerve stimulation—II: Stimulus waveform selection. *IEEE Trans. Biomed. Eng.* BME-26(2): 69–75.

Butler, E. D., and T. D. Gant (1978). Electrical injuries, with special reference to the upper extremities—A review of 182 cases. *Am. J. Surg.* 134: 95–101.

Cabanes, J., and C. Gary (1981). La perception directe du champ electrique. Paper 233-08. *CIGRE Symp. 22-81,* Stockholm, pp. 1–6.

Caldwell, C. W., and J. B. Reswick (1975). A percutaneous wire electrode for chronic research use. *IEEE Trans. Biomed. Eng.* 22: 429–432.

Campbell, J. N., and R. A. Meyer (1983). Sensitization of unmyelinated nociceptive afferents in monkey varies with skin type. *J. Neurophysiol.* 49(1): 98–110.

Campbell, J. N., R. A. Meyer, and R. H. LaMotte (1979). Sensitization of myelinated afferents that innervate monkey hand. *J. Neurophysiol.* 42(6): 1669–1679.

Carmeliet, E. (1977). Repolarization and frequency in cardiac cells. *J. Physiol.* (Paris). 73: 903–923.

Carter, A. O., and R. Morley (1969a). Electrical current flow through human skin at power frequency voltages. *Br. J. Ind. Med.* 26: 217–223.

Carter, A. O., and R. Morley (1969b). Effects of power frequency voltages on amputated human limb. *Br. J. Ind. Med.* 26: 224–230.

Caruso, P. M., J. A. Pearce, and D. P. DeWitt (1979). Temperature and current density distributions at electrosurgical dispersive electrode sites. *Proc. 7th N. Engl. Bioeng. Conf.,* Troy, NY, March 22–23, 1979, pp. 373–376.

Cattell, M. K., and R. W. Gerard (1935). The 'inhibitory' effect of high frequency stimulation and excitation state of nerve. *J. Physiol.* 83: 407–415.

Chakravarti, K., and G. J. Pontrelli (1976). The measurement of carpet static. *Textile Res. J.* 46(2): 129–134.

Chatterjee, I., D. Wu, and O. P. Gandhi (1986). Human body impedance and threshold currents for perception and pain for contact hazard analysis in the VLF-MF band. *IEEE Trans. Biomed. Eng.* BME-33(5): 486–494.

Chen, J. Y., and O. P. Gandhi (1988). Thermal implications of high SAR's in the body extremities at the ANSI—recommended MF-VHF safety levels. *IEEE Trans. Biomed. Eng.* 35(6): 435–441.

Chen, P., G. H. Myers, V. Parsonnet, K. Chatterjee, and P. Katz (1975). Relationship between pacemaker fibrillation thresholds and electrode area. *Med. Instrum.* 9(4): 165–170.

Chilbert, M. A., A. Sances, J. B. Mykelbust, T. Swiontek, and T. Prieto (1983). Post-mortem resistivity studies at 60 Hz. *J. Clin. Eng.* 8(3): 219–224.

Chilbert, M., D. Maiman, A. Sances Jr., J. Myklebust, T. E. Prieto, T. Swiontek, M. Heckman, and K. Pintar (1985a). Measure of tissue resistivity in experimental electrical burns. *J. Trauma* 25(3): 209–215.

Chilbert, M. T. Swiontek, T. Prieto, A. Sances, J. Myklebust, J. Ackmann, C. Brown, and J. Szablya (1985b). Resistivity changes of tissue during the application of injurious 60 Hz currents. In J. E. Bridges, G. L. Ford, I.A. Sherman, and M. Vainberg (eds.), *Electrical Shock Safety Criteria*, Pergamon Press, New York, pp. 193–201.

Chilbert, M., D. J. Moretti, T. Swiontek, J. B. Myklebust, T. Prieto, A. Sances Jr., and C. Leffingwell (1988). Instrumentation design for high-voltage electrical injury studies. *IEEE Trans. Biomed. Eng.* 35(7): 565–568.

Chilbert, M., T. Swiontek, C. Leffingwell, T. Prieto, J. Myklebust, A. Sances Jr., and T. Schneider (1989). Fibrillation induced at high current levels. *IEEE Trans. Biomed. Eng.* 36(8): 864–868.

Chiu, M. C. (1982). Fuel ignition by high voltage capacitive discharges. The Johns Hopkins University Applied Physics Laboratory, JHU PPSE T-18.

Chiu, S. Y., and J. M. Ritchie (1981). Evidence for the presence of potassium channels in the paranodal region of acutely demyelinated mammalian single nerve fibers. *J. Physiol.* 313: 415–437.

Chiu, S. Y., J. M. Ritchie, R. B. Rogart, and D. Stagg (1979). A quantitative description of membrane currents in rabbit myelinated nerve. *J. Physiol.* 292: 149–166.

Chizeck, H. J., R. Kobetic, E. B. Marsolais, J. J. Abbas, I. H. Donner, and E. Simon (1988). Control of functional neuromuscular stimulation systems for standing and locomotion in paraplegics. *Proc. IEEE* 76(9): 1155–1165.

Chokroverty, S., and J. DiLullo (1989). Percutaneous magnetic stimulation of the human lumbosacral spinal column: Physiological mechanism and clinical application. *Neurology.* 39: 376.

Christensen, J. A., R. T. Sherman, G. A. Balis, and J. D. Wuamett (1980). Delayed neurologic injury secondary to high-voltage current, with recovery. *J. Trauma* 20(2): 166–168.

Clairmont, B. A., G. B. Johnson, L. E. Zafanella, and S. Zelingher (1989). The effect of HVAC-HVDC line separation in a hybrid corridor. *IEEE Trans. Pwr. Del.* 4(2):1338–1350.

Clar, E. J., C. P. Her, and C. G. Sturelle (1975). Skin impedance and moisturization. *J. Soc. Cosmet. Chem.* 26: 337–353.

Clark, W. C., and S. B. Clark (1980). Pain responses in Nepalese porters. *Science* 209: 410–412.

Clerc, L. (1976). Direction differences of impulse spread in trabecular muscle from mammalian heart. *J. Physiol.* (London). 255: 335–346.

Coers, C. (1955). Les variations structurelles normales et pathologiques de la jonction neuromusculaire. *Acta Neurol. Psychiatr. Belg.* 55: 741.

Cohen, M. S., R. M. Weisshoff, R. R. Rzedzian, and H. C. Kantor (1990). Sensory stimulation by time-varying magnetic fields. *Magnetic Resonance in Medicine* 14: 409–414.

Cooley, J. W., and F. A. Dodge (1966). Digital computer solutions for excitation and propagation of the nerve impulse. *Biophys. J.* 6: 583–599.

Cowburn, J., and R. H. Fox (1974). A technique for studying thermal perception. *J. Physiol.* (London). 239: 77–78.

Cracco, R. Q. (1987). Evaluation of conduction in central motor pathways: Techniques, pathophysiology, and clinical interpretation. *Neurosurgery* 20: 199–203.

Crago, P. E., P. H. Peckham, J. T. Mortimer, and J. P. Van Der Meulen (1974). The choice of pulse duration for chronic electrical stimulation via surface, nerve, and intramuscular electrodes. *Annals Biomed. Eng.* 2: 252–264.

Crago, P. E., P. H. Peckham, and G. B. Thorpe (1980). Modulation of muscle force by recruitment during intramuscular stimulation. *IEEE Trans. Biomed. Eng.* 27: 679–684.

Craig, K. D., and S. M. Weiss (1971). Vicarious influences on pain-threshold determinations. *J. Personality and Social Psychol.* 19: 53–59.

Cranefield, P. F., B. F. Hoffman, and A. A. Siebens (1957). Anodal excitation of cardiac muscle. *Am. J. Physiol.* 190(2): 383–390.

Cranefield, P. F., H. O. Klein, and B. F. Hoffman (1971). Conduction of the cardiac impulse: I. Delay, block, and one-way block in depressed Purkinje fibers. *Circ. Res.* 28: 199–219.

Critchley, M. (1934). Neurological effects of lightning and of electricity. *Lancet.* 1: 68–72.

Crochetiere, W. J., L. Vodovnik, and J. B. Reswick (1967). Electrical stimulation of skeletal muscle—A study of muscle as an actuator. *Med. Biol. Eng.* 5: 111–125.

Dalziel, C. F. (1938). Danger of electric shock. *Electrical West* 80(4): 30–31.

Dalziel, C. F. (1943). Effect of wave form on let-go currents, *AIEE Trans.* 62: 739–744.

Dalziel, C. F. (1953). A study of the hazards of impulse current. *Trans. AIEE, Pt. III,* 72: 1032–1043.

Dalziel, C.F. (1954). The threshold of perception current. *Trans. AIEE, Pt. III* B: 990–996.

Dalziel, C. F. (1959). The effects of electric shock on man. *IRE Trans. Med. Elect.* PGME-5: 44–62.

Dalziel, C. F. (1960). Threshold 60-cycle fibrillating currents. *Trans. AIEE, Pt. III,* 79: pp. 667–673.

Dalziel, C. F. (1968). Reevaluation of lethal electric currents. *IEEE Trans. Ind. Appl.* IGA-4(5): 467–476.

Dalziel, C. F. (1972). Elecric shock hazard. *IEEE Spectrum* (9): 41–50.

Dalziel, C. F. (1978). Recent developments in ground fault circuit interrupters and ground fault receptacles. *Profession. Safety,* November, pp. 31–40.

Dalziel, C. F., and W. R. Lee (1968). Reevaluation of lethal electric currents. *IEEE Trans. Ind. Appl.* IGA-4 (5): 467–476.

Dalziel, C. F., and W. R. Lee (1969). Lethal electric currents. *IEEE Spectrum,* Feb., pp. 44–50.

Dalziel, C. F., and T. H. Mansfield (1950a). Perception of electric currents. *Electrical Eng.* 69: 794–800.

Dalziel, C. F., and T. H. Mansfield (1950b). Effects of frequency on perception currents. *AIEE Trans.* 69: 1162–1168.

Dalziel, C. F., and F. P. Massoglia (1956). Let-go currents and voltages. *AIEE Trans. P. II, Appl. Ind.* 75: 49–56.

Dalziel, C. F., J. B. Lagen, and J. L. Thurston (1941). Electric shock. *AIEE Trans.* 60: 1073–1079.

Dalziel, C. F., E. Ogden, and C. E. Abbott (1943). Effect of frequency on let-go currents. *AIEE Trans.* 62: 745–750.

Darien-Smith, I. (1982). Touch in primates. *Ann. Rev. Psychol.* 33: 155–194.

Darien-Smith, I., K. O. Johnson, C. LaMotte, P. Kenins, Y. Shigenaga, and V. C. Ming (1979). Coding of incremental changes in skin temperature by single warm fibers in the monkey. *J. Neurophysiol.* 42: 1316–1331.

Davey, K., K. C. Kalaitzakis, and C. Epstein (1988). Transcranial magnetic stimulation of the cerebral cortex. *Proc. 10th Ann. Conf. IEEE-EMBS,* pp. 922–923.

Davis, H. (1923). The relationship of the "chronaxie" of muscle to the size of the stimulating electrode. *Proc. Physiol. Soc.,* July 7, pp. lxxxi–lxxxii.

Dean, D., and P. D. Lawrence (1983). Application of phase analysis of the Frankenhaeuser-Huxley equations to determine threshold stimulus amplitudes. *IEEE Trans. Biomed. Eng.* BME-30(12): 810–818.

Dean, D., and P. D. Lawrence (1985). Optimization of neural stimuli based upon a variable threshold potential. *IEEE Trans. Biomed. Eng.* 32(1): 8–14.

Déléze, J. (1970). The recovery of resting potential and input resistance in sheep heart injured by knife or laser. *J. Physiol.* (London) 208: 547–562.

de Mello, W. C. (1972). The healing-over process in cardiac and other muscle fibres. In *Electrical Phenomena of the Heart,* Academic Press, New York, pp. 323–325.

Deno, D. W. (1975a). Calculating electrostatic effects of overhead transmission lines. *IEEE Trans. Pwr. Apparat. Sys.* PAS-93(5): 1458–1471.

Deno, D.W. (1975b). Electrostatic effect induction formulae. *IEEE Trans. Pwr. Apparat. Sys.* PAS-94(5): 1524–1536.

Deno, D. W. (1977). Currents induced in the human body by high voltage transmission line electric field—measurement of calculation of distribution and dose. *IEEE Trans. Pwr. Apparat. Sys.* PAS-96(5): 1517–1527.

Deno, D. W. (1978). Electrostatic and electromagnetic effects of ultrahigh-voltage transmission lines. Report EL-802. Electric Power Research Institute, Palo Alto, CA.

Deno, D. W., and L. E. Zaffanella (1982). Field effects of overhead transmission lines and stations. In *Transmission Line Reference Book*, Electric Research Institute, Palo Alto, CA, pp. 329–419.

DiFrancesco, D., and D. Noble (1985). A model of cardiac electrical activity incorporating ionic pumps and concentration changes. *Phil. Trans. R. Soc. London,* B: 307.

Dixon, G. F. (1983). The evaluation and management of electrical injuries. *Crit. Care Med.* 11: 384–387.

Dominguez, G., and H. A. Fozzard (1970). Influence of extracellular K^+ concentration on cable properties and excitability of sheep cardiac Purkinje fibers. *Circ. Res.* 26: 565–574.

Donly, K. J., and A. J. Nowak (1988). Oral electrical burns: Etiology, manifestations, and treatment. *Gen. Dent.* 36(2): 103–107.

Dowse, C. M., and C. E. Iredell (1920). The effective resistance of the human body to high frequency currents. *Arch. Radiol. Electrother.* 25(2): 34–46.

Drouhard, J. P., and F. A. Roberge (1982a). A simulation study of the ventricular myocardial action potential. *IEEE Trans. Biomed. Eng.* BME-29(7): 494–502.

Drouhard, J. P., and F. A. Roberge (1982b). The simulation of repolarization events of the cardiac Purkinje fiber action potential. *IEEE Trans. Biomed. Eng.* BME-29(7): 481–493.

DuBois, E. F., and D. DuBois (1916). Formula to estimate appoximate surface area if height and weight be known. *Arch. Intern. Med.* 17: 863.

Durney, C. H., C. C. Johnson, and H. Massoudi (1975). Long wavelength analysis of plane wave irradiation of a prolate spheroid model of a man. *IEEE Trans. Microwave Theory,* MTT-23(2): 246–253.

Eaton, H. A. C. (1992). The electric field induced in a spherical volume conductor from arbitrary coils: application to magnetic stimulation and MEG. *Med. Biol. Eng. Comp.* (in press).

Ebihara, L., and E. A. Johnson (1980). Fast sodium current in cardiac muscle. *Biophys. J.* 32: 779–790.

Edelberg, R. (1971). Electrical properties of the skin. In H. R. Elden (ed.), *Biophysical Properties of the Skin,* Wiley-Interscience, New York, pp. 513–550.

Edwards, R. H. T. (1981). Human muscle function and fatigue. In R. Porter and J. Whelan (eds.), *Human Muscle Fatigue: Physiological Mechanisms,* Pitman Medical, London, pp. 1–18.

Edwards, R. H. T., D. K. Hill, D. A. Jones, and P. A. Merton (1977). Fatigue of long duration in human skeletal muscle after exercise. *J. Physiol.* 272: 769–778.

Eifler, W. J., and R. Plonsey (1975). A cellular model for the simulation of activation in the ventricular myocardium. *J. Electrocardiol.* 8(2): 117–128.

Elden, H. R. (ed.) (1971). *Biophysical Properties of the Skin,* Wiley-Interscience, New York.

Elharrar, V., P. R. Forster, T. L. Jirak, W. E. Gaum, and D. P. Zipes (1977). Alterations in canine myocardial excitability during ischemia. *Circ. Res.* 40: 98–105.

El-Sherif, N., B. J. Scherlag, and R. Lazzara (1975). Electrode catheter recordings during malignant ventricular arrhythmias following experimental acute myocardial ischemia. *Circulation,* 51: 1003–1014.

Engelmann, T. W. (1875). Leitung und Erregung im Herzmuskel, *Pflugers Arch.* 11: 465–480.

Epstein, B. R., and K. R. Foster (1983). Anisotropy in the dielectric properties of skeletal muscle. *Med. Biol. Eng. Comp.* 21: 51–55.

Evans, B. A., W. J. Litchy, and J. R. Daube (1988). The utility of magnetic stimulation for routine peripheral nerve conduction studies. *Muscle & Nerve,* 11: 1074–1078.

Fang, Z. P., and J. T. Mortimer (1987). A method for attaining natural recruitment order in artificially activated muscles. *Proc. Ninth IEEE-EMBS Conf.,* pp. 657–658.

Ferris, L. P., B. G. King, P. W. Spence, and H. B. Williams (1936). Effect of electric shock on the heart. *AIEE Trans.* 55: 498–515.

Fitzhugh, R. (1962). Computation of impulse initiation and saltatory conduction in a myelinated nerve fiber. *Biophys. J.* 2: 11–21.

Fitzhugh, R. (1966). Theoretical effects of temperature on threshold in the Hodgkin-Huxley nerve model. *J. Gen. Physiol.* 49: 989–1005.

Forbes, T. W., and A. L. Bernstein (1935). The standardization of sixty-cycle electric shock for practical use in psychological experimentation. *J. Gen. Psychol.* 12/13: 436–441.

Foster, K. R., and H. P. Schwan (1986). Dielectric properties of tissues. In C. Polk and E. Postow (eds.), *CRC Handbook of Biological Effects of Electromagnetic Fields*, CRC Press Inc, Boca Raton, FL, pp. 27–96.

Fozzard, H. A. (1979). Conduction of the action potential. In R. M. Berne, N. Sperelakis, and S. Geiger (eds.). *The Cardiovascular System. Handbook of Physiology*, vol. I, sec. 2, American Physiological Society, Bethesda, MD, pp. 335–356.

Fozzard, H. A., and M. F. Arnsdorf (1986). Cardiac electrophysiology. In H. A. Fozzard, E. Haber, R. B. Jenings, A. M. Katz, and H. E. Morgan (eds.), *The Heart and Cardiovascular System*, Raven Press, New York, pp. 1–30.

Fozzard, H. A., and M. Schoenberg (1972). Strength-duration curves in cardiac Purkinje fibers: Effects of liminal length and charge distribution. *J. Physiol.* 226: 593–618.

Frankenhaeuser, B., and A. F. Huxley (1952). A quantitative description of membrane currents and its application to conduction and excitation in nerve. *J. Physiol.* 117: 500–544.

Frankenhaeuser, B., and A. F. Huxley (1964). The action potential in the myelinated nerve fiber of Xenopus laevis as computed on the basis of voltage clamp data. *J. Physiol.* 171: 302–315.

Freiberger, H. (1934). *Der elektrische Widerstand des menschlichen Körpers gegen technischen Gleich-und Wechselstrom*, Springer-Verlag, Berlin. Translation TR 79-45. The electrical resistance of the human body to commercial direct and alternating currents. Translated from German by Allen Translation Service, Maplewood, NJ, Item no. 9005, Published by Bell Laboratories, Maplewood, NJ.

Friedli, W. G., and M. Meyer (1984). Strength-duration curve: A measure for assessing sensory deficit in peripheral neuropathy. *J. Neurol. Psychiat.* 47: 184–189.

Furman, S., B. Parker, and D. J. W. Escher (1967). Endocardial electrical threshold of human cardiac response as a function of electrode surface area. *Digest 7th Int. Conf. Med. and Biol. Eng.*, Stockholm, Aug. 14–19, p. 71.

Furman, S., J. Garvey, and P. Hurzeler (1975). Pulse duration variation and electrode size as factors in pacemaker longevity. *J. Thorac. Cardiovas. Surg.* 69(3): 382–389.

Galvani, L. (1791). *Commentary on the Effect of Electricity on Muscular Motion*, translated by R. M. Green. Waverly Press, Baltimore, MD, 1953.

Gandhi, O. P., J. F. DeFord, and H. Kanai (1984). Impedance method for calculation of power deposition patterns in magnetically induced hyperthermia. *IEEE Trans. Biomed. Eng.* BME-31(10): 644–651.

Garrey, W. E. (1914). The nature of fibrillary contractions. Its relation to tissue mass and form. *Am. J. Physiol.* 33: 397–402.

Gauger, J. R. (1985). Household appliance magnetic field survey. *IEEE Trans. Pwr. Apparat. Sys.* PAS-104(9): 2436–2444.

Gaylor, D. C. (1989). Physical mechanism of cellular injury in electrical injury. In R. C. Lee (ed.), *Electric Trauma: Biophysical Mechanisms of Tissue Injury and Clinical Concepts*, Chicago, Ill., July 13–14.

Gaylor, D. C., K. Prakah-Asante, and R. C. Lee (1988). Significance of cell size and tissue structure in electrical trauma. *J. Theor. Biol.* 133: 223–237.

Geddes, L. A. (1972). *Electrodes and the Measurement of Bioelectric Events.* Wiley-Interscience, New York.

Geddes, L. A. (1985). The conditions necessary for the electrical induction of ventricular fibrillation. In J. E. Bridges, G. L. Ford, I. A. Sherman, and M. Vainberg (eds.),

Electrical Shock Safety Criteria, Pergamon Press, New York, pp. 45–59.

Geddes, L. A. (1987). Optimal stimulus duration for extracranial cortical stimulation. *Neurosurgery.* 20: 94–99.

Geddes, L. A., and H. Antoni (panel chairmen) (1985). Panel meeting on physiology of electrical shocks. In J. E. Bridges, G. L. Ford, I. A. Sherman, and M. Vainberg (eds.), *Electrical Shock Safety Criteria*, Pergamon Press, New York, pp. 89–112.

Geddes, L. A., and L. F. Baker (1967). The specific resistance of biological material—A compendium for the biomedical engineer and physiologist. *Med. Biol. Eng.* 5: 271–293.

Geddes, L. A., and L. E. Baker (1971). Response to passage of electric current through the body. *J. Assoc. Adv. Med. Instrum.* 5(1): 13–18.

Geddes, L. A., and L. E. Baker (1975). *Principles of Applied Biomedical Instrumentation*, 2nd ed. John Wiley, New York.

Geddes, L. A., and J. D. Bourland (1985). Tissue stimulation: Theoretical considerations and practical applications. *Med. Biol. Eng. Comp.* 23: 131–137.

Geddes, L. A., L. E. Baker, A. G. Moore, and T. W. Coulter (1969). Hazards in the use of low frequencies for the measurement of physiological events by impedance. *Med. Biol. Eng.* 7: 289–296.

Geddes, L. A., L. E. Baker, P. Cabler, and D. Brittain (1971). Response to passage of sinusoidal current through the body. In R. N. Wolfson and A. Sances (eds.), *The Nervous System and Electric Currents*, Vol. II, Plenum Press, New York.

Geddes, L. A., N. Morehouse, and A. Surawiez (1972). Effect of premature depolarization on the duration of action potentials in Purkinje and ventricular fibers of moderator band of the pig heart. *Circ. Res.* 30: 55–66.

Geddes, L. A., P. Cabler, A. G. Moore, J. Rosborough, and W. A. Tacker (1973). Threshold 60-Hz current required for ventricular fibrillation in subjects of various body weights. *IEEE Trans. Biomed. Eng.* BME-20: 465–468.

Geddes, L. A., M. J. Niebauer, C. F. Babbs, and J. D. Bourland (1985). Fundamental criteria underlying the efficacy and safety of defibrillating current waveforms. *Med. Biol. Eng. Comp.* 23: 122–130.

Geddes, L. A., J. D. Bourland, and G. Ford (1986). The mechanism underlying sudden death from electric shock. *Med. Instrum.* 20(6): 303–315.

Geddes, L. A., W. D. Voorhees, J. D. Bourland, and C. E. Riscili (1990). Optimum stimulus frequency for contracting the inspiratory muscles with chest-surface electrodes to produce artificial respiration. *Annals Biomed. Eng.*, 18: 103–108.

Geldard, F. A. (1972). *The Human Senses*. John Wiley, New York.

Gescheider, G. A., and J. H. Wright (1968). Effects of sensor adaptation on the form of the psychophysical magnitude function for cutaneous vibration. *J. Exp. Psychol.* 77: 308–313.

Gibson, R. H. (1968). Electrical stimulation of pain and touch. In D. R. Kenshalo (ed.), *The Skin Senses*, Charles C Thomas, Springfield, IL, pp. 223–261.

Gilbert, T. C. (1939). What is a safe voltage? *The Electrical Review* 119(3062): 145–146.

Girvin, J. P., L. E. Marks, J. L. Antunes, D. O. Quest, M. D. O'Keefe, P. Ning, and W. H. Dobelle (1982). Electrocutaneous stimulation—I. The effects of stimulus parameters on absolute threshold. *Perception & Psychophys.* 32(6): 524–528.

Glenn, W. W. L., and M. L. Phelps (1985). Diaphragm pacing by electrical stimulation of the phrenic nerve. *Neurosurgery* 17: 974–984.

Goff, G. D., B. S. Rosner, T. Detre, and D. Kennard (1965). Vibration perception in normal man and medical patients. *J. Neurol. Neurosurg. Psychiatr.* 18: 503–509.

Gorman, P. H., and J. T. Mortimer (1983). The effect of stimulus parameters on the recruitment characteristics of direct nerve stimulation. *IEEE Trans. Biomed. Eng.* BME-30(7): 407–414.

Gracely, R. H., R. Dubner, and P. A. McGrath (1979). Narcotic analgesia: Fentanyl reduces the intensity but not the unpleasantness of painful tooth pulp sensations, *Science.* 203(3): 1261–1263.

Grandjean, P. A., and J. T. Mortimer (1986). Recruitment properties of monopolar and bipolar epimysial electrodes. *Annals Biomed. Eng.* 14: 53–66.

Graupe, D. (1989). EMG pattern analysis for patient-responsive control of FES in paraplegics for walker-supported walking. *IEEE Trans. Biomed. Eng.* 36(7): 711–719.

Gray, H. (1985). *Gray's Anatomy*, 30th ed., C. D. Clemente (ed.), Lea & Febiger, Philadelphia.

Gray, H. (1989). *Gray's Anatomy*, 37th ed., P. L. Williams, R. Warwick, M. Dyson, and L. H. Bannister (eds.), Churchill Livingstone, New York.

Grayson, A. C. (1931). Treat low voltages with respect. *National Safety News.* 23: 32, 34, 64.

Green, B. G. (1977). The effect of skin temperature on vibrotactile sensitivity. *Perception & Psychophys.* 21: 243–248.

Green, D. M. (1960). Psychoacoustics and detection theory. *J. Acoust. Soc. Am.* 32: 1189–1203.

Green, D. M., and J. A. Swets (1966). *Signal Detection Theory and Psychophysics.* John Wiley, New York.

Green, H. L., E. B. Rafferty, and I. C. Gregory (1972). Ventricular fibrillation threshold of healthy dogs to 50 Hz current in relationship to earth leakage currents of electromedical equipment. *Biomed. Eng.*, 7: 408–414.

Green, H. L., J. Ross, and P. Kurn (1985). Danger levels of short (1 msec. to 15 msec.) electrical shocks from 50-Hz supply. In J. F. Bridges, G. L. Ford, I. A. Sherman, and M. Vainberg (eds.), *Electrical Shock Safety Criteria*, Pergamon Press, New York, pp. 259–272.

Greenberg, A. W. (1940). Experimental radiological observations on the action of electrical current upon the respiratory and circulatory organs. I. Respiratory organs. *J. Ind. Hyg. Toxicol.* 22: 104–110.

Grimnes, S. (1983a). Dielectric breakdown of human skin *in vivo*. *Med. Biol. Eng. Comp.* 21: 379–381.

Grimnes, S. (1983b). Electrovibration, cutaneous sensation of microampere currents. *Acta Physiol. Scand.* 118(1): 19–25.

Guinard, J. P., R. Chiolero, E. Buchser, A. Delaloye-Bischof, M. Payot, A. Grbic, S. Krupp, and J. Freeman (1987). Myocardial injury after electrical burns: Short and long term study. *Scand. J. Plast. Reconstr. Surg. Hand Surg.* 21(3): 301–302.

Gumbel, E. J. (1958). *Statistics of Extremes.* Columbia University Press, New York.

Gybels, J., H. O. Handwerker, and J. N, Hees (1979). A comparison between the discharges of human nociceptive nerve fibers and the subject's ratings of his sensations. *J. Physiol.* 292: 193–206.

Haberal, M. (1986). Electrical burns: A five-year experience—1985 Evans lecture. *J. Trauma* 26(2): 103–109.

Hahn, J. F. (1958). Cutaneous vibratory thresholds for square-wave electrical pulses. *Science* 127: 879–880.

Hallgren, R. (1973). Inductive Neural Stimulator. *IEEE Trans. Biomed. Eng.* BME-20(6): 470–472.

Hammond, J. S., and C. G. Ward (1988). High-voltage electrical injuries: Management and outcome of 60 cases. *South. Med. J.* 81(11): 1351–1352.

Han, J., and G. K. Moe (1964). Nonuniform recovery of excitability in ventricular muscle. *Circ. Res.* 14: 44–60.

Han, J., G. DeJalon, and G. K. Moe (1966). Fibrillation threshold of premature ventricular responses. *Circ. Res.* 18: 18–25.

Handa, Y., N. Hoshimiya, Y. Iguchi, and T. Oda (1989). Development of percutaneous intramuscular electrode for multichannel FES system. *IEEE Trans. Biomed. Eng.* 36(7): 705–710.

Hardy, J. D., H. G. Wolff, and H. Goodbell (1952). *Pain Sensations and Reactions.* Hafner, New York.

Hare, P. R. (1968). Detection threshold for electric shock in psychopaths. *J. Abnormal Psychol.* 73(3): 268–272.

Harkins, S. W., and C. R. Chapman (1976). Detection and decision factors in pain perception in young and elderly men. *Pain* 2: 253–264.

Harkins, S. W., and C. R. Chapman (1977). The perception of induced dental pain in young and elderly women. *J. Gerontol.* 32: 428–435.

Harkness, R. D. (1971). Mechanical properties of skin in relation to its biological function and its chemical components. In H. R. Elden (ed.), *Biophysical Properties of the Skin.* Wiley-Interscience, New York, pp. 393–436.

Harris, R. (1971). Chronaxy. In S. Licht (ed.), *Electrodiagnosis and Electromyography,* 3rd ed., E. Licht, New Haven, CT, chap. 9.

Hart, W. F. (1985). A five-part resistor-capacitor network for measurement of voltage and current levels related to electric shock and burns. In J. E. Bridges, G. L. Ford, I. A. Sherman, and M. Vainberg (eds.), *Electrical Shock Safety Criteria,* Pergamon Press, New York, pp. 183–192.

Hauf, G., K. Haap, M. Lay, and H. Antoni (1977). Beziehungen zwischen der Richtung des Stromdurchgangs und der Flimmerschwelle bei perfundierten Tierherzen. In R. Hauf (ed), *Beiträge Zur Ersten Hilfe,* Forschungsstelle fur Elektropathologie, Freiburg i. Br., pp. 146–163.

Hauf, R. (1986). *Beiträge zur Ersten Hilfe und Behandlung von Unfällen durch elektrischen Strom,* Proceedings of conference on electropathology, Sept. 11–13, 1986, Forschüngsstelle für Elektropathologie, Freiburg, Germany.

Hawkes, G. R. (1962). Effect of skin temperature on absolute threshold for electrical current. *J. Appl. Physiol.* 17: 110–112.

Hawkes, G. R., and J. S. Warm (1960). The sensory range of electrical stimulation of the skin. *Am. J. Psychol.* 73: 485–487.

Heckmann, J. R. (1972). Excitability curve: A new technique for assessing human peripheral nerve excitability *in vivo. Neurology* 22: 224–230.

Heft, M. (1982). Conjoint measurement analysis of verbal category judgments for electrocutaneous stimulation. Ph.D. dissertation, American University, Washington, DC.

Heinz, M., and F. Lippay (1928). Über die beziehungen zwischen der unterscheidsempfindlichkeit und der zahl der erregten sinneselemente: I. *Pflugers Arch. ges. Physiol. Menschen Tiere* 218: 437–447.

Henneman, E., and L. M. Mendell (1981). Functional organization of motoneuron pool and its inputs. In V. B. Brooks (ed.), *Handbook of Physiology, Section 1: The Nervous System. Vol. II. Motor Control, Part II,* American Physiological Society, Bethesda, MD, pp. 423–507.

Henneman, E., G. Somjen, and D. O. Carpenter (1965a). Functional significance of cell size in spinal motoneurons. *J. Neurophys.* 28: 581–598.

Henneman, E., G. Somjen, and D. O. Carpenter (1965b). Excitability and inhibitability of motoneurons of different sizes. *J. Neurophys.* 28: 599–620.

Henriques, F. C. (1947). Studies of thermal injury v. the predictibility and the significance of thermally induced rate processes leading to irreversible epidermal injury. *Arch. Pathol.,* 43: 489–502.

Henriques, F. C., and A. R. Moritz (1947). Studies of thermal injury I. The conduction of heat to and through skin and the temperatures attained therein. A theoretical and experimental investigation. *Am. J. Pathol.* 23: 531–549.

Henriquez, C. S., and R. Plonsey (1987). Effect of resistive discontinuities on waveshape and velocity in single cardiac fiber. *Med. Biol. Eng. Comput.* 25: 428–438.

Hensel, H. (1973). Cutaneous thermoreceptors.. In A. Iggo (ed.), *Handbook of Sensory Physiology: Somatosensory System,* vol. 2, Springer-Verlag, New York, pp. 79–110.

Hess, C. W., K. R. Mills, N. M. F. Murray, and T. N. Schriefer (1987). Magnetic Brain stimulation: Central motor conduction studies in multiple sclerosis. *Annals Neurol.* 22: 744–752.

Higgins, J. D., B. Tursky, and G. E. Schwartz (1971). Shock-elicited pain and its reduction by concurrent tactile stimulation. *Science* 172: 866–867.

Hill, A. V., R. S. Fullerton, B. Katz, and D. Y. Solandt (1937). Nerve excitation by alternating current. *Proc. R. Soc. London, Ser. B* 121: 74–132.

Hille, B. (1984). *Ionic Channels in Excitable Membranes*, Sinauer Associates, Sunderland, MA.

Hodgkin, A. L., and A. F. Huxley (1952). A quantitative description of membrane current and its application to conduction and excitation in nerve. *J. Physiol.* 117: 500–544.

Hodgkin, A. L., and B. Katz (1949). The effect of temperature on the electrical activity of the giant axon of the squid. *J. Physiol* (London). 109: 240–249.

Hoffman, B. F., and P. F. Cranefield (1960). *Electrophysiology of the Heart.* McGraw-Hill, New York.

Hohnloser, S., S. Weirich, and H. Antoni (1982). Influence of direct current on the electrical activity of the heart and on its susceptibility to ventricular fibrillation. *Basic Res. Cardiol.* 77: 237–249.

Holle, J., M. Frey, H. Gruber, H. Kern, H. Stohr, and H. Thoma (1984). Functional electrostimulation of paraplegics: Experimental investigations and first clinical experience with an implantable stimulation device. *Orthop.* 7: 1145–1155.

Hooker, D. R., W. B. Kouwenhoven, and O. R. Langworthy (1932). The effect of alternating currents on the heart. *Am. J. Physiol.* 103: 444–454.

Hooshmand, H., F. Radfar, and E. Beckner (1989). The neurophysiological aspects of electrical injuries. *Clin. Electroencephalogr.* 20(2): 111–120.

Hoque, M., and O. P. Gandhi (1988). Temperature distributions in the human leg for VLF–VHF exposures at the ANSI-recommended safety levels. *IEEE Trans. Biomed. Eng.* 35(6): 442–449.

Hoshimiya, N., A. Naito, M. Yajima, and Y. Handa (1989). A multichannel FES system for the restoration of motor functions in high spinal cord injury patients: A respiration-controlled system for multijoint upper extremity. *IEEE Trans. Biomed. Eng.* 36(7): 754–760.

Howatson, A. M. (1965). *An Introduction to Gas Discharges.* Pergamon Press, Oxford.

HSRI (1977). Anthropometry of infants, children, and youths to age 18 for product safety design. Final Report, May 31, 1977, prepared for the U.S. Consumer Product Safety Commission by the Highway Safety Research Institute, University of Michigan, Contract CPSC-C-75-0068.

Hunt, J. L. (1979). The use of technetiun-99m stannous pyrophosphate scintigraphy to identify muscle damage in acute electric burns. *J. Trauma.* 19(6): 409–413.

Hunt, J. L., R. M. Sato, and C. R. Baxter (1980). Acute electric burns. Current diagnostic and therapeutic approaches to management. *Arch. Surg.* 115: 434–438.

Huxley, H. E., and J. Hanson (1954). Changes in the cross-atriation of muscle during contraction and stretch and their structural interpretation. *Nature* 173: 973–976.

Hylten-Cavallius, N. (1975). Certain ecological effects of high voltage power lines. Institut de recherche de l'Hydro-Quebec (IREQ), IREQ-1160.

IAEI News Bulletin (Anon.) (1940). Oregon's first death from an electric fence. Vol. 12, p. 70.

IEC (1982). Report of Technical Committee 64: Electrical installation of buildings. International Electrotechnical Commission, Geneva, Switzerland.

IEC (1984). Effects of current passing through the human body, Part 1: General Aspects. Publication 479-1, International Electrotechnical Commission, Geneva, Switzerland.

IEC (1985). Meeting Minutes TC74/WG5, Everett, Washington, April 2, 3, 4, 1985. International Electrotechnical Commission, Geneva, Switzerland, July 16, 1985.

IEC (1986). Safety of information technology equipment including electical business equipment. Publication 950 (including Amendment No. 1, Nov., 1988).

IEC (1987). Effects of current passing through the human body, Part 2: Special Aspects. Publication 479-2, International Electrotechnical Commission, Geneva, Switzerland.

IEC (1990a). Safety requirements for electrical equipment for measurement, control, and laboratory use. Publication 1010-1, International Electrotechnical Commission, Geneva, Switzerland.

IEC (1990b). Methods of measurement of touch current and protective conductor current. Report of TC74/WG5, Publication 990, International Electrotechnical Commission, Geneva, Switzerland.

IEEE (1986). Guide for safety in ac substation grounding. ANSI/IEEE Std 80-1986, Institute of Electrical and Electronics Engineers, New York.

IEEE (1990). National Electrical Safety Code. ANSI C2-1990, Institute of Electrical and Electronic Engineers, New York.

Inancsi, W., and T. L. Guidotti (1987). Occupation-related burns: Five-year experience of an urban burn center. *J. Occup. Med.* 29(9): 730–733.

Irnich, W. (1973). Physikalische Überlegungen zur Elektrostimulation (Physical consideration on electrostimulation). *Biomed. Technik.* 18: 97–104.

Irnich, W. (1980). The chronaxie time and its practical importance. *PACE*, 3: 292–301.

Irwin, D., S. Rush, R. Everling, E. Lepeschkin, D. Bruce Montgomery, and R. J. Weggel (1970). Stimulation of cardiac muscle by a time-varying magnetic field. *IEEE Trans. Magnet.* MAG-6 (2): 321–322.

ITT (1979). *Reference Data for Radio Engineers.* Howard W. Sams, New York.

Jack, J. J. B., D. Noble, and R. W. Tsien (1983). *Electric Current Flow in Excitable Cells.* Clarendon Press, Oxford.

Jackson, J. D. (1962). *Classical Electrodynamics*, John Wiley, New York.

Jackson, T. A., and B. F. Riess (1934). Electric shock with different size electrodes. *J. Gen. Psychol.* 45: 262–266.

Jacobsen, J., S. Buntenkutter, and H. J. Reinhard (1975). Experimentelle untersuchungen an schweinen zur frage der mortalität durch sinusformige, phasengeschnittene sowie gleichgerichtete elektrische ströme. *Biomed. Technik* 20: 99–107.

Jaeger, R. J., G. M. Yarkony, and R. M. Smith (1989). Standing the spinal cord injured patient by electrical stimulation: Refinement of a protocol for clinical use. *IEEE Trans. Biomed. Eng.* 36(7): 720–728.

Jalife, J., and G. K. Moe (1976). Effect of electrotonic potentials on pacemaker activity of canine Purkinje fibers in relation to parasystole. *Circ. Res.* 39: 801–808.

Jalife, J., and G. K. Moe (1981). Excitation, conduction, and reflection of impulses in isolated bovine and canine cardiac Purkinje fibers. *Circ. Res.* 49: 233–247.

Janse, M. J., F. J. L. Van Capelle, H. Morsink, A. G. Kleber, F. Wilms-Schopman, R. Cardinal, C. Naumann D'Alnoncourt, and D. Durrer (1980). Flow of "injury" current and patterns of excitation during early ventricular arrhymias in acute regional myocardial ischemia in isolated porcine and canine hearts. *Circ. Res.* 47: 151–165.

Janzekovic, Z. (1975). The burn wound from the surgical point of view. *J. Trauma* 15: 42.

Jex-Blake, A. J. (1913) *Br. Med. J.* 1: 425.

Johna, R. (1989). Elektrophysiologische Eigenschaften gekoppelter Herzmuskelzellen in Zellkultur. Dissertation, Universität Freiburg i. Br., Germany.

Johnston, F. E., and R. M. Malina (1966). Age changes in the composition of the upper arm in Philadelphia children. *Human Biol.* 38: 1–21.

Jones, M., and L. A. Geddes (1977). Strength-duration curves for cardiac pacemaking and ventricular fibrillation. *Cardiovasc. Res. Bull.* 15(4): 101–112.

Jones, R. O., J. C. Wright, W. T. Jones, and A. Berger (1983). A case study of high voltage electrical injury. *J. Am. Podiatry Assoc.* 73: 638–642.

Joyner, R. W., F. R. Ramon, and W. Moore (1975). Simulation of action potential propagation in an inhomogeneous sheet of electrically coupled excitable cells. *Circ. Res.* 36: 654–661.

Kalmijn, A. J. (1990). Transduction of nanovolt signals: Limits of electric-field detection. *Bioelectromagnetics Soc. Newsl.* Jan./Feb. issue, no. 92, pp. 1–7.

Kandel, E. R., and J. B. Schwartz (1981). *Principles of Neural Science.* Elsevier/North-Holland, New York.

Kanitkar, S., and A. H. Roberts (1988). Paraplegia in an electrical burn: A case report. *Burns Incl. Therm. Ini.* 14(1): 49–50.

Kaplan, E. B. (1984). *Kaplan's Functional and Surgical Anatomy of the Hand*, 3rd ed. Morton Spinner, Philadelphia.

Katims, J. J., E. H. Naviasky, M. S. Randell, K. Y. Lorenz, and M. L. Bleecker (1987). Constant current sinewave transcutaneous nerve stimulation for the evaluation of peripheral neuropathy. *Archives Phys. Med. Rehab.* 68: 210–213.

Kato, M., S. Ohta, T. Kobayashi, and G. Matsumoto (1986). Response of sensory receptors of the cat's hindlimb to a transient, step function dc electric field. *Bioelectromagnetics.* 7: 395–404.

Kato, M., S. Ohta, K. Shimizu, Y. Ysuchida, and G. Matsumoto (1989). Detection-threshold of 50-Hz electric fields by human subjects. *Bioelectromagnetics* 10: 319–327.

Katz, B. (1939). Nerve excitation by high-frequency alternating current. *J. Physiol.* 96: 202–224.

Katz, B. (1966). *Nerve, Muscle, and Synapse.* McGraw-Hill, New York.

Kaune, W. T. (1981). Power frequency electric fields averaged over the body surfaces of grounded humans and animals. *Bioelectromagnetics* 2: 403–406.

Kaune, W. T., and W. C. Forsythe (1985). Current densities measured in human models exposed to 60-Hz electric fields. *Bioelectromagnetics* 6(1): 13–32.

Kaune, W. T., and M. F. Gills (1981). General properties of the interaction between animals and ELF electric fields. *Bioelectromagnetics* 2: 1–11.

Kaune, W. T., and R. D. Phillips (1980). Comparison of grounded humans, swine and rats to vertical, 60-Hz electric fields. *Bioelectromagnetics* 1: 117–129.

Kaune, W. T., R. G. Stevens, N. J. Callahan, R. K. Steverson, and D. B. Thomas (1987). Residential magnetic and electric fields. *Bioelectromagnetics* 8: 315–335.

Kenshalo, D. R. (1979). Aging effects on cutaneous and kinesthetic sensibilities. In S. S Han and D. H. Coon (eds.), *Special Senses in Aging: A Current Biological Assessment*, Institute of Gerontology, University of Michigan, Ann Arbor, MI, pp. 189–217.

Kiang, N. Y. S. (1965). *Discharge patterns of single fibers in the cat's auditory nerve.* M.I.T. Press, Cambridge, MA.

Kieback, D. (1988). International comparison of electrical accident statistics. *J. Occup. Accidents* 10: 95–106.

Kinnunen, E., M. Ojala, H. Taskinen, and E. Matikainen (1988). Peripheral nerve injury and Raynaud's syndrome following electric shock. *Scand. J. Work Environ. Health* 14 (5): 332–333.

Kirk, R. E. (1982). *Experimental design: Procedures for the behavioral sciences.* 2nd ed. Brooks/Cole, Belmont, CA.

Kiselev, A. P. (1963). Threshold values of safe current at commercial frequency (in Russian). *Vopr. Elektoborud, Elekt-snabzh, i Elekt. Izmerenii Sob. MITT* 17: pp. 47–58. CEGB Information Services, Translation no. 1167.

Kloss, D. A., and E. L. Carstensen (1982). Effects of ELF electric fields on isolated frog heart. *IEEE Trans. Biomed. Eng.* BME-30 (6): 347–348.

Klump, D., and M. Zimmerman (1980). Irreversible differential block of A- and C-fibers following local nerve heating in the cat. *J. Physiol.* (London) 298: 471–482.

Knickerbocker, G. G. (1973). Fibrillating parameters of direct and alternating (20-Hz) currents separately and in combination—An experimental study. *IEEE Trans. Comm.* COM-21 (9): 1015–1027.

Koniarek, J. P. (1989). Mechanical and electrical effects of high-frequency and high-intensity stimulation of muscle. *Bioelectromagnetics* 10: 335–345.

Koning, G., H. Schneider, and A. J. Hoelen (1975). Amplitude-duration relation for direct ventricular defibrillation with rectangular current pulses. *Med. Biol. Eng.*, May, pp. 388–395.

Kopeliowitch, J. (1946). The physiological effects of an alternating current and the danger of shock in ac electrical installations. *Assoc. Eng. Architects Palestine J.* (Tel Aviv, Palestine) 7 (6): 2–8.

Kouwenhoven, W. B. (1949). Effects of electricity on the human body. *Elec. Eng*, 68: 199–203.

Kouwenhoven, W. B. (1956). Effect of capacitor discharges on the heart. *AIEE Trans. Pwr Apparat. Sys.* 75, part 3, no. 23.

Kouwenhoven, W. B. (1964). The effects of electricity on the human body. *Bull. Johns Hopkins Hosp.* 114: 425.

Kouwenhoven, W. B., and O. R. Langworthy (1931). Effects of electric shock—II. *Trans. AIEE* 50: 1165–1171.

Kouwenhoven, W. B., D. R. Hooker, and O. R. Langworthy (1932). Heart injury from electric shock. *Trans. AIEE* 242.

Kouwenhoven, W. B., D. R. Hooker, E. L. Lotz (1936). Electric shock effects of frequency. *AIEE Trans.* 55: 384–386.

Kouwenhoven, W. B., et al. (1957). The effects of high-voltage, low capacitance electrical discharges in the dog. *IRE Trans. Med. Electron.* PGME 11: 41.

Kouwenhoven, W. B., and W. R. Milnor (1958). The effects of high-voltage, low-capacitance electrical discharges in the dog. *IRE Trans. Biomed. Eng.* PGME 11: 41–45.

Kouwenhoven, W. B., G. G. Knickerbocker, R. W. Chestnut, W. R. Milnor, and D. J. Sass (1959). AC shocks of varying parameters affecting the heart. *Trans. AIEE* 73, part III, pp. 163–169.

Kralj, A., T. Bajd, R. Turk, J. Krajnik, and H. Benko (1983). Gait restoration in paraplegic patients: A feasibility demonstration using multichannel surface electrode FES. *J. Rehabil. R & D* 20: 3–20.

Kralj, A., T. Bajd, R. Turk, and H. Benko (1986). Posture switching for prolonging functional electrical stimulation standing in paraplegic patients. *Paraplegia* 24: 221–230.

Kugelberg, J. (1976). Electrical induction of ventricular fibrillation in the human heart. *Scand. J. Cardiovasc. Surg.* 10: 237–240.

LaCourse, J. R., W. T. Miller, M. Vogt, and S. M. Selibowitz (1985). Effect of high-frequency current on nerve and muscle tissue. *IEEE Trans. Biomed. Eng.* BME-32 (1): 82–86.

Lamb, J. F., C. G. Ingram, I. A. Johnston, and R. M. Pitman (1984). *Essentials of Physiology.* Blackwell, Oxford.

LaMotte, R. H., and J. N. Campbell (1978). Comparison of warm and nociceptive C-fiber afferents in monkey with human judgments of thermal pain. *J. Neurophysiol.* 41 (6): 509–528.

LaMotte, R. H., H. E. Torebjörk, C. J. Robinson, and J. G. Thalhammer (1984). Time-intensity profiles of cutaneous pain in normal and hyperalgesic skin: A comparison with C-fiber nociceptor activities in monkey and human. *J. Neurophysiol.* 51 (6): 1434–1450.

Lane, J. F., and T. J. Zebo (1967). Volume potential fields developed in cats' limbs during the passage of constant current pulses. *Digest 7th Int. Conf. Med. and Biol. Eng.*, Stockholm, Aug. 14–19, p. 207.

Lapicque, L. (1907). Recherches quantitatives sur l'excitation électrique des nerfs traitée comme une polarization. *J.Physiol. Paris* 9: 620–635.

Larkin, W. D., and J. P. Reilly (1984). Strength/duration relationships for electrocutaneous sensitivity: Stimulation by capacitive discharges. *Perception Phychophys.* 36 (1): 68–78.

Larkin, W. D., and J. P. Reilly (1986). Electrocutaneous sensitivity: Effect of skin temperature. *Somatosensory Res.* 3 (3): 261–271.

Larkin, W. D., J. P. Reilly, and L. B. Kittler (1986). Individual differences in sensitivity to transient electrocutaneous stimulation. *IEEE Trans. Biomed. Eng.* BME-33 (5): 494–504.

Lazzara, R., N. El-Sherif, and B. J. Scherlag (1975). Disorders of the cellular electrophysiology produced by ischemia of the canine His bundle. *Circ. Res.* 36: 444–454.

Lee, R. C., and M. S. Kolodney (1987a). Electrical injury mechanisms: Electrical breakdown of cell membranes. *Plast. Reconstr. Surg.* 80 (5): 672–679.

Lee, R. C., and M. S. Kolodney (1987b). Electrical injury mechanisms: Dynamics of the thermal response. *Plast, Reconstr. Surg.* 80 (5): 663–671.

Lee, R. C., D. C. Gaylor, D. Bhatt, and D. A. Israel (1988). Role of cell membrane rupture in the pathogenesis of electrical trauma. *J. Surg. Res.* 44 (6): 709–719.

Lee, W. R. (1961). A clinical study of electrical accidents. *B. J. Ind. Med.* 18: 260–269.

Lee, W. R. (1964). Electrophysiology. In *Proc. Int. Symp. on Electrical Accidents,* International Labour Office, Geneva, Switzerland, Chap. 2.

Lee, W. R. (1966). Death from electric shock. *Proc. IEEE* 111 (1): 144–148.

Lee, W. R. and S. Zoledziowski (1964). Effects of electric shock on respiration in the rabbit. *B. J. Ind. Med.* 21: 135–144.

Levitt, H. (1971). Transformed up-down methods in psychoacoustic. *J. Acoust. Soc. Am.* 49: 467–477.

Levy, W. J. (1987). Clinical experience with motor and cerebellar evoked potential monitoring. *Neurosurgery* 20: 169–182.

Lewis, T. H., H. S. Feil, and W. D. Stroud (1920). Observations upon flutter and fibrillation. *Heart* 7: 191–233.

Leyden, J. G. (1990). Death in the hot seat: A century of electrocutions. *The Washington Post,* Aug. 5, p. D5.

Li, C. L., and A. Bak (1976). Excitability characteristics of the A- and C-fibers in a peripheral nerve. *Exp. Neurol.* 50: 67–79.

Liberson, W. T. (1971). Progressive and alternating currents. In S. H. Licht (ed.), *Electrodiagnosis and Electromyography,* E. Licht, New Haven, CT, pp. 272–285.

Libet, B., Alberts, W. W., Wright, E. W., Jr., DeLattre, L., Levin, G., and Feinstein, V. (1964). Production of threshold levels of conscious sensation by electrical stimulation of human somatosensory cortex. *J. Neurophysiol.* 27: 546–578.

Licht, S. H. (1971). History of electrodiagnosis. In S. H. Licht (ed.), *Electrodiagnosis and Electromyography,* E. Licht, New Haven, CT, pp. 1–23.

Lindermans, F. W., and J. J. Danier van der Gon (1978). Current thresholds and liminal size in excitation of heart muscle. *Cardiovasc. Res.* 12: 477–485.

Lochner, J. P. A., and J. F. Burger (1961). Form of the loudness function in the presence of masking noise. *J. Acoust. Soc. Am.* 33: 1705–1707.

Lord, F. M., and M. R. Novick (1968). *Statistical Theories of Mental Test Scores.* Addison-Wesley, Reading, MA.

Lövsund, P., and P. A. Öberg (1982). ELF magnetic fields in electrosteel and welding industries. *Radio Sci.* 17 (55): 355–385.

Lövsund, P., P. A. Öberg, S. A. Nilson, and T. Reuter (1980a). Magnetophosphenes: a quantitative analysis of thresholds. *Med. Biol. Eng. Comput.* 18: 326–334.

Lövsund, P., P. A. Öberg, and S. E. Nilson (1980b). Magneto- and electrosphosphenes: A comparative study. *Med. Biol. Eng. Comput.* 18: 785–764.

Luce, E. A., and S. E. Gottlieb (1984). True high-tension electrical injuries. *Ann. Plast. Surg.,* 12: 321–326.

Luce, E. A., W. L. Dowden, C. T. Su, and J. E. Hoopes (1978). High tension electrical injury of the upper extremity. *Surg. Gynecol. Obstet.* 147: 38.

Lundborg, G. L. (1988). *Nerve Injury and Repair.* Churchill Livingstone, New York.

Lykken, D. T. (1971). Square-wave analysis of skin impedance. *Psychophysiology* 7 (2): 262–275.

Maccabee, P. J., V. E. Amassian, R. Q. Cracco, and J. A. Cadwell (1988a). Analysis of peripheral motor stimulation in humans using the magnetic coil. *EEG Clin. Neurophysiol.* 70: 524–533.

Maccabee, P. J., V. E. Amassian, R. Q. Cracco, J. B. Cracco, and B. J. Anziska (1988b). Intracranial stimulation of facial nerve in humans with the magnetic coil. *EEG Clin. Neurophysiol.* 70: 350–354.

Malina, R. M. (1975). *Growth and Development: The First Twenty Years in Man.* Burgess Publishing Co., Minneapolis, MN.

Marks, L. E. (1974). *Sensory Processes: The New Psychophysics.* Academic Press, New York.

Marsolais, E. B., and R. Kobetic (1987). Functional electrical stimulation for walking in paraplegia. *J. Bone Joint Surg.* 69A: 728–733.

Martin, J. H. (1985). Receptor physiology and submodality coding in the somatic sensory system. In E. R. Kandel and J. H. Schwartz (eds.), *Principles of Neural Science,* 2nd edition, Elsevier, New York.

Mason, J. L., and N.A.M. Mackay (1976). Pain sensations associated with electrocutaneous stimulation. *IEEE Trans. Biomed. Eng.* BME-23 (5): 405–409.

Massello, W., 3d (1988). Lightning deaths. *Med. Leg. Bull.* 37 (1): 1–7.

McAllister, R. E., D. Noble, and R. W. Tsien (1975). Reconstruction of the electrical activity of cardiac Purkinje fibres. *J. Physiol.* 251: 1–59.

McCarroll, G. D., and B. A. Rowley (1979). An investigation of the existence of electrically located acupuncture points. *IEEE Trans. Biomed. Eng.* BME-26 (3): 177–181.

McConville, J. T., T. D. Churchill, I. Kaleps, C. E. Clauser, and J. Cuzzi (1980). *Anthropometric Relationships of Body and Body Segment Moments of Inertia.* Air Force Aerospace Medical Research Lab., Wright-Patterson Air Force Base, Ohio.

McCreery, D. B., and W. F. Agnew (1990). Mechanisms of stimulation-induced neural damage and their relation to guidelines for safe stimulation. In W. F. Agnew and D. B. McCreery (eds.), *Neural Prostheses: Fundamental Studies,* Prentice-Hall, Englewood Cliffs, NJ, pp. 297–317.

McNeal, D. R. (1976). Analysis of a model for excitation of myelinated nerve. *IEEE,Trans. Biomed, Eng.* BME-23: 329–337.

McNeal, D. R. (1977). 2000 years of electrical stimulation. In T. F. Hambrecht and J. B. Reswick (eds.), *Functional Electrical Stimulation.* Marcel Dekker, New York, pp. 3–35.

McNeal, D. R., and L. L. Baker (1988). Effects of joint angle, electrodes and waveform in electrical stimulation of the quadriceps and hamstrings. *Annals Biomed. Eng.* 16: 299–310.

McNeal, D. R., and B. R. Bowman (1985a). Peripheral neuromuscular stimulation. In J. B. Mykleburst, J. F. Cusick, A. Sances, and S. J. Larsons (eds.), *Neural Stimulation,* vol. II, CRC Press, Boca Raton, FL, pp. 95–118.

McNeal, D. R., and B. R. Bowman (1985b). Selective activation of muscles using peripheral nerve electrodes. *Med. Biol. Eng. Comput.* 23: 249–253.

McNeal, D. R., and D. A. Teicher (1977). Effect of electrode placement on threshold and initial site of excitation of a myelinated nerve fiber. In T. F. Hambrecht and J. B. Reswick (eds.), *Functional Electrical Stimulation,* Marcel Decker, New York: 405–412.

McNeal, D. R., B. R. Bowman, and W. L. Momsen (1973). Peripheral block on motor activity. In M.Gavrilovic and A. B. Wilson (eds.), *Advances in External Control of Human Extremities,* Yugoslav Committee for Electronics and Automation, Belgrade, pp. 575–583.

McNeal, D. R., R. Waters, and J. Reswick (1977). Experience with implanted electrodes. *Neurosurgery,* 1: 228–229.

McRobbie, D., and M. A. Foster (1984). Thresholds for biological effects of time-varying magnetic fields. *Clin. Phys. Physiol. Measurements* 5 (2): 67–78.

McRobbie, D., and M. A. Foster (1985). Cardiac response to pulsed magnetic fields with regard to safety in NMR imaging. *Phys. Med. Biol.* 30 (7): 695–702.

Melzak, R., and P. D. Wall (1965). Pain mechanisms: A new theory. *Science* 150 (3699): 971–979.

Merton, P. A., and H. B. Morton (1986). A magnetic stimulation for the human motor cortex. *J. Physiol.* 381: 10P.

Meyer, R. A., and J. N. Campbell (1981). Myelinated nociceptive afferents accounts for the hyperalgesia that follows a burn to the hand. *Science* 213: 1527–1529.

Mills, K. R., and N. M. F. Murray (1985). Corticospinal tract conduction time in multiple sclerosis. *Annals Neurol.* 18: 601–610.

Mines, G. R. (1914). On circulating excitations in the heart muscle and their possible relation to tachycardia and fibrillation. *Trans. R. Soc. Can.* 8: 43.

Mogul, D. J., N. V. Thakor, J. R. McCullogh, G. A. Meyers, R. E. Teneick, and D. H. Siniger (1984). Modified Beeler-Reuter model yields improved simulation of myocardial action potentials. *Proc. IEEE Conf., Computers in Cardiology*, Salt Lake City, Utah, Sept. 18–21, pp. 159–162.

Moncrief, J. A., and B. A. Pruitt, Jr. (1971). Hidden damage from electrical injury. *Geriatrics* 26 (4): 84–85.

Monster, A. W., and H. Chan (1977). Isometric force produced by motor units of extensor digitorum communis muscle in man. *J. Neurophysiol.* no. 40: 1432–1443.

Montagu, M. F. A. (1960). *A Handbook of Anthropometry.* Charles C Thomas, Springfield, IL.

Moriarty, B. J., and J. N. Char (1987). Electrical injury and cataracts—An unusual case," *West Indian Med. J.* 36(2): 114–116.

Mortimer, J. T. (1981). Motor prostheses. In J. M. Brookhart, V. B. Mountcastle, V. B. Brooks, and S. R. Geiger (eds.), *Handbook of Physiology, Section 1: The Nervous System. Vol. II. Motor Control, Part I*, Am. Physiol. Soc., Bethesda, MD.

Morton, D. J. (1944). *Manual of Human Cross Section Anatomy.* Williams & Wilkins, Baltimore.

Morton, R., and K. A. Provins (1960). Finger numbness after acute local exposure to cold. *J. Appl. Physiol.* 15: 149–154.

Moskowitz, H. R., B. Scharf, and J. C. Stevens (1974). *Sensation and Measurement.* D. Reidel, Boston.

Motz, H., and F. Rattay (1986). A study of the application of the Hodgkin-Huxley and the Frankenhaeuser-Huxley model for electrostimulation of the acoustic nerve. *Neuroscience* 18: 699–712.

Mouchawar, G. A., L. A. Geddes, J. D. Bourland, and J. A. Pearce (1989). Ability of the Lapicque and Blair strength-duration curves to fit experimentally obtained data from the dog heart. *IEEE Trans. Biomed. Eng.* 36(9): 971–974.

Mueller, E. F., R. Loeffel, and S. Mead (1953). Skin impedance in relation to pain threshold testing by electrical means. *J. Appl. Physiol.* 5: 746–752.

Myklebust, J., A. Sances, M. Chilbert, T. Prieto, and T. Swiontek (1985). Capacitive discharge studies. In J. E. Bridges, G. L. Ford, I. A. Sherman, and M. Vainberg (eds.), *Electrical Shock Safety Criteria*, Pergamon Press, New York, pp. 71–76.

Nahin, P. J. (1987). *Oliver Heaviside, Sage in Solitude.* IEEE Press, New York.

Naples, G. G., J. T. Mortimer, A. Scheiner, and J. D. Sweeney (1988). A spiral nerve cuff electrode for peripheral nerve stimulation. *IEEE Trans. Biomed. Eng.* BME-35: 905–916.

Naples, G. G., J. T. Mortimer, and T. G. H. Yuen (1990). Overview of peripheral nerve electrode design and implantation. In W. F. Agnew and D. B. McCreery (eds.), *Neural Prostheses: Fundamental Studies*, Prentice-Hall, Englewood Cliffs, NJ, pp. 107–145.

National Radiological Protection Board (1983). Revised guidance on acceptable limits of exposure during nuclear magnetic resonance clinical imaging. *Brit. J. Radiol.* 56: 974–977.

Nethken, R. T., and M. A. Bulot (1967). Threshold of electrical signals on the upper arm. *Proc. IEEE Region III Convention*, Jackson, MI, pp. 81–83.

Newman, A. L. (1984). Self-injurious behavior inhibiting system. *John Hopkins APL Tech. Digest* 5(3): 290–295.

NFPA (1990). *National Electrical Code.* ANSI/NFPA 70-1990, National Fire Protection Association.

Nicholson, P. W. (1965). Specific impedance of cerebral white matter. *Exp. Biol.* 13: 386–401.

Noble, D. (1962). A modification of the Hodgkin-Huxley equations applicable to

Purkinje fibre action and pace-maker potentials. *J. Physiol* 160: 317–353.

Noble, D. (1984). The surprising heart: A review of recent progress in cardiac electrophysiology. *J. Physiol.* 353: 1–50.

Noble, D., and S. J. Noble (1984). A model of sino-atrial node electrical activity based on a modification of the DiFrancesco-Noble (1984) equations. *Proc. R. Soc. London* B222: 295–304.

Noble, D., and R. B. Stein (1966). The threshold conditions for initiation of action potentials by excitable cells. *J. Physiol.* 187: 129–162.

Notermans, S. L. H. (1966). Measurement of the pain threshold determined by electrical stimulation and its clinical application, Part I: Method and factors possibly influencing the pain threshold. *Neurology* 16: 1071–1086.

Notermans, S. L. H. (1967). Measurement of the pain threshold determined by electrical stimulation and its clinical application, Part II: Clinical applications in neurological and neurosurgical patients. *Neurology* 17: 58–73.

Notermans, S. L. H., and M. M. W. A. Tophofff (1975). Sex differences in pain tolerance and pain apperception. In M. Weisenberg (ed.), *Pain: Clinical and Experimental Perspectives*, C. V. Mosby, St. Louis MO, pp. 111–116.

Nute, R. (1985). Dynamic aspects of body impedance. In J. E. Bridges, G. L. Ford, I. A. Sherman, and M. Vainberg (eds.), *Electrical Shock Safety Criteria*, Pergamon Press, New York.

Oester, Y. T., and S. H. Licht (1971). Routine electrodiagnosis. In S. H. Licht (ed.), *Electrodiagnosis and Electromyography*, E. Licht, New Haven, CT, pp. 201–217.

Olsen, R. W., L. J. Hayes, E. H. Wissler, H. Nikaidoh, and R. C. Eberhart (1985). Influence of hypothermia and circulatory arrest on cerebral temperature distributions. *ASME J. Biomech. Eng.* 107: 354–360.

Omura, Y. (1977). Critical evaluation of the methods of measurement of "tingling threshold" and "pain tolerance" by electrical stimulation. *Acupuncture & Electro-therapy Res. Int. J.* 2: 161–236.

Osypka, P. (1963). Quantitative investigation of current strength, duration, and routing in ac electrocution accidents involving human beings and animals. *Elektromedizen* 8, Sonderdruck Fachverlag Schiele and Schön, Berlin, Translation by SLA Translations Center, TT 66-1588 & TT-11470.

Paintal, A. S. (1967). A comparison of the nerve impulses of mammalian non-medullated nerve fibers with those of the smallest diameter medulled fibers. *J. Physiol.* 193: 523–533.

Paintal, A. S. (1973). Conduction in mammalian nerve fibrs. In J. E. Desmedt (ed.), *New Developments in Electromyography and Clinical Neurophysiology*, vol. 2, Karger, Basel, Switzerland, pp. 19–41.

Palin, W. E., Jr., A. M. Sadove, J. E. Jones, W. F. Judson, and H. D. Stambaugh (1987). Oral electrical burns in a pediatric population. *J. Oral Med.* 42(1): 17–21, 34.

Papoulis, A. (1965). *Probability, Random Variables, and Stochastic Processes*. McGraw-Hill, New York.

Parry, C. B. W. (1971). Strength-duration curves. In S. Licht (ed.), *Electrodiagnosis and Electromyography*, E. Licht, New Haven, CT, pp. 241–271.

Parshley, P. F., J. Kilgore, J. F. Pulito, P. W. Smiley, and S. H. Miller (1985). Aggressive approach to the extremity damaged by electric current. *Am. J. Surg.* 150(1): 78–82.

Pearce, J. A., J. D. Bourland, W. Neilsen, L. A. Geddes, and M. Voelz (1982). Myocardial stimulation with ultrashort duration current pulses. *PACE* 5: 52–58.

Peckham, P. H. (1983). Restoration of upper extremity function. *EMBS Magazine* 2(3): 30–32.

Peckham, P. H. (1987). Functional electrical stimulation: Current status and future prospects of applications to the neuromuscular system in spinal cord injury. *Paraplegia* 25: 279–288.

Peleska, B. (1963). Cardiac arrhythmias following condenser discharges and their dependence upon the strength of current and phase of cardiac cycle. *Circ. Res.* 13: 21–32.

Peleska, B. (1965). Cardiac arrhythmias following condenser discharges led through an inductance. *Circ. Res.* 16: 11–18.

Pennes, H. H. (1948). Analysis of issue and arterial blood temperatures in the resting human forearm. *J. Appl. Physiol.* 1: 93–122.

Peter, J. B., R. J. Barnard, V. R. Edgerton, C. A. Gillespie, and K. E. Stempel (1972). Metabolic profiles of three fiber types of skeletal muscle in guinea pigs and rabbits. *Biochemistry* 11: 2627–2633.

Pethig, R. (1979). *Dielectric and Electronic Properties of Biological Materials*. John Wiley, Chichester, U.K.

Petrofsky, J. S. (1978). Control of the recruitment and firing frequencies of motor units in electrically stimulated muscles in the cat. *Med. Biol. Eng. Comput.* 16: 302–308.

Petrofsky, J. S., and C. A. Phillips (1983). Computer controlled walking in the paralyzed individual. *J. Neurol. Orthoped. Surg.* 4: 153–164.

Pette, D., and G. Vrobova (1985). Invited review: Neural control of phenotypic expression in mammalian muscle fibers. *Muscle and Nerve* 8: 676–689.

Pfeiffer, E. A. (1968). Electrical stimulation of sensory nerves with skin electrodes for research, diagnosis, communication, and behavioral conditioning: A survey. *Med. Biol. Eng.* 6: 637–651.

Plonsey, R. (1969). *Bioelectric Phenomena*. McGraw-Hill, New York.

Plonsey, R., and R. C. Barr (1986). A critique of impedance measurements in cardiac tissue. *Annals Biomed. Eng.* 14: 307–322.

Plonsey, R., and R. C. Barr (1988). *Bioelectricity*. Plenum Press, New York.

Polk, C. (1986). Introduction. In C. Polk and E. Postow (eds.), *CRC Handbook of Biological Effects of Electromagnetic Fields*, CRC Press, Boca Raton, FL, pp. 1–24.

Polson, M. J. R., A. T. Barker, and I. L. Freeston (1982a). Stimulation of nerve trunks with time-varying magnetic fields. *Med. Biol. Eng. Comput.* 20: 243–244.

Polson, M. J. R., A. T. Barker, and S. Gardiner (1982b). The effect of rapid rise-time magnetic fields on the ECG of the rat. *Clin. Physiol. Measurements* 3: 231–234.

Ponten, B., U. Erikson, S. H. Johansson, and L. Olding (1970). New observations on tissue changes along the pathway of the current in an electrical injury. Case report. *Scand. J. Plast. Reconstr. Surg.* 4(1): 75–82.

Prieto, T., A. Sances, J. Myklebust, and M. Chilbert (1985). Analysis of cross-body impedance at household voltage levels. In J. E. Bridges, G. L. Ford, I. A. Sherman, and M. Vainberg (eds.), *Electric Shock Safety Criteria*, Pergamon Press, New York, pp. 151–160.

Procacci, B. G. (1968). A study on the cutaneous pricking pain threshold in normal man. In A. Soulairac, J. Cahn, and J. Charpentier (ed.), *Pain*, Academic Press, London.

Procacci, P., M. Zoppi, M. Maresca, and S. Romano (1974). Studies on the pain threshold in man. In J. J. Bonica (ed.), *Advances in Neurology*, vol. 4, *International Symposium on Pain*, Raven Press New York.

Project UHV (1982). *Transmission Line Reference Book, 345 kV and Above*, 2nd ed. The Electric Power Research Institute, Palo Alto, CA.

Provins, K. A., and R. Morton (1960). Tactile discrimination and skin temprature. *J. App. Physiol.* 15: 155–160.

Pruna, S., C. Ionescu-Tirogoviste, E. Popa, and I. Mincu (1989). Measurement of perception threshold to an electrical stimulus using a phase-sensitive technique in normal and diabetic subjects. *Med. Biol. Eng. Compt.* 27: 111–116.

Quiring, D. P. (1944). Surface area determination. In O. Glasser (ed.), *Medical Physics*, vol. 1, Year Book Publishers, Chicago, pp. 1490–1494.

Rafferty, E. B., H. L. Green, and I. C. Gregory (1975a). Disturbances of heart rhythms produced by 50 Hz leakage currents in dogs. *Cardiovasc. Res.* 9: 256–262.

Rafferty, E. B., H. L. Green, and M. H. Yacoub (1975b). Disturbances of heart rhythm by 50 Hz leakage currents in human subjects. *Cardiovasc. Res.* 9, pp. 263–265.

Rall, W. (1977). Core conductor theory and cable properties of neurons. In *Handbook of Physiology: A Critical, Comprehensive Presentation of Physiological Knowledge and Concepts*, vol. 1, American Physiological Society Bethesda, MD, pp. 39–97.

Ranck, J. B. (1963). Specific impedance of rabbit cerebral cortex. *Exp. Neurol.* 7: 144–152.

Ranck, J. B. (1975). Which elements are excited in electrical stimulation of mammalian central nervous system: A review. *Brain Res.* 98: 417–440.

Rasch, P. J. (1989). *Kinesiology and Applied Anatomy*, 7th ed., Lea & Febiger, Philadelphia.

Rattay, F. (1986). Analysis of models for external stimulation of axons. *IEEE Trans. Biomed. Eng.* 33: 974–977.

Rattay, F. (1988). Modeling the excitation of fibers under surface electrodes. *IEEE Trans. Biomed. Eng.* 35(3): 199–202.

Reilly, J. P. (1978a). Electric field induction on sailboats and vertical poles. *IEEE Trans. Pwr. Apparat. Sys.* PAS-97(4): 1373–1381.

Reilly, J. P. (1978b). Electric and magnetic coupling from high voltage AC power transmission lines—Classification of short-term effects on people. *IEEE Trans. Pwr. Apparat. Sys.* PAS-97 (6): 2243–2252.

Reilly, J. P. (1979a). Electric field induction on long objects—A methodology for transmission line impact studies. *IEEE Trans. Pwr. Apparat. Sys.* PAS-98(6): 1841–1849.

Reilly, J. P. (1979b). An approach to the realistic-case analysis of electric field induction from AC transmission lines. Third Int. Symp. High Voltage Engineering, Milan, Italy.

Reilly, J. P. (1980). Spark discharge characteristics of vehicles energized by AC electric field. JHU PPSE T-16, The Johns Hopkins University Applied Physics Laboratory, Laurel, MD.

Reilly, J. P. (1982). Characteristics of spark discharges from vehicles energized by ac electric fields. *IEEE Trans. Pwr. Apparat. Sys.* PAS-101(9): 3178–3186.

Reilly, J. P. (1988). Electrical models for neural excitation studies. *Johns Hopkins APL Tech. Digest* 9(1): 44–58.

Reilly, J. P. (1989). Peripheral nerve stimulation by induced electric currents: Exposure to time-varying magnetic fields. *Med. Biol. Eng. Comput.* 27: 101–110.

Reilly, J. P (1991). Magnetic field excitation of peripheral nerves and the heart: A comparison of thresholds. *Med. Biol. Eng. Comput.* 29(6): 571–579.

Reilly, J. P., and R. H. Bauer (1987). Application of a neuroelectric model to electrocutaneous sensory sensitivity: Parameter variation study. *IEEE Trans. Biomed. Eng.* BME-34(9): 752–754.

Reilly, J. P., and M. Cwiklewski (1978). A realistic-case analysis of electric field induction on vehicles near AC transmission lines. IEEE Can. Conf. Communications and Power, Montreal.

Reilly, J. P., and M. Cwiklewski (1981). Rain gutters near high-voltage power lines: A study of electric field induction. *IEEE Trans. Pwr. Apparat. Sys.* PAS-100(4): 2068–2076.

Reilly, J. P., and W. D. Larkin (1983). Electrocutaneous stimulation with high voltage capacitive discharges. *IEEE Trans. Biomed. Eng.* BME-30: 631–641.

Reilly, J. P., and W. D. Larkin (1984). Understanding electric shock. *Johns Hopkins APL Tech. Digest* 5(3): 296–304.

Reilly, J. P., and W. D. Larkin (1985a). Human reactions to transient electric currents—Summary report. The Johns Hopkins University Applied Physics Laboratory, Rep. PPSE T-34 (NTIS No. PB 86-117280/AS), Laurel, MD.

Reilly, J. P., and W. D. Larkin (1985b). Mechanisms for human sensitivity to transient electric currents. In J. E. Bridges, G. L. Ford, I. A. Sherman, and M. Vainberg (eds.), *Electrical Shock Safety Criteria*, Pergamon Press, New York, pp. 241–249.

Reilly, J. P., and W. D. Larkin (1987). Human sensitivity to electric shock induced by power frequency electric fields. *IEEE Trans. Electromagnetic Compatibility* EMC-29(3): 221–232.

Reilly, J. P., W. Larkin, R. J. Taylor, and V. T. Freeman (1982). Human reactions to transient electric currents, annual report, July 1981–July 1982. CPE-8203 (NTIS No. PB83 204628), The Johns Hopkins University Applied Physics Laboratory, Laurel, MD.

Reilly, J. P., W. Larkin, R. J. Taylor, V. T. Freeman, and L. B. Kittler (1983). Human reactions to transient electric currents, annual report, July 1982–June 1983. Report CPE-8305 (NTIS No. PB84-112895), The Johns Hopkins University Applied Physics Laboratory, Laurel, MD.

Reilly, J. P., W. D. Larkin, L. B. Kittler, and V. T. Freeman (1984). Human reactions to transient electric currents—Annual report, July 1983–June 1984. Report CPE-8313 (NTIS No. PB84-231463), The Johns Hopkins University Applied Physics Laboratory, Laurel, MD.

Reilly, J. P., V. T. Freeman, and W. D. Larkin (1985). Sensory effects of transient electrical stimulation—Evaluation with a neuroelectric model. *IEEE Trans. Biomed. Eng.* BME-32 (12): 1001–1011.

Robblee, L. S., and T. L. Rose (1990). Electrochemical guidelines for selection of protocols and electrode materials for neural stimulation. In W. F. Agnew and D. B. McCreery (eds.), *Neural Prostheses: Fundamental Studies*, Prentice-Hall, Englewood Cliffs, NJ, pp. 25–66.

Rollman, G. B. (1969). Electrocutaneous stimulation: Psychometric functions and temporal integration. *Perception Psychophys.* 5(5): 289–293.

Rollman, G. B. (1974). Electrocutaneous stimulation. In F. A. Geldard (ed.), *Conference on Cutaneous Communication Systems and Devices*, Monterey, CA, 1973, The Psychonomic Society, Austin, TX, pp. 38–51.

Rollman, G. B. (1975). Behavioral assessment of peripheral nerve function. *Neurology* 26: 339–342.

Rollman, G. B., and G. Harris (1987). The detectability and perceived magnitude of painful electrical shock. *Perception Phychophys.* 42(3): 257–268.

Rosenberg, D. B., and M. Nelson (1988). Rehabilitation concerns in electrical burn patients: a review of the literature. *J. Trauma* 28(6): 808–812.

Rosenblueth, A., and J. Garcia Ramos (1947). Studies on flutter and fibrillation. II. The influence of artificial obstacles on experimental, auricular flutter. *Am. Heart J.* 33: 677.

Rösler, K. M., C. W. Hess, R. Heckmann, and H. P. Ludin (1989). Significance of shape and size of the stimulating coil in magnetic stimulation of the human motor cortex. *Neurosci. Lett.* 100: 347–352.

Rosner, B. S. (1961). Neural factors limiting cutaneous spatio temporal discrimination. In W. A. Rosenblith (ed.), *Sensory Communication*, M.I.T. Press, Cambridge, MA, pp. 725–737.

Rosner, B. S., and W. R. Goff (1967). Electrical response of the nervous system and subjective scales of intensity. In W. D. Neff (ed.), *Contributions to Sensory Physiology*, vol. 2, Academic Press, New York.

Roth, B. (1989). Interpretation of skeletal muscle four-electrode impedance measurements using spatial and temporal frequency-dependent conductivities. *Med. Biol. Eng. Comput.* 27: 491–495.

Rothberger, C. J., and H. Winterberg (1941). Über Vorhofflimmern und Vorhofflattern. *Pflügers Arch.* 160: 42–90.

Rouse, R. G., and A. R. Dimick (1978). The treatment of electrical injury compared to burn injury: A review of pathophysiology and comparison of patient management protocols. *J. Trauma* 18: 43–47.

Roy, O. Z. (1980). Summary of cardiac fibrillation thresholds for 60-Hz currents and voltages applied directly to the heart. *Med. Biol. Eng. Comput.* 18: 657–659.

Roy, O. Z., J. R. Scott, and G. C. Park (1976). 60-Hz ventricular fibrillation and pump failure thresholds versus electrode area. *IEEE Trans. Biomed. Eng.* BME-23(1): 45–48.

Roy, O. Z., G. C. Park, and J. R. Scott (1977). Intracardiac catheter fibrillation thresholds as a function of the duration of 60-Hz current and electrode area. *IEEE Trans. Biomed. Eng.* BME-24(5): 430–435.

Roy, O. Z., A. J. Mortimer, B. J. Trollope, and E. J. Villeneuve (1985). Electrical stimulation of the isolated rabbit heart by short duration transients. In J. E. Bridges, G. L. Ford, I. A. Sherman, and M. Vainberg (eds.), *Electrical Shock Safety Criteria*, Pergamon Press, New York, pp. 77–86.

Roy, O. Z., J. R. Scott, and B. J. Trollope (1986). 60 Hz ventricular fibrillation thresholds for large-surface-area electrodes. *Med. Biol. Eng. Comput.* 24: 471–474.

Roy, O. Z., B. J. Trollope, and J. R. Scott (1987). Measurement of regional cardiac fibrillation thresholds. *Med. Biol. Eng. Comput.* 25: 165–166.

Ruch, T. C. (1979). Somatic sensation: Receptors and their axons. In T. C. Ruch and H. D. Patton (eds.), *Physiology and Biophysics*, W. B. Saunders, Philadelphia, pp. 157–200.

Ruch, T. C., and H. D. Patton (1979). *Physiology and Biophysics*. W. B. Saunders, Philadelphia.

Ruch, S., J. A. Abildskov, and R.McFee (1963). Resistivity of body tissues at low frequencies. *Circ. Res.* 12: 40–50.

Ruch, T. C., H. D. Patton, J. W. Woodbury, and A. L. Towe (1968). *Neurophysiology*. W. B. Saunders, Philadelphia.

Sachs, R. M., J. D. Miller and K. W. Grant (1980). Perceived magnitude of multiple electrocutaneous pulses. *Perception Psychophys.* 28: 255–262.

Sandove, A. M., J. E. Jones, T. R. Lynch, and P. W. Sheets (1988). Appliance therapy for perioral electrical burns: A conservative approach. *J. Burn Care Rehabil.* 9(4): 391–395.

Sagan, P. M., M. E. Stell, G. K. Bryan, and W. R. Adey (1987). Detection of 60-Hertz vertical electric fields by rats. *Bioelectromagnetics* 8: 303–313.

Sances, A., Jr., S. J. Larson, J. Myklebust, and J. F. Cusick (1979). Electrical injuries. *Surg. Gynecol. Obstet.* 149(1): 97–108.

Sances, A., Jr., J. B. Myklebust, S. J. Larson, J. C. Darin, T. Swiontek, T. Prieto, M. Chilbert, and J. F. Cusick (1981a). Experimental electrical injury studies. *J. Trauma* 21(8): 589–597.

Sances, A., Jr., J. B. Myklebust, J. F. Szablya, T. J. Swiontek, S. J. Larson, M. Chilbert, T. Prieto, and J. F. Cusick (1981b). Effects of contacts in high voltage injuries. *IEEE Trans. Pwr. Apparat. Sys.* PAS-100(6): 2987–2992.

Sances, A., Jr., J. F. Szyblya, J. D. Morgan, J. B. Myklebust, and S. J. Larson (1981c). High voltage powerline injury studies. *IEEE Trans. Pwr. Apparat. Sys.* PAS-100(2): 552–558.

Sances, A., Jr., J. B. Myklebust, J. F. Szablya, T. J. Swiontek, S. J. Larson, M. Chilbert, T. Prieto, J. F. Cusick, D. J. Maiman, and K. Pintar (1983). Current pathways in high-voltage injuries. *IEEE Trans. Biomed. Eng.* BME-30(2): 118–124.

Sasyniuk, G. I., and C. Mentez (1971). A mechanism for reentry in canine ventricular tissue. *Circ. Res.* 28: 3–15.

Sato, M., and J. Ushiyama (1950). On the relation of strength-frequency curve in excitation by low frequency A. C. to the minimal gradient of the nerve fiber. *Jpn. J. Physiol.* 1: 141–146.

Saunders, F. A. (1974). Electrocutaneous displays. In F. A. Geldard (ed.), *Conference on Cutaneous Communication Systems and Devices*, Monterey, CA, 1973, Psychonomic Society, Austin, TX, pp. 20–26.

Schaefer, D. J., (1992). Doimetry and effects of MR exposure to RF and switched magnetic fields. In R. L. Magin, R. P. Liburdy, and B. Persson (eds.), *Biological Effects and Safety Aspect of Nuclear Resonance Imaging and Spectroscopy*, New York Academy of Sciences, New York.

Scherf, D. (1947). Studies on auricular tachycardia caused by aconitine administration. *Proc. Soc. Exp. Biol. Med.* 64: 233–239.

Schludermann, E., and J. P. Zubeck (1962). Effect of age on pain sensitivity. *Perceptual and Motor Skills.* 14: 295–301.

Schmid, E. (1961). Temporal aspects of cutaneous interaction with two-point electrical stimulation. *J. Exp. Psychol.* 67: 191–192.

Schmidt, R. (ed.) (1978). *Fundamentals of Sensory Physiology.* Springer-Verlag, New York.

Schmidt-Nielsen, K. (1984). *Scaling: Why Is Animal Size so Important?* Cambridge Univ. Press, Cambridge.

Schriefer, T. N., K. R. Mills, N. M. Murray, and C. W. Hess (1988). Evaluation of proximal facial nerve conduction by transcranial magnetic stimulation. *J. Neurol. Neurosurg. Psychiatr.* 51: 60–66.

Schwan, H. P. (1954). Die elektrischen Eigenschaften von muskelgewebe bie Niederfrequenz. *Z. Naturforsche,* 96: 245–251.

Schwan, H. P. (1966). Alternating current electrode polarization. *Biophysik.* 3: 181–201.

Schwan, H. (1968). Electrical impedance of the human body. Technical Report TR-2199, U. S. Naval Weapons Laboratory, Dahlgren, VA, NTIS No. AD 842306.

Scott, J. P., W. P. Lee, and S. Zoledziowski (1973). Ventricular fibrillation thresholds for a.c. shock of long duration in dogs with normal acid-base states. *Br. J. Ind. Med.* 30: 155.

Scott, W. T. (1966). *The Physics of Electricity and Magnetism.* John Wiley, New York.

Sharma, M., and A. Smith (1978). Paraplegia as a result of lightning injury. *Br. Med. J.* 12: 1464–1465.

Sharp, G. H., and R. W. Joyner (1980). Stimulated propagation of cardiac action potential. *Biophys. J.* 31: 403–424.

Shields, C. B., H. L. Edmonds, M. Paloheimo, J. R. Johnson, and R. T. Holt (1988). Intraoperative use of transcranial magnetic motor-evoked potentials. *Proc. 10th Ann. Conf. IEEE-EMBS*, pp. 926–927.

Silny, J. (1986). The influence of threshold of the time-varying magnetic field in the human organism. In J. H. Bernhardt (ed.), *Biologial Effects of Static and Extremely Low Frequency Magnetic Fields,* MMV Medzin Verlag, Munchen.

Silva, M., N. Hummon, D. Ruttor, and C. Hooper (1989). Power frequency magnetic fields in the home. *IEEE Trans. Pwr. Del.* 4(1): 465–478.

Silverglade, D. (1983). Splinting electrical burns utilizing a fixed splint technique: A report of 48 cases. *ASDC J. Dent. Child.* 50(6): 455–458.

Silversides J. (1964). The neurological sequelae of electrical injury. *Calif. Med. Assoc. J.* 91: 195–204.

Skoog, T. (1970). Electrical injuries. *J. Trauma.* 10: 816–830.

Smith, L. (1990). Electrocutions involving consumer products. Memorandum of March 30, 1990, U. S. Consumer Product Safety Commission, Washington, DC.

Smoot, A. L. (1985). The seventh meeting of IEC TC74/WGS. Memorandum of April 11, 1985, Underwriters Laboratories, Inc.

Smoot, A., and J. Stevenson (1968a). Investigation of reaction current. Technical Report of April 1968, Underwriters Laboratories, Inc.

Smoot, A., and J. Stevenson (1968b). Report on investigation of reaction currents. Technical Report of Dec. 3, 1968, Underwriters Laboratories, Inc.

Smyth, P. D., P. P. Tarjan, E. Chernof, and N, Baker (1976). The significance of electrode surface area and stimulating thresholds in permanent cardiac pacing. *J. Thorac. Card. Surg.* 71(4): 559–565.

Snyder, R. G., M. L. Spencer, C. L. Owings, and L. W. Schneider (1975). Anthropometry of US infants and children. SAE Automotive Engineering Congress and Exposition, Paper No. 750423, as Referenced by A. F. Roche and R. M. Malina (1983), *Manual of Physical Status and Performance in Childhood, Vol. 1B: Physical Status,* Plenum Press, New York.

Solem, L., R. P. Fischer, and R. G. Strate (1977). The natural history of electrical injury. *J. Trauma.* 17(7): 487–492.

Solomonow, M., E. Eldred, J. Lyman, and J. Foster (1983). Control of muscle contractile force through indirect high-frequency stimulation. *Am. J. Phys. Med.* 62: 71–82.

Song, W. J., S. Weinbaum, and L. M. Jiji (1988). A combined macro and microvascular model for whole limb heat transfer. *J. Biomed. Eng.* 110(4): 259–268.

Spach, M. S., W. T. Miller, D. B. Geselowitz, R. C. Barr, J. M. Kootsey, and E. A. Johnson (1981). The discontinuous nature of propagation in normal canine cardiac muscle. Evidence for recurrent discontinuities of intracellular resistance that affect the membrane currents. *Circ. Res.* 48: 39–54.

Spach, M. S., W. T. Miller, P. C. Dolber, J. M. Kootsey, J. R. Sommer, and C. E. Mosher (1982). The functional role of structural complexities in the propagation of depolarization in the atrium of the dog. *Circ. Res.* 50: 175–191.

Spiegel, R. J. (1976). Magnetic coupling to a prolate spheroid model of man. *IEEE Trans. Pwr. Apparat. Sys.* PAS-96(1): 208–212.

Starmer, C. F., and Whalen, R. E. (1973). Current density and electrically induced ventricular fibrillation. *Med. Instrum.* 7(1): 3–6.

Stein, R. B. (1980). *Nerve and Muscle.* Plenum Press, New York.

Sten-Knudsen, O. (1960). Is muscle contraction initiated by internal current flow? *J. Physiol.* 151: 363–384.

Stern, S., and V. G. Laties (1985). 60 Hz electric fields: Detection by female rats. *Bioelectromagnetics* 6: 99–103.

Stern, S., V. G. Laties, C. V. Stancompiano, C. Cox, and J. O. deLorge (1983). Behavioral detection of 60-Hz electric fields by rats. *Bioelectromagnetics.* 4: 215–247.

Sternbach, R. A., and B. Tursky (1964). On the psychophysical power function in electric shock. *Psychol. Sci.* 1: 217–218.

Sternbach, R. A., and B. Tursky (1965). Ethnic differences among housewives in psychophysical and skin potential responses to electric shock. *Psychophysiology* 1: 241–246.

Stevens, J. C. (1980). Thermo-tactile interactions: Some influences of temperature on touch. In D. R. Kenshalo (ed.), *Sensory Functions of the Skin of Humans,* Plenum, Press, New York, pp. 207–222.

Stevens, J. C., J. D. Mack, and S. S. Stevens (1960). Growth of sensation on seven continua as measured by force of handgrip. *J. Exp. Psychol.* 59: 60–67.

Stevens, J. C., L. E. Marks, and D. C. Simonson (1974). Regional sensitivity and spatial summation in the warmth sense. *Physiol. Behav.* 13: 825–836.

Stevens, J. C., B. G. Green, and Krimsley (1977). Punctuate pressure sensitivity: Effects of skin temperature. *Sensory Processes* 1: 238–243.

Stevens, S. S. (1959). Cross-modality validations of subjective scales for loudness, vibration, and electric shock. *J. Exp. Psychol.* 57: 201–209.

Stevens, S. S. (1966). Matching functions between loudness and ten other continua *Perception Psychophys.* 1: 5–8.

Stevens, S. S. (1975). *Psychophysics: Introduction to Its Perceptual Neural, and Social Prospects.* John Wiley, New York.

Stevens, W. G. S. (1963). The current-voltage relationship in human skin. *Med. Electron. Biol. Eng.* 1: 389–399.

Stevenson, J. (1969). Progress report on investigation of reaction current. Technical Report Nov. 18, 1969, Underwriters Laboratories, Inc.

Stevenson, J. (1971). Reaction to leakage currents. *UL Lab Data* 2(1): 16–19. Underwriters Laboratories Inc.

Stoy, R. D., K. R. Foster, and H. P. Schwan (1982). Dielectric properties of mammalian tissues from 0.1 to 100 MHz: A summary. *Phys. Med. Biol.* 27(4): 501–513.

Strasser, E. J., R. M. David, and M. J. Mehshey (1977). Lightning injuries. *J. Trauma.* 17(4): 315–319.

Stuchly, M. A., and D. W. Lecuyer (1989). Exposure to electromagnetic fields in arc welding. *Health Phys.* 56(3): 297–302.

Stuchly, M. A., and S. S. Stuckly (1980). Dielectric properties of biological substances—tabulated. *J. Mic. Pwr.* 15(1): 19–26.

Suchi, T. (1954). Experiments on electrical resistance of the human epidermis. *Jpn. J. Physiol.* 5: 75–80.

Sugimoto, T., S. F. Schaal, and A. G. Wallace (1967). Factors determining vulnerability to ventricular fibrillation induced by 60-cps alternating current. *Circ. Res.* 21: 601–608.

Sunderland, S. (1978). *Nerves and Nerves Injuries*, Churchill Livingstone, New York.

Sweeney, J. D., J. T. Mortimer, and D. Durand (1987). Modeling of mammalian myelinated nerve for functional neuromuscular stimulation. *Proc. Ninth Ann. Int. Conf. of the IEEE-EMBS*, pp. 1577–1578.

Sweeney, J. D., K. Deng, E. Warman, and J. T. Mortimer (1989). Modeling of electric field effects on the excitability of myelinated motor nerve. *Proc Eleventh Ann. Int. Conf. of the IEEE-EMBS*, pp. 1281–1282.

Sweeney, J. D., D. Ksienski, and J. T. Mortimer (1990). A nerve cuff technique for selective activation of peripheral nerve trunk regions. *IEEE Trans. Biomed. Eng.* BME-31: 706–715.

Szeto, A. Y., and F. A. Saunders (1982). Electrocutaneous stimulation for sensory communication in rehabilitation engineering. *IEEE Trans. Biomed. Eng.* BME-25(4): 300–308.

Takagi, T., and T. Muto (1971). Influence upon human bodies and animals of electrostatic induction caused by 500 kV transmission lines (English transl.), Tokyo Electric Power Co.

Takemoto-Hambleton, R. M., W. J. Dunseath, and W. T. Joines (1988). Electromagnetic fields induced in a person due to devices radiating in the 10 Hz to 100 kHz range. *IEEE Trans. Elect. Compat.* 30(4): 529–537.

Tanner, J. A. (1962). Reversible blocking of nerve conduction by alternating current excitation. *Nature* 195: 712–713.

Tasaki, I. (1953). *Nervous Transmission*. Charles C. Thomas, Springfield, IL.

Tasaki, I. (1982). *Physiology and Electrochemistry of Nerve Fibers*. Academic Press, New York

Tasaki, I., and M. Sato (1951). On the relation of the strength-frequency curve in excitation by alternating current to the strength-duration and latent addition curves of the nerve fiber. *J. Gen. Physiol.* 34: 373–388.

Tashiro, T., and A. Higashiyama (1981). The perceptual properties of electrocutaneous stimulation: Sensory quality, subjective intensity, and intensity-duration relation. *Perception Psychophys.* 30(6): 579–586.

Taussig, H. (1968). Death from lightning and the possibility of living again. *Ann. Intern. Med.* 68: 1345–1349.

Taylor, R. J. (1985). Body impedance for transient high voltage currents. In J. E. Bridges, G. L. Ford, I. A. Sherman, and V. Vainberg (eds.), *Electric Shock Safety Criteria*, Symposium on Electric Shock Safety Criteria, Pergamon Press, pp. 251–258.

Teghtsoonian, R. (1973). Range effects in psychological scaling and a revision of Stevens' law, *Am. J. Psychol.* 86: 3–27.

Teicher, D. A., and D. R. McNeal (1978). Comparison of a dynamic and steady-state model for determining nerve fiber threshold. *IEEE Trans. Biomed. Eng.* BME-25(1): 105–107.

Tenforde, T.S. (1989). Electroreception and magnetoreception in simple and complex organisms. *Bioelectromagnetics* 10: 215–221.

Thalen, H. J. T., J. V. P. Berg, J. N. Heide, and J. Nieveen (1975). *The Artificial Cardiac Pacemaker*. Van Gorum, Assem, The Netherlands.

Thompson, G. (1933). Shock threshold fixes appliance insulation resistance. *Electrical World* 101: 793–795.

Torebjörk, H. E., and R. G. Hallin (1973). Perceptual changes accompanying controlled preferential blocking of A and C fibre responses in intact human skin nerves. *Exp. Brain Res.* 16: 321–322.

Trautwein, W., and J. Dudel (1954). Actionspotential und Mechanogramm des Warmblüterherzmuskels als Funktion der Schlagfrequenz. *Pflügers Arch. ges. Physiol.* 260: 24–39.

Treagear, R. T. (1966). *Physical Functions of Skin.* Acadamic Press, London, U. K.

Tsuji, S., Y. Murai, and M. Yarita (1988). Somatosensory potentials evoked by magnetic stimulation of lumbar roots, cauda equina, and leg nerves. *Annals Neurol.* 24: 568–573.

Tucker, R. D., and O. H. Schmitt (1978). Tests for human perception of 60 Hz moderate strength magnetic field. *IEEE Trans. Biomed. Eng.* BME-25(6): 509–518.

Tursky, B., and P. D. Watson (1964). Controlled physical and subjective intensities of electric shock.*Psychophysiology* 1(2): 151–162.

Ueno, S., K. Harada, C. Ji, and Y. Oomura (1984). Magnetic nerve stimulation without interlinkage between nerve and magnetic flux. *IEEE Trans. Magnet.* MAG-20(5): 1660–1662.

Ueno, S., T. Tashiro, and K. Harada (1988). Localized stimulation of neural tissues in the brain by means of a paired configuration of time-varying magnetic fields. *J. Appl. Phys.* 64: 5862–5864.

UL (1945). Measurement of electric shock hazard in radio equipment. *Bulletin of Research No. 33,* Underwriters Laboratories.

UL (1975). Method of development—Revision—Implementation—Standards for safety. Publication F 200-46 3M575, Underwriters Laboratories.

UL (1981). Development of test equipment and methods for measuring potentially lethal and otherwise damaging current levels. Consumer Product Safety Commission, Contract Number CPSC-C-79-1034, Underwriters Laboratories.

UL (1983). Standard for Safety, double insulation systems for use in electrical equipment. UL 1097, second edition, Underwriters Laboratories.

UL (1985). Standard for safety, ground-fault circuit interrupters. UL943, 2nd ed., Underwriters Laboratories.

UL (1987). Standard for safety, telephone equipment. UL1459, 2nd ed., December 18, 1987, Underwriters Laboratories.

UL (1988). Electric shock—A safety seminar on theory and prevention. Underwriters Laboratories.

UL (1990). Report to the instrumentation committee of ANSI C101, Preliminary report by A. W. Smoot, J. Stevenson, W. Myrick, and W. Tuthill. Underwriters Laboratories.

Vallbo, A. B., K. A. Olsson, K. G. Westberg, and F. J. Clark (1984). Microstimulation of single tactile afferents from the human hand. *Brain* 107(3): 727–749.

van Boxtel, A. (1977). Skin resistance during square-wave electrical pulses of 1 to 10 mA. *Med. Biol. Eng. Comp.* 15: 679–687

van den Honert, C., and J. T. Mortimer (1979a). The response of the myelinated nerve fiber to short duration biphasic stimulating currents. *Annals Biomed. Eng.* 7: 117–125.

van den Honert, C., and J. T. Mortimer (1979b). Generation of unidirectionally propagated action potentials in peripheral nerve by brief stimuli. *Science* 206: 1311–1312.

Veltink, P. H., J. A. van Alste, and H. B. K. Boom (1988). Influences of stimulation conditions on recruitment of myelinated nerve fibers: A model study. *IEEE Trans. Biomed. Eng.* BME-35: 917–924.

Veltink, P. H., B. K. van Veen, J. J. Struijk, J. Holsheimer, and H. B. K. Boom (1989a). A modeling study of nerve fascicle stimulation. *IEEE Trans. Biomed. Eng.* BME-36(7): 683–692.

Veltink, P. H., J. J. Hermens, and J. A. van Alste (1989b). Multielectrode intrafascicular and extraneural stimulation. *Med. Biol. Eng. Comput.* 27: 19–24.

Verrillo, R. T. (1979). Comparison of vibrotactile threshold and suprathreshold responses in men and women, *Perception Psychophys.* 26: 20–24.

Verrillo, R. T. (1982). Effects of aging on the suprathreshold responses to vibration. *Perception Psychophys.* 32: 61–68.

Verillo, R. T., and G. A. Gesheider (1979). Psychophysical measurements of enhancement, suppression, and surface gradient effects in vibrotaction. In D. R. Kenshalo (ed.), *Sensory Functions of the Skin of Humans*, Plenum Press, New York, pp. 153–181.

Vodovnik, L., W. J. Crochetiere, and J. B. Reswick (1967). Control of a skeletal joint by electrical stimulation of antagonists. *Med. Biol. Eng.* 5: 97–109.

Vodovnik, L., T. Bajd, F. Gracanin, A. Kralj, and P. Strojnik (1981). Functional electrical stimulation for control of locomotor systems. *CRC Crit. Rev. Bioeng.* 6: 63-132.

Volta, A. (1800). On the electricity excited by the mere contact of conducting substances of different kinds. *Phil. Trans. R. Soc. London* 90: 403–431.

Voorhes, W. D., K. S. Foster, L. A. Geddes, and C. F. Babbs (1983). Safety factor for transchest pacing. *Proc., 36th ACEMB Conf.*, Columbus, Ohio, Sept. 12–14, p. 19.

Walthard, K. M., and M. Thicaloff (1971). Motor points. In S. Licht (ed.), *Electrodiagnosis and Electromyography*, E. Licht, New Haven, CT, pp. 153–170.

Wang, X. W., and W. H. Zoh (1983). Vascular injuries in electrical burns—The pathologic basis for mechanism of injury. *Burns Incl. Therm. Inj.* 9(5): 335–338.

Wang, X. W., C. S. Lu, N. Z. Wang, H. C. Lin, H. Su, J. N. Wei, and W. Z. Zoh (1984), High tension electrical burns of upper arms treated by segmental excision of necrosed humerus. An introduction of a new surgical method. *Burns Incl. Therm. Inj.* 10(4): 271–281.

Wang, X. W., B. B. Roberts, R. L. Zapata-Sirvent, W. A. Robinson, J. P. Waymack, E. J. Law, B. G. MacMillan, and J. W.Davies (1985). Early vascular grafting to prevent upper extremity necrosis after electrical burns. Commentary on indications for surgery. *Burn Incl. Therm. Inj.* 11(5): 359.

Wang, X. W., E. J. Bartle, and B. B. Roberts (1987). Early vascular grafting to prevent upper extremity necrosis after electric burns: Additional commentary on indications for surgery. *J. Burn Care Rehabil.* 8(5): 391–394.

Warren, R. M. (1981). Measurement of sensory intensity. *Behav. Brain Sci.* 4: 175–223.

Waters, R. L., D. R. McNeal, and J. Perry (1975). Experimental correction of footdrop by electrical stimulation of the peroneal nerve. *J. Bone Joint Surg.* 57-A: 1047–1054.

Waters, R. L., D. R. McNeal, W. Faloon, and B. Clifford (1985). Functional electrical stimulation of the peroneal nerve for hemiplegia. *J. Bone Joint Surg.* 67-A: 792–793.

Watson, A. B., J. S. Wright, and J. Loughman (1973). Electrical thresholds for ventricular fibrillation in man. *Med. J. Austral.* 1, June 16, pp. 1179–1182.

Wedensky, N. (1884). Wie Rasch ermudet der Nerv. *Z. Med. Wissen.* 22: 65–68.

Wégria, R., G. K. Moe, and C. J. Wiggers (1941). Comparison of the vulnerable periods and fibrillation threshold of normal and idioventricular beats. *Am. J. Physiol.* 132: 651–657.

Weigel, R. J., R. A. Jaffe, D. L. Lunstorm, W. C. Forsythe, and L. E. Anderson (1987). Stimulation of cutaneous mechanoreceptors by 60-Hz electric field. *Bioelectromagnetics* 8: 337–350.

Weinstein, S. (1963). The relationship of laterality and cutaneous area to breast-sensitivity in sinistrals and dextrals. *Am. J. Psychol.* 76: 475–479.

Weinstein, S. (1968). Intensive and extensive aspects of tactile sensitivity as a function of body part, sex, and laterality. In D. R. Kenshalo (ed.), *The Skin Senses*, Charles C Thomas, Springfield, IL, pp. 195–218.

Weinstein, S. (1978). New methods for the in-vivo assessment of skin smoothness and skin softness. *J. Soc. Cosmet. Chem.* 29: 99–115.

Weirich, J., S. Hohnloser, and H. Antoni (1983). Factors determining the susceptibility of the isolated guinea pig heart to ventricular fibrillation induced by sinusoidal alternating current at frequencies from 1 to 1000 Hz. *Basic Res. Cardiol.* 78: 604–615.

Weirich, J., K. Haverkampf, and A. Antoni (1985). Ventricular fibrillation of the heart induced by electric current. *Rev. Gen. de l'Elect.*, no. 11, pp. 833–843.

Weiss, G. (1901). Sur la possibilité de rendre comparables entre eux les appareils servant a l'excitation électrique. *Arch. Ital. Biol.* 35: 413–446.

Werner, G., and V. B. Mountcastle (1968). Quantitative relations between mechanical stimuli to the skin and neural response evoked by them. In D. R. Kenshalo (ed.), *The Skin Senses*, Charles C Thomas, Springfield, IL, pp. 112–138.

Wessale, J. L., J.D. Bourland, W. A. Tacker, and L. A. Geddes (1980). Bipolar catheter defibrillation in dogs using trapezoidal waveforms of various tilts. *J. Electrocardiol.* 13: 359–366.

Wetherill, G. B. (1963). Sequential estimation of quantal response curves. *J. R. Stat. Soc.* B25: 1–48.

Whitaker, H. B. (1939). Electric shock hazard as it pertains to the electric fence. *Underwriters Laboratories, Bull. Res.* 14: 3–56.

Wiggers, C. J., and R. Wégria (1939). Ventricular fibrillation due to single, localized induction and condenser shocks applied during the vulnerable phase of ventricular systole. *Am. J. Physiol* 128: 500–505.

Williams, D. O., B. J. Scherlag, R. R. Hope, N. El-Sherif, and R. Lazzara (1974). The pathophysiology of malignant ventricular arrhythmias during acute myocardial ischemia. *Circulation* 50: 1163–1172.

Willis, R. J., and W. M. Brooks (1984). Potential hazards of NMR imaging. No evidence of the possible effects of static and changing magnetic fields on cardiac function of the rat and guinea pig. *Magnetic Resonance Imaging* 2: 89-95.

Wilson, C. M., J. D. Allen, J. B. Bridges, and A. A. Adgey (1988). Death and damage caused by multiple direct current shocks: Studies in an animal model. *Eur. Heart J.* 9(11): 1257–1265.

Wit, A. L., P. E. Cranefield, and B. F. Hoffman (1972). Slow conduction and reentry in the ventricular conducting system. II. Single and sustained circus movement in networks of canine and bovine Purkinje fibers. *Circ. Res.* 30: 11–22.

Wolff, B. B., and S. Langley (1975). Cultural factors and the response to pain. In M. Weisenberg (ed.), *Pain: Clinical and Experimental Perspectives*, C. V. Mosby, St. Louis, pp. 144–151.

Woodbury, J. W. (1968). Action potential: Properties of excitable membranes. In T. C. Ruch, H. D. Patton, J. W. Woodbury, and A. L. Towe (eds.), *Neurophysiology*, 2nd ed., W. B. Saunders, Philadelphia, pp. 26–53.

Woodbury, J. W., A. M. Gordon, and J. T. Conard (1966). Muscle. In T. C. Ruch and H. D. Patton (eds.), *Physiology and Biophysics*, W. B. Saunders, Philadelphia, pp. 113–152.

Woodrow, K. W., G. D. Friedman, A. P. Siegelaub, and M. F. Collen (1975). Pain tolerance: Differences according to age, sex, and race. In M. Weisenberg (ed.), *Pain: Clinical and Experimental Perspectives*, C. V. Mosby, St. Louis, pp. 133–140. (Reprinted from *Psychosomatic Medicine*, 1972, vol. 34, pp. 548–556.)

Woodworth, R. S., and H. Schlosberg (1954). *Experimental Psychology*. Holt, Rinehart and Winston, New York.

Worthen E. F. (1982). Surgical treatment of electrical burns of the scalp and skull: Past and present. *Clin. Plast Surg.* 9(2): 161–165.

Wu, Y. C. (1979). Electrical injuries—A literature review. *Natl. Bur. Std.* NBSIR, pp. 79–1710.

Wuerker, R. B., A. M. McPhedran, and E. Henneman (1965). Properties of motor units in a heterogeneous fast muscle (M. Gastrocenmius) of the cat. *J. Neurophys* 28: 85–99.

Wynn Parry, C. B. (1971). Strength-duration curves. In S. Licht (ed.), *Electrodiagnosis and Electromyography*, 3rd ed., E. Licht, New Haven, CT.

Wyss, A. M. (1963). Die Reizwirkung sinuförger Wechselströme, untersucht, bis zur oberen Grenze der Niederfrequenz (1000Hz). *Helv. Physiol. Pharmacol.* 21: 419–443.

Yamamoto, T., and Y. Yamamoto (1977). Analysis for the change of skin impedance. *Med. Biol. Eng. Comput.* 15: 219–227.

Yamamoto, T., Y. Yamamoto, and A. Yoshida (1986). Formative mechanisms of current concentration and breakdown phenomena dependent on direct current flow through the skin by a dry electrode. *IEEE Trans. Biomed. Eng.* BME-33(4): 396–404.

Young, J. S., P. E. Burns, A. M. Bowen, and R. McCutchen (1982). *Spinal Cord Injury Statistics. Experience of the Regional Spinal Cord Injury Systems,* Good Samaritan Medical Center, Phoenix, AZ.

Young, J. W., R. F. Chandler, C. C. Snow, K. M. Robinette, G. F. Zehner, and M. S. Lofberg (1983). *Anthropometric and Mass Distribution Characteristics of the Adult Female.* FAA-AM-83-16 revised ed.

Younossi, V. K., H. Z. Rüdiger, K. Haap, and H. Antoni (1973). Untersuchungen über die Flimmerschwelle dis isolierten Meerchweinchen-Herzens für Gleichstrom und sunusförmigen Wechselstrom. *Basic Res. Cardiol.* 68: 551–568.

Yuen, T. G. H., W. F. Agnew, L. A. Bullara, and D. B. McCreery (1990). Biocompatibility of electrodes and materials in the central nervous system. In W. F. Agnew and D. B. McCreery eds., *Neural Prostheses: Fundamental Studies,* Prentice-Hall Englewood Cliffs, NJ, pp. 197–223.

Zborowski, M. (1952). Cultural components in response to pain. *J. Social Issue* 8: 16–30.

Zelt, R. G., R. K. Daniel, P. A. Ballard, Y. Brissette, and P. Heroux (1988). High-voltage electrical injury: Chronic wound evolution. *Plast. Reconstr. Surg.* 82(6): 1027–1041.

Zipes, D. P. (1975). Electrophysiological mechanisms involved in ventricular fibrillaton, *Supplement III to Circulation,* vols. 51 & 52, pp. III-120 to III-130.

Zoll, P. M., R. H. Zoll, R. H. Flak, J. E. Clinton, D. R. Eitel, and E. M. Antman (1985). External noninvasive temporary cardiac pacing: Clinical trial. *Circulation* 71(5): 937–944.

Index